Production, Growth, and the Environment

An Economic Approach

WILLIAM L. WEBER

CRC Press
Taylor & Francis Group
Boca Raton London New York

CRC Press is an imprint of the
Taylor & Francis Group, an **informa** business

CRC Press
Taylor & Francis Group
6000 Broken Sound Parkway NW, Suite 300
Boca Raton, FL 33487-2742

First issued in paperback 2017

Version Date: 20140709

ISBN 13: 978-1-4822-4305-5 (hbk)
ISBN 13: 978-1-138-07509-2 (pbk)

Visit the Taylor & Francis Web site at
http://www.taylorandfrancis.com

and the CRC Press Web site at
http://www.crcpress.com

For Mary Jeanne, Michael, and Julia

Contents

List of Figures

List of Tables

Acknowledgments

In the late 1990s former provost of Southeast Missouri State University, Chuck Kupchella, urged the faculty to start a program in environmental science. I had the good fortune to be asked to prepare a class in environmental economics. The program has prospered under the leadership of Steve Overmann who has been director of Environmental Science since its inception. Steve has provided encouragement and sent numerous good students to take my class for which I am thankful. Over the years Rolf Färe and Shawna Grosskopf tried to steer me in the right direction about how to think and model jointly produced desirable and undesirable by-products. In preparing this book I benefited from discussions with Moriah Bostian, Mike Devaney, Bruce Domazlicky, Hirofumi Fukuyama, Hamner Hill, Stephanie Shreve, and Rebecca Summary as they read various chapters and made numerous comments and suggestions. Thanks to Irma Britton, Jill Jurgensen, and Robin Lloyd-Starkes of CRC Press for helping to make this book possible. Of course, I retain the property rights to any undesirable ideas presented. My final thanks go to my wife, Mary Jeanne. Without her support and encouragement, I would still be writing.

1

Introduction

And did those feet in ancient time,
Walk upon Englands mountains green:
And was the holy Lamb of God,
On Englands pleasant pastures seen!

And did the Countenance Divine,
Shine forth upon our clouded hills?
And was Jerusalem builded here,
Among these dark Satanic Mills?

Bring me my Bow of Burning gold;
Bring me my Arrows of desire:
Bring me my Spear: O clouds unfold!
Bring me my Chariot of fire!

I will not cease from Mental Fight,
Nor shall my Sword sleep in my hand:
Till we have built Jerusalem,
In Englands green and Pleasant Land.

William Blake, "And did those feet in ancient time"

1.1 Economic Growth and the Environment

The start of the Industrial Revolution, sometime around 1750, caused per capita income in the Western world to increase 15 to 23 times by the start of the twenty-first century. This growth in income was also accompanied by a change in the industrial process that substituted the use of non-renewable energy derived from coal and oil for renewable energy human, draught animals, wind, water, and wood. From 1820 to 2001, the world's population is estimated to have grown from around 1 billion people to more than 6 billion

[154]. Does economic growth necessitate environmental degradation? Will the world's population outgrow the ability to feed itself? Are we running out of resources? Will man's dominance over earth's ecosystem cause widespread extinction? What are the best institutions to foster a rising standard of living for all people? We will address these questions and more in this book.

E.J. Hobsbawm [110, p. 10] remarked that a traveler through England's countryside in 1750 "would be immediately struck by the greenness, tidiness, the apparent prosperity of the countryside, and by the apparent comfort of 'the peasantry'." However, the "apparent comfort of the peasantry" did not stop the great rural to urban migration made possible by a rising agricultural surplus. Today, more of the world's population lives in an urban rather than rural area. In developed countries, like the United Kingdom and the United States, more than 80% of the population lives in a city. London's rising population tended to be crowded into tenements with high crime rates and "'common sewers' which flowed through or past these crowded communities." [83, p. 77] Hobsbawm tells us that upon arriving in London, the traveler would encounter a rather "gloomy" city that was neither clean nor well-lit. The "dark Satanic mills" of Birmingham and Manchester in the 1800s, and Kansas City, Pittsburgh, and Chicago in the late 1800s, filled the skies with smoke and soot, not unlike many Chinese cities today.

Would a reverse migration and return to a more pastoral way of life enhance human well-being? Although Hobsbawm's depiction of the tranquil English countryside entices us, Dorothy George [83] tells us that the ditches outside London were "full of nastiness and stinking dirt" so that "many persons who have occasion to go in or come out of town are forced to stop their noses to avoid the ill-smell occasioned by it." [83, p. 105] Robert J. Gordon cautions that even in the late 1880s, women in rural areas bore the costs of a lack of running water and indoor plumbing. Water for drinking and cooking had to be carried in and waste water carried out. "The average North Carolina housewife in 1885 had to walk 148 miles per year while carrying 35 tons of water" [91, p.7]. In addition, a lack of electricity meant cooking was done over an open hearth and dwellings were dark and smoky. In 1990, only 27% of the world's rural population had access to improved sanitation facilities, rising to 47% in 2010. In contrast, 76% of the world's urban population had access to improved sanitation in 1990, rising to 79% in 2010 [234]. Given that people migrate to cities because they think it offers a higher standard of living than rural areas, such a reverse migration might very well lower living standards.

For 1000 years from AD 1 to 1000 AD, world per capita GDP fell from around $467 to $450 in 1990 inflation adjusted dollars. By 1820, per capita GDP was $667, for an average growth rate of 0.05% per year for 820 years. Western Europe and its offshoots (America and the Caribbean) grew slightly faster at 0.13% per year from 1000 to 1820. Then, from 1820 to 2003 growth in the West accelerated to 1.63% per year. The rising standard of living allowed the "common sewers" to be closed reducing disease and death. Life expectancy in the West soared from an average of 24 years in the year 1000 to 36 years in

1820 to 76 years in 2003. In the non-Western world life expectancy increased to 67 years [155, p. 70–72].

Life in the "Bad Old Days"

The urban streets of the 1870s and 1880s were full not just of horses but pigs, which were tolerated because they ate garbage. In Kansas City, the stench of the patrolling hogs was so penetrating that Oscar Wilde observed, "They made granite eyes weep." The increasing production of animal waste caused pessimistic observers to fear that American cities would disappear like Pompeii—but not under ashes. Added to that was acrid industrial smog, sidewalks piled high with kitchen slops, coal dust, and dumped merchandise, which became a liquid slime after a rain. All of this was made worse in the summer, which was almost as unbearable outdoors as inside, especially with the heavy clothes of the day. Rudyard Kipling said of Chicago, "Having seen it, I desire urgently never to see it again. Its air is dirt." Added to putrid air was the danger of spoiled food—imagine meat and poultry hung unrefrigerated for days, spoiled fruit, bacteria-infected milk, and so on. Epidemics included yellow fever, scarlet fever, and smallpox. Many hospitals were deathtraps.

Before the invention of electricity, urban streets were a chaotic jungle of horse-drawn conveyances of all types, made even more congested in winter by horse-drawn snowplows that did little more than move the snow out of the way of trolleys by dumping it on the sidewalks. Rural life was marked by isolation, loneliness, and the drudgery of fireplace cooking and laundry done by muscle power. Travel between cities on railroads was surprisingly dangerous; in 1890, railroad-connected accidents caused 10,000 deaths.

In 1882 only 2 percent of New York City's houses had water connections. Urban apartments were crowded, damp, airless, and often firetraps. Even middle-class apartment buildings were little more than glorified tenements. In the slums as many as eight persons shared a single small room.

Coal miners, steel workers, and many others worked 60-hour weeks in dirty and dangerous conditions, exposed to suffocating gas and smoke. Danger was not confined to mines or mills; in 1890 one railroad employee was killed for every 300 employed. Sewing in a sweatshop might have been the most oppressive occupation for women, but was not as dangerous as soap-packing plants or the manual stripping of tobacco leaves.

From Robert J. Gordon, "Does the 'New Economy' Measure up to the Great Inventions of the Past?" [90, p. 58]

The Industrial Revolution provided inexpensive cotton underwear, paper for cleaning, and soap to the common man and woman helping reduce the transmission of disease from the privy [136]. The economic growth of the last 200 years has made people in the West and increasingly in Asia better off, but parts of the world, such as Africa and southeast Asia, have not experienced such a rising standard of living.

Although income has increased by a factor of 15–23 in the West and 10 in the rest of the world it has been accompanied by increased greenhouse gas emissions of carbon dioxide, nitrogen oxides, and methane. Such emissions might cause significant climate change that could dampen or reverse future economic growth if continued. Thus, several challenges exist. Should we stabilize or reduce greenhouse gas emissions and if so, how and when? Should some countries be allowed to increase emissions as other countries cut emissions? Table 1.1 shows that renewable energy sources make up less than 10% of total world energy consumption and without hydroelectric, the renewable energy share falls to 2%. Is it realistic to design and promote policies that substitute renewable energy for non-renewable energy? Here, the choices we make will impact both current and future generations and might redistribute wealth between countries. How can we increase the standard of living for the 2.4 billion people who, in 2008, still lived on less than $2 per day and the 26% of the world's population still without access to electricity [234]? People in the developed world burn cheap fossil fuels to enhance their welfare. However, further increases in fossil fuel use will likely contribute to climate change. Are the 2.4 billion low income people destined to remain poor so that the detrimental effects of climate change can be minimized? Or, are those costs too great to deny a higher living standard to the world's poorest people?

Table 1.1: Fuel Shares of Energy Consumption, 2010

Fuel	World	Non-OECD	OECD
Liquids	33.2%	28.7%	38.5%
Natural Gas	22.4%	21.1%	23.7%
Coal	28.6%	37.3%	18.3%
Nuclear	5.3%	1.8%	9.5%
Hydroelectric and renewables	10.5%	11.1%	10.0%

Source: Energy Information Administration.

1.2 Brief History of Environmental Economics

Concern over environmental degradation in the US can be traced back at least 375 years. In 1637, amid concerns over declining herring runs in local streams, the town of Plymouth, Massachusetts began requiring millers to allow water to flow freely at certain times of the year so that spawning could occur. Game laws came into effect in Rhode Island Colony in 1639 and in 1668, the Governor of New York Colony forbid people from polluting the harbor. Two hundred years later, the River and Harbors Act of 1899 became the first federal law prohibiting the dumping of refuse in navigable waters [182].

Economists have long recognized the interaction of humans, the economy, and the environment. In 1776, Adam Smith recognized the joint production of various goods, such as skins as a by-product of meat production. In 1848, John Stuart Mill pointed out that coal gas (methane) is a jointly produced by-product of coal production. Analyzing joint production is important if one is to understand the trade-off between the production of desirable goods such as electricity and its undesirable by-products such as nitrogen, sulfur, and carbon emissions and we address this topic in Chapter 5. Mill also understood that natural resources have amenity values: people value natural resources in their undeveloped state.

Thomas Malthus [161] was concerned about population problems stemming from his observation that any increase in the aggregate economic surplus tended to cause population to increase. At the time, Malthus observed that since food production tended to increase linearly, while population grew exponentially, the bulk of the population would be forced to live at the subsistence level. David Ricardo examined the consequences for land rents and wages due to diminishing returns to land in agriculture. As the population grew and the demand for food increased, farmers would be forced to bring less productive lands into production causing the marginal cost of food to increase. Although landowners would reap higher rents on the most productive pieces of land, workers would be forced to pay higher food prices, thus reducing their wages to the subsistence level.

John Stuart Mill recognized that technological progress might offset the effects of diminishing returns in agriculture, much like today's advances in biotechnology. For instance, advances in biotechnology that have allowed some varieties of cotton, corn, and potatoes to contain the *Bacillus thuringiensis* (Bt) gene have reduced the need for insecticides [28]. Scientists have also learned to transfer genes that increase quantities of vitamin A into rice varieties, helping to reduce the incidence of blindness. Advances in biotechnology which have increased yields have also simultaneously reduced the amount of land under cultivation. However, France and other countries, concerned about so-called "Franken-foods," have resisted the introduction of genetically modified varieties, potentially slowing the pace of progress and hastening the effects of diminishing returns.

In 1865 Stanley Jevons [118] sounded the alarm that Britain might run out of coal. Of course such alarm has been subsequently refuted. In 2012, known coal reserves were still enough to power the planet for 112 to 130 more years. In addition, known reserves depend on prices. If coal or any resource were to become more scarce, its price would rise providing incentives for exploration among suppliers and conservation among consumers. However, Jevons raised a more subtle issue that still has legs. Jevons's Paradox is the idea that technologies that provide greater energy efficiency have the paradoxical effect of increasing energy use. As an example, Savery's steam engine, which was designed to pump water out of coal mines in 1698, wasted so much heat that it was uneconomical to operate in spite of access to cheap coal. Once the machine was improved with the addition of a piston, the greater efficiency allowed it to consume more coal [4].

Nineteenth and early twentieth century economist Alfred Marshall deserves special mention for recognizing the importance of "sustainable" development and for his concern "with many aspects of the quality of life, generally considered: the quality of the air man breathes, and of the water he drinks, and that of the 'environment' (physical, social, cultural) he work and lives in" [33, p. 533]. Marshall seemed to be especially concerned with Londoner's lack of fresh air and lack of "bright sunshine and green fields which lowers the physical constitution."

In 1920, A.C. Pigou [196] formalized the idea of an externality which is either a benefit conferred to or cost imposed on a third-party for which there is no compensation. Pigou thought that external costs, such as pollution, could be internalized by the market process via taxes equal to the marginal damage borne by the third-party. In contrast to Pigou, Ronald Coase [43], in his famous paper "The Problem of Social Cost" argued that private parties sometimes negotiate amongst themselves to internalize externalities without government intervention. If costs are already internalized, any government intervention moves the economy away from its efficient point and in the process reduces welfare.

In 1952, the independent research organization, Resources for the Future (RFF), was established. Much of the early concern of RFF was with whether the supply of natural resources would be sufficient to allow continued economic growth [195]. An influential book commissioned by RFF and written by Harold Barnett and Chandler Morse [18] appeared in 1963 and examined the limits to economic growth posed by Malthus and Ricardo. As resources become more scarce their costs should rise. However, using data from 1870–1957 Barnett and Morse found that increased economic activity did not result in higher unit costs of production. Instead, technological progress, substitution between resources, and the discovery of new resources tended to cause costs to decline.

The year 2012 marked the fiftieth anniversary of the publication of Rachel Carson's *Silent Spring* [36]. Carson, a biologist, focused on the problems posed by synthetic chemicals, particularly on the pesticide DDT, used to keep mosquitoes at bay in tropical climates. The insecticide had been found to cause a thinning and cracking of the eggs of tropical songbirds. Carson claimed that like certain radioactive compounds, various chemical insecticides might also be linked to increased risks of cancer, especially in children where the "rapidly growing tissues of a child would also afford conditions most suitable for the development of malignant cells" [36, p. 234].

Kenneth Boulding [26] was an early advocate of sustainability—conserving resources today so as not to detract from the welfare of future generations. Boulding used the metaphor of a "spaceship Earth" to argue that there are limits to growth since there are finite supplies of energy, water, and natural resources. These limits require that society focus not on consumption, but on maintaining the stock of natural assets by switching to solar power and recycling and reusing materials. Other researchers focused on the problems posed by population growth, economic growth, and food shortages. In 1968, biologist Paul Ehrlich wrote *The Population Bomb* in which he argued that the American lifestyle was driving the world to the brink of ecological destruction. Ehrlich argued that because only finite amounts of natural resources exist, the continued pace of consumption was not only undesirable, it was also unsustainable. In 1972, The Club of Rome published *The Limits to Growth*, which looked at what they saw as a set of inter-related problems including poverty, environmental degradation, urban sprawl, overconsumption, and overpopulation [15].

Of course, trying to link over-population and food shortages is not a recent endeavor, but dates at least from the Reverend Thomas Malthus's warning in *An Essay on the Principle of Population.* Malthus based his essay on two ideas [69]. The first idea is that "our ability to produce children will always exceed our ability to secure food for their survival." In essence, it is easier to double the population than it is to double agricultural productivity. Because of this theory, Malthus thought that population has to somehow be checked, either through preventive checks, such as birth control, abstinence, delayed marriage, or non-procreative sex, or, by positive checks such as premature death, infanticide, abortion, famine, or disease. While the upper classes tend to rely more on preventive checks, Malthus thought that the lower classes would be more susceptible to positive checks. Malthus's second theory suggests that any increase in the food supply removes one of the checks on population, resulting in an expansion in population to a new level where it must be checked again.

Malthus's vision suggests that there is little that can be done to overcome the population problem. In his view, structural reforms, the spread of democracy or socialism, industrialization or globalization, will do nothing to overcome the problem of population running up to its maximum limit. However, Nobel Prize winner Amartya Sen [214] presented evidence that the link between over-population and food shortages disappears when the institutions that are chosen to allocate resources are taken into account. According to Sen, no country with a working democracy and a free press has ever experienced a famine. Furthermore, famines, such as those in North Korea, occur not because of food shortages, but because of dictators and their policies. When a famine begins, it is usually because a group of people living at the subsistence level experiences a decline in income due to changing relative prices. In countries with democracies and a free press, the plight of those people suffering deprivation becomes quickly known and action is taken to alleviate the problem. In countries without those democratic institutions, knowledge of the problem does not become public, no one takes action, and people die.

In a counter to the neo-Malthusians economist Julian Simon argued that human ingenuity, not natural resources, is the limit to economic growth and living standards. In 1980, Paul Ehrlich and Simon agreed to a bet to test their hypotheses. The two parties agreed to purchase \$1000 of five metals. Simon allowed Ehrlich to choose the five metals and Ehrlich chose copper, chrome, nickel, tin, and tungsten. A rise in the inflation adjusted metal prices would indicate greater relative scarcity and Simon would owe Ehrlich the difference between the five metal's prices and \$1000. On the other hand, a fall in the prices of the five metals would indicate that the metals were less scarce, in which case Ehrlich would owe Simon the difference between \$1000 and the prices of the five metals. To some observers it was a sucker's bet. The most that Simon could win was \$1000 and Ehrlich had unlimited upside potential. Even though the world's population grew by 800 million people from 1980 to 1990, Simon won the bet and received a check from Ehrlich in the amount of \$576.07 as the price of the metals fell.

1.3 Tragedy of the Commons

Common property resources such as open access fisheries, aquifers, migratory animals, and oil pools are non-excludable and rival. A resource in non-excludable when it is prohibitively costly to exclude people from using the resource and it is rival in that one person's use reduces the amount available to others. The prisoners' dilemma game is a useful starting point for analyzing common property resources. Two shepherds share a common pasture where they graze their flocks. The carrying capacity of the pasture might be 10 sheep, so if each shepherd grazes 5 sheep the pasture attains its maximum yield and each shepherd receives a payoff of 5, represented by the northwest cell in Table 1.2. Each cell in the table represents the payoffs to the two shepherds, where the first number in the cell represents the payoff to Shepherd B when they play the strategy in the particular row and Shepherd A plays the strategy in the particular column, and the second number represents the payoff to Shepherd A. If one shepherd decides to double his flock size and graze 10 sheep while the other shepherd continues to graze only 5 sheep, the pasture's yield decreases. Although the shepherd with the larger flock receives a higher payoff when he increases flock size, the total payoff to the two shepherds falls and is only 9 instead of 10. In this game the shepherd who doubles flock size receives 2/3 of the now reduced yield while the other shepherd receives the remaining 1/3.

What strategy will each shepherd choose if they do not account for the effects of their choice on the other shepherd? Shepherd A has the incentive to increase flock size to 10 sheep regardless of whether Shepherd B grazes 5 sheep or grazes 10 sheep. Similarly, Shepherd B also has the incentive to increase flock size. Clearly, when both shepherds increase flock size they both receive a smaller payoff than when they both graze only 5 sheep. This equilibrium corresponds to over-exploitation of the pasture: the tragedy of the commons.

Table 1.2: Common Property Resource Game

		Shepherd A	
		Graze 5 Sheep	**Graze 10 Sheep**
Shepherd B	**Graze 5 Sheep**	5,5	3,6
	Graze 10 Sheep	6,3	4,4

Can the tragedy of the commons be avoided? One way might be for government to impose rules that limit the number of sheep each shepherd can graze. Alternatively, the pasture might be privatized so that responsibility for both the benefits and the costs are assigned to a private party. Nobel Prize winner Elinor Ostrom [191] spent much of her academic life analyzing the common property resource problem. Even in the absence of formal government restrictions, Ostrom found that under certain conditions, the users of the common property resource are sometimes able to cooperate and adhere to formal or informal rules that help prevent the tragedy of the commons. We examine some of Ostrom's results in Chapter 8.

1.4 Scarcity and the Laws of Thermodynamics

The fundamental problem of economics is the scarcity of resources relative to our unlimited wants. Because resources are scarce, individuals and societies must make choices. These choices entail opportunity costs which represent the subjective value of foregone alternatives. From an environmental perspective examples of scarcity and opportunity cost abound. A decision by a city to preserve open space is simultaneously a decision to forego a new shopping center or subdivision. National and global efforts to reduce the effects of global warming will require less energy consumption. Efforts to expand habitat for the Northern spotted owl means that land previously used to grow trees for houses be taken out of production.

The first two laws of thermodynamics are useful for our understanding of scarcity. These laws are:

1st Law of thermodynamics—Matter cannot be created or destroyed.

2nd Law of thermodynamics—In a closed system, the use of matter-energy causes a one way flow from low entropy resources to high entropy resources, from order to disorder.

As an energy resource is used, the amount of work that energy can do is diminished and waste is generated. Thus, the production of any desirable product or service, such as the production of iron from coal and iron ore, simultaneously generates an undesirable or wasteful by-product. In the conversion of coal and iron ore into iron the jointly produced undesirable by-products consist of carbon dioxide emissions and slag waste [20]. Furthermore, the second law ensures that there is a degree of irreversibility to production. Once the genie is let out of the bottle it cannot be put in again.

Nicholas Georgescu-Roegen was an influential economist in the realm of environmental economics. In *The Entropy Law and the Economic Process*, Georgescu-Roegen used an "entropy hourglass analogy" to describe the first two laws of thermodynamics that is described by Herman Daly [57, p.29–30].

> First, the hourglass is an isolated system: no sand enters, and no sand exits.
>
> Second, within the glass there is neither creation nor destruction of sand, the amount of sand in the glass is constant. This of course is the analog of the first law of thermodynamics-conservation of matter-energy.
>
> Third, there is a continuing running down of sand in the top chamber, and an accumulation of sand in the bottom chamber. Sand in the bottom chamber, since it has used up its potential to fall and thereby do work is high-entropy or unavailable matter/energy. This is the second law of thermodynamics: entropy increases in an isolated system. The hourglass analogy is particularly apt since entropy is time's arrow in the physical world.

While the work of laws of thermodynamics [52, 57, 84] originally caused economists to think that entropy would limit growth, today most economists think that the limits to growth are derived from the 1st law of thermodynamics. Since matter can neither be created nor destroyed, the accumulation of waste over time may constrain growth. It is important to note, though, that the earth is not a closed system since we import matter from the sun in the form of solar energy and it might even be feasible to export matter away from earth; for example, by shooting waste into space.

1.5 Circular Flow of Income and Spending

Figure 1.1 depicts a "näive" circular flow of income and spending between households and business firms. Households are shown as suppliers of resources (land, labor, capital, and entrepreneurship) and demanders of goods and services. Business firms are the demanders of resources and the suppliers of goods and services to households. Households earn income (rent, wages, interest, and profits) from their continual supply of resources to business firms. Businesses are willing and able to pay households since they receive a flow of revenue in exchange for the goods and services they provide.

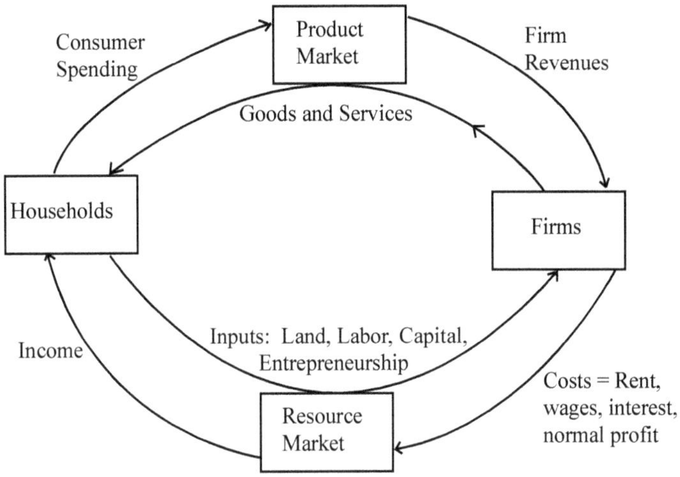

Figure 1.1: Näive Circular Flow of Income and Spending.

We need to modify the näive circular flow to account for various environmental factors. First, not all resources are privately owned, so that some of the costs of using the resource are borne by third parties. Second, undesirable outputs are jointly produced by-products of the production of desirable goods and services. Third, there is a degree of irreversibility in some actions that needs to be accounted for.

A stock variable can be measured at a point in time: the amount of money in your checking account on January 1, 2013 at 10 a.m. On the other hand, the process of generating income and then consumption of goods and services are flows which are measured between two points in time. Since time moves in only one direction, it is important to note that today's choices affect tomorrow's opportunities. This fact is especially important since some choices today might be irreversible and must be addressed within a closed system. Using irrigation in many western states has caused some farmland to become saline and unusable for generations to come. Generating electricity through nuclear fission results in nuclear waste with a half-life of hundreds of thousands of years. The destruction of wildlife habitat and entire ecosystems through mining and timber harvesting might cause extinction of some species.

In Figure 1.2 we modify the circular flow of income and spending model to account for the first two laws of thermodynamics. The production process transforms inputs (x) into desirable outputs (y) and jointly produced undesirable outputs (b). The environment in this model provides life support services and amenities for consumption, energy, and materials for production, and serves as a waste sink for the undesirable outputs. Production and consumption simultaneously determines available environmental services. The size of the boxes for environmental amenities, land and natural resources, and the waste sink can shrink or grow in response to human activity and might also overlap. Producers use inputs (x) that consist of both privately owned resources such as labor and capital and publicly owned environmental inputs. Consumers derive utility from desirable outputs (y) that consist of material goods and services and environmental amenities. Consumers receive disutility from undesirable outputs (b) that are generated as part of the production and consumption processes.

As a waste sink, the environment absorbs human caused and natural pollutants, such as volcanic ash, methane gases from animals and decaying plant life, and smoke from forest and grass fires. Over time, the environment has an assimilative capacity that makes the pollutants less harmful [99]. The stock of pollutants within the closed system in period t_0 (B_{t_0}) is a function of all past flows (b_t), and assimilated amounts (A_t):

$$B_{t_0} = \sum_{t<t_0} (b_t - A_t). \tag{1.1}$$

Clearly, if the flow of pollutants is greater than the amount assimilated, the stock of pollutants will increase. The resulting increase in toxicity can cause damage to humans and wildlife and change the ecological equilibrium. Some of the carbon dioxide released by burning fossil fuels is absorbed by oceans relatively quickly, but full assimilation can take thousands of years. Methane, another greenhouse gas, has an atmospheric lifetime of only twelve years, but has twenty-five times the global warming potential as carbon dioxide.

As a consequence of the laws of thermodynamics [20] it is also true that

$$\begin{aligned} \text{mass of inputs}(x) = &\text{mass of desirable outputs}(y) + \\ &\text{mass of undesirable outputs}(b). \end{aligned} \tag{1.2}$$

For instance, one ton of high entropy iron ore and one ton of low entropy coal (x) can be processed into low entropy iron (y) and high entropy greenhouse gases (b) with the constraint that the mass of y plus the mass of b equal two tons.

In today's world humans dominate the Earth's ecosystem which has consequences for land and water resources, biochemical and nitrogen cycles, the use of synthetic chemicals, and biotic changes [241]. Approximately 10–15% of the Earth's land surface is occupied by row-crops or urban and industrial activities. An additional 6–8% of land has been converted to pastures grazed by domestic animals. Further land fragmentation has been caused by highways,

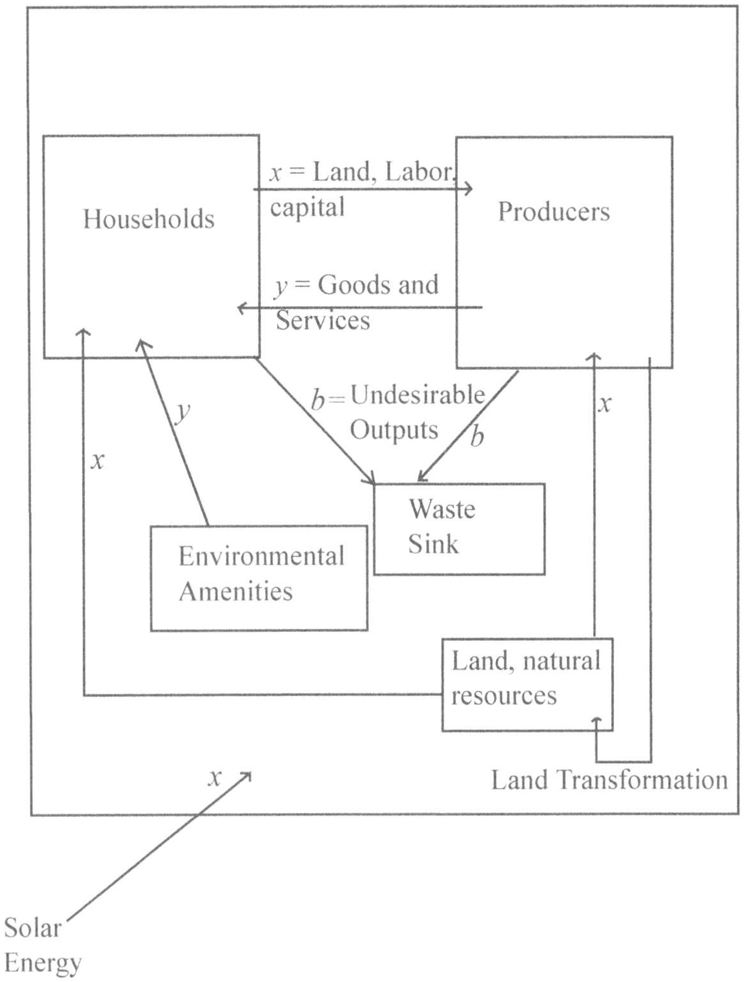

Figure 1.2: Human Activity and the Environment.

railways, and urbanization. In total, 39–50% of land has been transformed or degraded. Land transformation caused by fires and tilling contributes about 20% to carbon dioxide and other greenhouse gases such as methane and nitrous oxides. Oceans have also been transformed as 60% of the world's population lives within 100 km of a coast. While humans use only 8% of the primary production of oceans, they use 25–35% of up-swelling and shelf areas. Approximately 22% of marine fisheries are overexploited and 44% more are at their biological limits. Of freshwater resources, humans use one half of freshwater runoff, with 70% of that runoff used by agriculture. Only 2% of rivers in the United States are unimpeded by dams or diversions. The biochemical cycle has been altered with CO_2 emissions increasing from 280 $\pm$10 parts per million around 1750 to 391 ppm in 2011 as a consequence of fossil-fuel burning, timber harvesting, and plowing.

Human actions also alter the nitrogen cycle. For nitrogen to be of use to plants, it must be fixed (combined) with carbon, hydrogen, or oxygen. Too much nitrogen fixation leads to an increase in nitrogen oxides, acid rain and smog, and increases in fluxes of reactive nitrogen. Even though nitrogen fixation increases the productivity of an ecosystem, it tends to decrease biological diversity. For example, algal blooms caused by too much nitrogen fixation lead to oxygen depletion results in fish kills. Biotic changes caused by land transformation and invasive species have caused some species to go extinct.

As the world's economy industrialized there was a rapid decline in native species' diversity in large marine ecosystems [261]. Healthy ocean ecosystems offer biodiversity, provide ecosystem services, and help reduce various kinds of risk. Increased biodiversity is positively associated with species' recovery in overfished large marine ecosystems [261]. In addition, healthy ecosystems provide nursery habitats consisting of oyster reefs, seagrass beds, and wetlands. Suspension feeders, submerged vegetation, and wetlands help filter and detoxify the habitat. When these ecosystems are lost, water quality declines and there is an increased occurrence of algal blooms, fish kills, beach closures, and oxygen depletion. Finally, more diverse ecosystems reduce the risk that species within the system will collapse or go extinct. Diverse ocean ecosystems also enhance recovery capability as fishermen can more easily switch from one over-fished species to another. In economic terms, increased diversity provides a kind of portfolio diversification so that a collapse in one fish stock is partially offset by fishers switching to another fish stock. In addition, with a more diverse fishery average fish catch rates (productivity) increase and the variability of the catch decreases from year to year.

Since time moves in only one direction it is important to think about the preferences of future generations and how our actions today will affect their well-being. In 1995, The Convention on Highly Migratory and Straddling Stocks was the first international fisheries treaty advocating the precautionary principle. There is usually uncertainty about how human actions affect the environment. The **precautionary principle** is a stance of taking action in the face of uncertainty about possible environmental damage. For fisheries, the precautionary principle suggests controlling fishery access and fishing methods, such as the use of drift nets, before a tragedy of the commons occurs.

Environmentalists often argue that action should be taken before such uncertainty is resolved, since the costs of not taking action may be greater than the costs of preventative action taken now. Some people have gone so far as to argue that society should have as its objective the elimination of activities where uncertain environmental damages are involved. Alternatively, some, but not all economists argue that the costs of implementing the precautionary principle might be too large relative to its benefits. That is, some environmental policies are likely inefficient. Moreover, all decisions are made with some degree of uncertainty. Since all environmental policies have some degree of uncertainty, a strict adherence to the precautionary principle would result in doing nothing. To the extent that individual and society's demand for a clean environment and well-functioning ecological system depend on income, well-intentioned but inefficient policies might lead to environmental degradation. To implement any cost-benefit test it is necessary to accurately measure the costs and benefits accruing to current and future generations from environmental policy changes. If past trends continue future generations will be much richer than current generations and that increased wealth can be used to enhance environmental amenities.

1.6 Evaluating Environmental Policy

1.6.1 Efficiency, Equity, Growth, Stability, and Flexibility

An allocation of resources refers to how resources are used. Outputs are efficiently allocated when their value is maximized so that any change in the mix of outputs results in extra costs that are greater than the extra benefits. Alternatively, inputs are efficiently allocated when it is not possible to produce the same amount of output at a lower cost. Named for the late nineteenth and early twentieth century Italian economist Vilfredo Pareto, an allocation is **Pareto efficient** if no person can be made better off by reallocating resources without making at least one other person worse off. Any change in the allocation of inputs and outputs that makes at least one person better off and nobody else worse off is said to be **Pareto improving**. Very few policy changes are Pareto improving since almost all policies create winners and losers. An alternative way to evaluate policy is to use the **Pareto compensation principle**. Here, a policy is efficiency enhancing if the benefits received by one person or group are more than enough to offset the costs borne by the losers. A social welfare function that somehow aggregates individual utilities is often necessary to implement the Pareto compensation principle, a topic taken up in Chapter 11.

Efficiency is only one criterion for evaluating policy. Other criteria include the effects of environmental policies on equity, stability, growth, and flexibility. Marketable pollution permits have the potential to abate pollution at a lower cost than command and control regulations and lead to gains in efficiency. Greater efficiency allows larger amounts of pollution to be abated for a given budget. On the other hand, marketable pollution permits have been criticized because many older polluting factories are located in decaying neighborhoods where firms find it cheaper to buy permits rather than clean up toxic waste sites or reduce emissions. Thus, even if a larger amount of pollution is cleaned up in a higher income neighborhood, some people might argue against the policy on equity grounds. Similarly, other people have criticized corporations for building polluting manufacturing plants in minority and low income neighborhoods, even if the new manufacturing plant results in lower costs.

Future generations should also be accounted for when we take actions today. The notion of **sustainability** implies that our actions taken today should not leave future generations worse off. If one can appropriately sum up the welfare received by all future generations, then sustainability would mean that any action taken today does not reduce that welfare. Care should be taken in adopting norms regarding resource use since the value of resources tends to change over time as new technologies become available. As Barnett and Morse note "the particular resources with which one starts increasingly become a matter of indifference. The reservation of particular resources for later use, therefore, may contribute little to the welfare of future generations. ... Resource reservation, by limiting output, and thereby research, education, and investment, might even diminish the value of the social heritage" [18, p.11–12].

As economies grow, how will the flow and stock of pollutants change and how will individual and societal demands for environmental amenities change? When individuals have met their basic needs for food, clothing, and housing, will they demand more environmental amenities such as clean air, clean water, and biodiversity and demand fewer material goods and services or will their demand for material goods increase and lead to greater environmental degradation? In Chapter 4 we will examine the theory and evidence of the effects of economic growth on equity and the environment.

Policies should also be flexible enough to incorporate new information about the costs and benefits of human activity on the environment as it becomes available. Markets tend to be flexible in regard to exogenous changes. Prices incorporate information available to thousands of individuals. Rising prices send a signal to consumers that the good or service is becoming more scarce relative to their wants; information that can be immediately acted upon. On the other hand, government bureaucracies tend to respond slowly in incorporating new information into their decision making. However, should the whims of people via the market process be catered to? Such whims might include short-lived preferences that cause irreversible environmental damage. For example, high demands for bluefin tuna prized in making sushi have caused precipitous and perhaps irreversible declines in bluefin tuna stocks as fishers seek to take advantage of high prices. It might be that the reason we dedicate parcels of land as wilderness areas to be forever wild is because we don't want land use to change as prices change.

1.6.2 "Law" of Unintended Consequences

Quite often when individuals pursue their own self-interest, the effect is to simultaneously make other people better off. This idea is best represented by Adam Smith's metaphor of an "invisible hand" guiding resources toward their highest and best use. Markets work well when we don't have to worry about the effects of pollution, species' extinction, and the effects of one's actions on third parties, but they tend to break down when things like pollution or bio-diversity matter. On the other hand, well-intentioned government policies can also yield adverse unintended consequences.

Alex Tabarrok [232] writes that "the law of unintended consequences is what happens when a simple system tries to regulate a complex system. The political system is simple, it operates with limited information (rational ignorance), short time horizons, low feedback, and poor and misaligned incentives. Society in contrast is a complex, evolving, high-feedback, incentive-driven system."

Most environmentalists consider the Endangered Species Act (ESA) an important signal of society's intent to protect endangered species. However, by taking private land to protect endangered species, the ESA provides private landowners an incentive to not reveal that they have any endangered species on their land, potentially harming the species. For instance, suppose that you own land that harbors the endangered black-footed ferret and your plans are to develop a subdivision on the land. If you reveal your discovery of the ferret to wildlife authorities you have an increased probability of forfeiting your investment in the land so that the newly discovered ferret habitat can be protected. If the government does not compensate you for your loss, what will you do? In the face of such a Faustian bargain landowners who might like to be conservation-minded have adopted the strategy of "shoot, shovel, and shut-up." Clearly this kind of behavior is an unintended and adverse consequence of the ESA.

As another example, some cattle ranchers in Missouri allow their cattle to roam freely and wade and drink in Ozark streams and rivers, thereby leading to a decline in water quality. To prevent this activity the Missouri Department of Conservation decides to provide a subsidy to cattle ranchers who install windmills or other kinds of water pumps if they agree to keep their cattle out of the river. At first glance this policy might seem to enhance the environment. But what if the subsidy provides the incentive for existing landowners to become cattle ranchers to capture the subsidy? Now we have a case where the well-intentioned policy increases the number of cows, and although they are no longer wading in the river, the algal blooms caused from the runoff of manure leads to fish kills. This is the kind of off-setting behavior that policy makers must be wary of when formulating environmental policy.

A third example concerns US wildfire suppression efforts. When wildfire suppression is funded by federal and state governments, those efforts act as a kind of fire insurance for private developers who might not develop an area if they had to privately bear the costs of fire fighting and suppression. Researchers [189] have found that in the western United States, development is greater on lands benefiting from federal fire suppression efforts. Similarly, crop and flood insurance programs encourage farming and development in areas that might have been avoided were it not for a third party—the federal government—providing subsidized insurance.

These examples of unintended consequences are meant to illustrate the potential for harmful, as well as beneficial, effects when well-intentioned environmental policies are put in place. Almost all policies, even those that help protect the environment, involve trade-offs. Any policy for which there are no costs would have been implemented long ago, since no one would object to it. Good intentions are not enough to ensure good policy. Even though unintended consequences occur with new policies, they are not necessarily unpredictable. If we keep in mind the potential costs as well as the benefits of alternative environmental policies, the potential to improve societal welfare will be improved. Finally, we should recognize that policy makers must balance many trade-offs between alternative societal goals, of which concern over the environment is only one argument.

1.7 Limits to the Economic Approach

Thomas M. Power [202, p. 1087–1088] emphasizes the limits of the economic approach to evaluating policies related to the environment. First, "the realm of economics is limited to settings where trade-offs are considered culturally and ethically appropriate." For instance, in settings such as loyalty, religion, principles to live by, and basic human rights, it is difficult if not impossible to consider trading off one principle for another principle. Second, "economic values are relative values stated in barter or trade-off terms." This approach to determining values ignores metaphysical or philosophic values. Third, economists determine values at the margin, which might not be appropriate when examining issues such as the extinction of a species or the irreversible loss of habitat or policies that allow catastrophic climate events to occur. Fourth, economists are not so good at evaluating long-term events since preferences and institutions change over time. Fifth, public policy debates frequently focus on equity, while economists tend to be more comfortable in addressing efficiency issues. Sixth, "economics is self-consciously amoral. It accepts peoples' preferences as they are." However, some research suggests that individual preferences, at least those stated in public, can be molded by the group of people an individual identifies with. Thus, individuals like Paul Ehrlich, who called attention to what he perceived as environmental problems, helped to change individuals' preferences toward reducing environmental degradation. Finally, recent work in behavioral economics suggests that individuals might not be the rationally calculating machines that are assumed by neoclassical economic assumptions such as firms' pursuit of maximum profits and consumers' pursuit of utility [122].

These limits do not mean that economics has no role to play in evaluating economic growth and environmental policies. Somehow, various policies will be chosen and implemented, whether by a dictator, a bureaucrat, some form of democratic voting, by custom and tradition, or by a group of local citizens trying to pursue their own interests when those interests are affected by the group. Economics provides a way of systematically thinking about how various policies and events come about and how a different set of policies might make things better or worse off.

1.8 Questions to Ponder

1. How is the economist's notion of scarcity related to the 1st and 2nd laws of thermodynamics?
2. Describe the precautionary principle as it relates to the environment.
3. In 1990, a United Kingdom White Paper stated that the precautionary principle should only be applied "if the balance of likely costs and benefits justify it." Explain.
4. Why did policy makers at The Rio Summit adopt the precautionary principle and state that it should be applied to all countries, but only according to their capabilities?
5. A law is passed that attempts to encourage people to recycle to reduce the waste stream. The law stipulates that people may now throw away only one government approved can of trash each week. To help them, government also institutes a day of each week in which people can, at no money cost, set recyclables out for pickup. What "unintended consequences" might occur with such a policy?
6. Did the highest and best use of the land in northern Alaska dedicated as the Arctic National Wildlife Area change after the September 11, 2001 terrorist attacks?
7. When evaluating environmental policies we should take into account the efficiency of the policy, the likely equity effects of the policy, how the policy affects economic growth, and the likely effects of the policy on political and economic stability. Explain each of these policy criteria and tell why they are important in evaluating environmental policy.
8. After losing the bet with Julian Simon, several of Paul Ehrlich's

colleagues proposed another bet. In *The Ultimate Resource* 2, Simon offered this blanket wager to Ehrlich: "I'll bet that just about any environmental and economic trend pertaining to basic human material welfare (though not, of course, the progress of this group compared to that one) will show improvement in the long run." Paul Ehrlich and Steven Schneider offered a series of 15 items they were willing to bet would worsen over the next decade. The list of criteria included:

- The three years 2002–2004 will on average be warmer than 1992–1994.
- There will be more carbon dioxide in the atmosphere in 2004 than in 1994.
- There will be more nitrous oxide in the atmosphere in 2004 than 1994.
- The concentration of ozone in the lower atmosphere (the troposphere) will be greater in 2004 than in 1994.
- Emissions of the air pollutant sulfur dioxide in Asia will be significantly greater in 2004 than in 1994.
- There will be less fertile cropland per person in 2004 than in 1994.
- There will be less agricultural soil per person in 2004 than 1994.
- There will be on average less rice and wheat grown per person in 2002–2004 than in 1992–1994.
- In developing nations there will be less firewood available per person in 2004 than in 1994.
- The remaining area of virgin tropical moist forests will be significantly smaller in 2004 than in 1994.
- The oceanic fisheries harvest per person will continue its downward trend and thus in 2004 will be smaller than in 1994.
- There will be fewer plant and animal species still extant in 2004 than in 1994.
- More people will die of AIDS in 2004 than in 1994.
- Between 1994 and 2004, sperm counts of human males will continue to decline and reproductive disorders will continue to increase.
- The gap in wealth between the richest 10% of humanity and the poorest 10% will be greater in 2004 than in 1994.

Julian Simon declined the bet. What differences exist between the ways Simon and the ecologists think about what determines human welfare?

2

Review of Microeconomic Tools

Fundamentally, in a system where the knowledge of the relevant facts is dispersed among many people, prices can act to coördinate the separate actions of different people in the same way as subjective values help the individual to coördinate the parts of his plan.

F.A. Hayek [103, p.526]

2.1 Markets and Prices

Economists study the behavior of buyers and sellers and a key to understanding this behavior is that in the face of scarcity, people respond to incentives. We will assume that individuals try to maximize their well-being (utility) and that business firms try to maximize profits. Households gain utility from the goods and services they consume, including environmental amenities, but experience disutility from pollution and from supplying labor to firms. Business firms earn revenue from the sale of goods and services to households but incur costs in hiring inputs.

At the simplest level, an economic agent's behavior depends on the costs and benefits of any action. When the cost of an action increases, economic theory predicts that less of the action will be undertaken. Likewise, when the benefits of an action increase, economic theory predicts that more of the action will be undertaken. Building on the predicted response of economic agents to changes in costs or benefits leads to the "laws" of supply and demand. Households are demanders in the market for goods and services, but are suppliers of inputs such as land, labor, capital, and entrepreneurship. Likewise, business firms are demanders of inputs, but are suppliers of goods and services.

2.1.1 Utility Theory

To study consumer choices between alternative competing wants economists rely on utility theory. Utility measures the amount of satisfaction an individual receives from consuming a given bundle of goods and services and economic theory only requires utility to be ordinal; i.e., consumers can rank various bundles according to their preferences, but are not required to determine whether or not they like one bundle twice or three times as much as another bundle as would a cardinal utility function. Various assumptions are employed in utility theory: in general, more desirable goods are preferred to less and fewer undesirable goods (pollution) are preferred to more, substitution between two or more goods is possible, and preferences are subject to diminishing marginal rates of substitution.

Suppose an individual consumes two goods—an environmental good like water (good x) and other goods and services such as restaurant dinners, blue jeans, and housing (good y). The two goods have prices p_x and p_y and the quantities the individual consumes are represented by x and y. An indifference curve represents all the various quantities of the two goods that yield the same level of utility. In general, as an individual consumes more of one good they must consume less of the other good to remain at the same level of utility. For instance, for the utility function $u = x \times y$ the equation of the indifference curve is found by solving for the quantity of one of the goods, say good y: $y = \frac{u}{x}$. Table 2.1 gives the alternative quantities of the two goods that yield the same level of utility, say $u = 10$.

Table 2.1: Individual Indifference Set for $u = x \times y$

u	x	y	$MRS = \lvert\frac{\triangle y}{\triangle x}\rvert$
10	1	10	x
10	2	5	5.00
10	3	3.33	1.67
10	4	2.5	0.83
10	5	2	0.50
10	6	1.67	0.33
10	7	1.43	0.24
10	8	1.25	0.18
10	9	1.11	0.14
10	10	1	0.11

An individual's marginal rate of substitution (MRS) equals the amount of one good (y) the individual is willing to give up to obtain an additional unit of another good: $MRS = \lvert\frac{\triangle y}{\triangle x}\rvert$. The slope of the indifference curve equals the consumer's MRS. In general as an individual consumes more of one good, in this case water (x), their MRS falls. Figure 2.1 depicts an indifference map which shows indifference curves for various levels of utility. The indifference curves have a negative slope, are convex to the origin, and do not intersect. In addition, notice that the indifference curves are very steep when very small amounts of good x (water) are consumed. That is, since water is essential for life, individuals are willing to give up extremely large quantities of other goods to obtain the first few units of water. In contrast, the indifference curves flatten out as more of good x is consumed meaning that after consuming a lot of good x (water) the individual is only willing to give up a small amount of good y for another unit of good x.

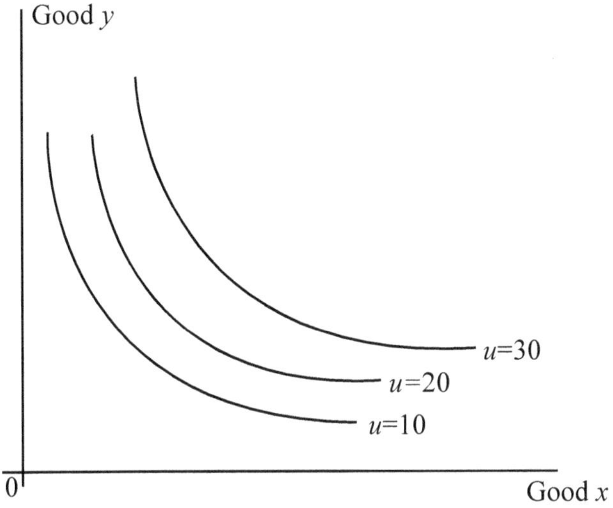

Figure 2.1: Indifference Map.

Individuals have to make choices because of scarcity. When individuals have an income (I) or budget to spend on water and other goods and services their choices must satisfy their budget constraint: $I \geq p_x x + p_y y$. In Figure 2.1 we arbitrarily placed good y on the vertical axis and good x (water) on the horizontal axis; so to graphically depict the budget set we solve the budget constraint for the quantities of good y that can be consumed given income (I), prices and quantities of good x: $y \leq \frac{I}{p_Y} - \frac{p_x}{p_y}x$. The vertical intercept, $\frac{I}{p_y}$, gives the quantity of good y that can be consumed if none of good x is consumed. The slope, $-\frac{p_x}{p_y}$, represents the quantity of good y that must be given up for one additional unit of good x. For instance, if $p_x = \$3$ and $p_y = \$1$ then the consumer must give up 3 units of good y for an additional unit of good x.

How do individuals choose among their alternative competing wants? In general, if the amount of good y an individual is willing to trade for another unit of good x (the MRS) is greater than the amount of good y that has to be given up in the marketplace ($\frac{p_x}{p_y}$) the individual chooses more x and less y. If instead, the $MRS < \frac{p_x}{p_y}$ the individual chooses less of good x and more of good y. The consumer maximizes utility subject to her budget constraint when $MRS = \frac{p_x}{p_y}$. Such an equilibrium is depicted in Figure 2.2 at point E where the indifference curve is tangent to the budget constraint.

A **numeraire** is a good or service in which all other goods are priced. If we let good y (other goods) be the numeraire and set its price equal to \$1, the price ratio becomes $\frac{p_x}{p_y} = \frac{p_x}{\$1} = p_x$. In addition, with good y as the numeraire, the

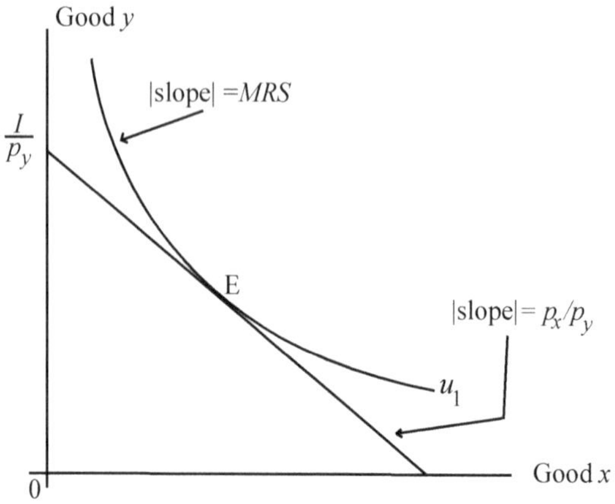

Figure 2.2: Utility Maximization.

marginal rate of substitution can be interpreted as the consumer's willingness to pay for another unit of good x or as a money measure of the marginal benefits (MB) they receive from an additional unit of good x. In the next section we introduce the consumer's demand function which is related to their willingness to pay and derived from the underlying utility function.

2.1.2 Demand

The "law" of demand states that at higher relative prices less quantity will be demanded than at lower relative prices, *ceteris paribus* (holding everything else constant). This law holds because of the income effect and the substitution effect, or alternatively, because individuals have diminishing marginal rates of substitution. The income effect of a price change means that as the relative price changes an individual's real purchasing power changes, enabling them to buy more or limiting them to less. The substitution effect implies that when the relative price increases, an individual will tend to buy less as they instead consume lower priced substitutes.

Society can choose to allocate scarce resources by custom and tradition, by government and the political process, or by markets and the price mechanism. In this chapter we want to think about how markets allocate resources using the price mechanism. Suppose two people, Alice and Bill, have to decide how to allocate a scarce resource like water. Each person wants to use the water in different ways and places different priorities on each alternative use. The different priorities are consistent with the idea of diminishing marginal rates of substitution. Alice and Bill would normally choose to consume water for higher priority uses before lower priority uses. Alice's and Bill's priorities and amounts of water wanted for each use are listed in Table 2.2.

If water is free, Alice would like to use 29.5 (=2+5+2+18+2.5) units of

Table 2.2: Uses and Priorities for Water

Alice			Bill		
Priority	Use	Amount	Priority	Use	Amount
1	Drinking	2	1	Drinking	4
2	Bathing	5	2	Wash Car	8
3	Water plants	2	3	Bathing	5
4	Water Grass	18	4	Maintain pool	10
5	Wash Dog	2.5			

water and Bill would like to use 27 (=4+8+5+10) units. Suppose the amount of water available to Alice and Bill is 26 units and the price of water is zero. Since Alice and Bill want 56.5 units but only 26 units are available, a way has to be found to decide how much water each person will receive. One possibility would be that each person could be given 13 units. If Alice chooses her highest priority uses first she would drink, bathe, water houseplants, and use her other six units of water to partially water the grass. With 13 units of water Bill would drink, wash the car, and take a sponge bath with his one remaining unit of water. Another possibility is that rules could be determined by a dictator or by some form of voting. One rule might be to disallow the use of water for watering houseplants and grass or maintaining pools. In this case, Alice would drink, bathe, and wash the dog, and Bill would drink, wash his car, and bathe. Another rule to limit consumption to 26 would be to disallow watering grass and washing cars and dogs. Still another way to limit use to 26 units would be to outlaw the use of water for bathing, washing cars and dogs, and maintaining pools. In fact, it is possible to choose many different rules to limit water use to 26 units and with each rule, Alice and Bill would receive different amounts of water.

As an alternative to rules, the available water might be allocated through a market by the price mechanism. The market process allocates water to the person who is willing to pay the most. We assume that higher priority uses elicit a higher willingness to pay than lower priority uses. Alice's and Bill's willingness to pay (WTP) per unit of water are listed in Table 2.3.

For Alice, drinking has a higher priority than bathing, so her WTP is higher for drinking than for bathing. Similar reasoning holds for Bill's WTP for alternative uses. The quantity of water each person demands is inversely related to price. Suppose the price of water is \$1. At that price, Alice would be willing to buy 27 total units of water: 2 units for drinking, an additional 5 units for bathing, and an additional 20 units for watering plants and grass. However, she would forego washing the dog since she is only willing to pay \$0.75 for that use. Similarly, at a price of \$1, Bill would be willing to buy 27 units of water: 4 units for drinking, 8 units for washing the car, 5 units for bathing, and 10 units for maintaining his pool. The market demand schedule

Table 2.3: Willingness to Pay for Water

Alice			Bill		
Priority	Amount	WTP	Priority	Amount	WTP
Drinking	2	\$5	Drinking	4	\$6
Bathing	5	\$4.5	Wash Car	8	\$5
Water Plants	2	\$3.8	Bathing	5	\$4
Water Grass	18	\$1	Maintain Pool	10	\$1
Wash Dog	2.5	\$0.75			

for water equals the sum of the individual demand schedules and is presented in Table 2.4. Together, Alice and Bill would be willing to buy up to 54 units of water at a price of \$1. If the price of water were to rise to \$4.50, Alice would demand 7 units and Bill would demand 12 units, for a total quantity demanded of 19. Thus, as prices rise each person foregoes some lower priority uses and the market quantity demanded falls.

Table 2.4: Market Demand for Water

Price	Quantity Demanded
\$0.75	56.5
\$1.00	54
\$3.80	26
\$4.00	24
\$4.50	19
\$5.00	14
\$6.00	4

Consider the case where each person receives an equal amount of water (13 units) when the total amount of water available is 26. At the margin, Alice would use the 13th unit of water for watering grass (perhaps only the front yard, but not the back) and Bill would use his 13th unit to take a sponge bath. Given the amounts of water each receives, Bill would be willing to pay up to \$4 for an additional unit while Alice only values her water at \$1 per unit and would sell it at any price above \$1. Since Bill's willingness to pay is greater than Alice's willingness to accept, a trade could take place. Suppose the trade takes place at a price of \$3.00. At that price, Bill receives water at \$1 less than he was willing to pay and Alice receives \$2 more for her water than the value she placed on it. Thus, the equal allocation of water is Pareto inefficient and the trade makes both Alice and Bill better off and corresponds to a Pareto improving reallocation of water.

A similar argument could be made for the rule which outlaws watering grass and maintaining pools. In this case, higher valued uses are foregone as Alice would use her 8th and 9th units of water for washing the dog which she values at only $0.75, while Bill uses his 13th unit of water to bathe which he values at $1.

2.1.3 Supply

So far we have assumed that the amount of water available is 26 units, but the quantity of water sellers are willing to offer for sale is positively related to its price, *ceteris paribus.* Table 2.5 lists the alternative sources of water that can be used to satisfy the demands of Alice and Bill. Sellers have an incentive to supply water from the lowest marginal cost source first. As the cheapest sources of water are exhausted, sellers have to turn to sources that are more expensive. Only if prices are high enough to justify the added expense will sellers tap those more expensive sources of water.

Table 2.5: Costs of Water from Alternative Sources

Water from:	Amount Available	Source Cost	Cumulative Amount	Marginal Cost/Unit
Aquifer (Upper)	10	$5.60	10	$0.56
Aquifer (Lower)	19	$14.25	29	$0.75
Lake (10 miles distant)	6	$4.86	35	$0.81
River	12	$11.16	47	$0.93
Lake (50 miles distant)	7	$7.00	54	$1.00
Lake (100 miles distant)	20	$24.00	74	$1.20
Desalinization Plant	280	$1120.00	354	$4.00

When markets are competitive sellers take prices as given and choose to supply a given quantity of water if the price is high enough to cover the marginal cost of production. For instance, if the price of water is $0.93 per unit, sellers will have the incentive supply 29 (=10+19) units of water from the aquifer, an additional 6 units of water from the lake that is 10 miles away, and an additional 12 units of water from the river, for a total of 47 units of water. If the price rises to $1, it will become efficient to produce an additional 7 units of water (for a total of 54) from the lake that is a distance of 50 miles. Table 2.6 presents the supply schedule and illustrates that as price increases a larger quantity of water is supplied. Alternatively, if sellers are to be induced to offer more water for sale they need to be compensated by a higher price since the marginal cost of producing additional units of water increases.

Table 2.6: Market Supply for Water

Price	Quantity Supplied
\$0.56	10
\$0.75	29
\$0.81	35
\$0.93	47
\$1.00	54
\$1.20	74
\$4.00	354

2.1.4 Trade Creates Value

When people use competitive markets to allocate scarce resources prices provide information to both the buyers and the sellers about relative scarcity. In Table 2.7 we see that at a price of \$0.75 per unit of water, quantity demanded is greater than quantity supplied. When a shortage exists, price tends to rise. As price rises, some consumers cut back on the amount of water they would use in lower priority uses. At the same time, as price rises, water that was once too costly to produce becomes profitable as the price received more than covers the marginal cost of production. Thus, a rising price for water works to reduce the shortage by rationing the available water among its competing uses and, simultaneously, signals producers to increase production. Alternatively, at a price of \$4.00, quantity demanded is less than quantity supplied and a surplus exists, leading price to fall. A lower price signals water producers that their resources are now more valuable in producing something else. At the same time, the lower price gives consumers the incentive to increase water consumption by engaging in activities, like washing the dog or maintaining the pool, that before were of too low a priority to justify the price. At a price of \$1, the water market depicted in Table 2.7 reaches an equilibrium quantity of 54 units.

Voluntary trade creates value for both the buyer and the seller. Trades that create value are Pareto improving since at least one party to the trade is made better off (and usually both parties to the trade) and the other party is made no worse off. Consumer's surplus is the value created by the trade that accrues to the consumer. Formally, consumer's surplus equals the difference between willingness to pay and price summed over all trades: $CS = \sum_{trades} (WTP - P)$. For instance, referring back to Table 2.3 when water has a price of \$1 per unit,

Table 2.7: Determining the Equilibrium Price and Quantity

Quantity Demanded	Price	Quantity Supplied
	\$0.56	10
56.5	\$0.75	29
	\$0.81	35
	\$0.93	47
54	\$1.00	54
	\$1.20	74
26	\$3.80	
24	\$4.00	354
19	\$4.50	
14	\$5.00	
4	\$6.00	

Alice and Bill's consumer's surplus is:

$$CS = \underbrace{(6-1)\times 5}_{\text{B drinks}} + \underbrace{(5-1)\times(8+2)}_{\text{B's car wash, A drinks}} + \underbrace{(4.5-1)\times 5}_{\text{A bathes}} + \underbrace{(4-1)\times 5}_{\text{B bathes}} + \underbrace{(3.8-1)\times 2}_{\text{A waters plants}} + \underbrace{(1-1)\times(10+18)}_{\text{B's pool, A's grass}} = \$98.1. \quad (2.1)$$

Figure 2.3 depicts the market demand schedule and consumer's surplus at a price of \$1. Higher prices reduce consumer's surplus and lower prices increase consumer's surplus. The quantities for Alice's and Bill's uses of water are discrete. As the change in quantity for each alternative use of water approaches zero, consumer's surplus is approximately equal to the area under the demand curve but above price. That is, in the context of Figure 2.3, each of the non-shaded triangles would also become part of consumer's surplus.

Producer's surplus is the value of the trade(s) that accrue to the seller and equals the difference between price and the marginal cost of production summed over all trades: $PS = \sum_{trades} (P - MC)$. When the price of water equals \$1, producer's surplus is:

$$PS = \underbrace{(1-0.56)\times 10}_{\text{Upper Aquifer}} + \underbrace{(1-0.75)\times 19}_{\text{Lower Aquifer}} + \underbrace{(1-0.81)\times 6}_{\text{Lake 10 miles}} + \underbrace{(1-0.93)\times 12}_{\text{River}} + \underbrace{(1-1)\times 7}_{\text{Lake 50 miles}} = \$11.13. \quad (2.2)$$

Figure 2.4 depicts the supply/marginal cost curve of water and producer's surplus. Higher prices increase producer's surplus and lower prices reduce pro-

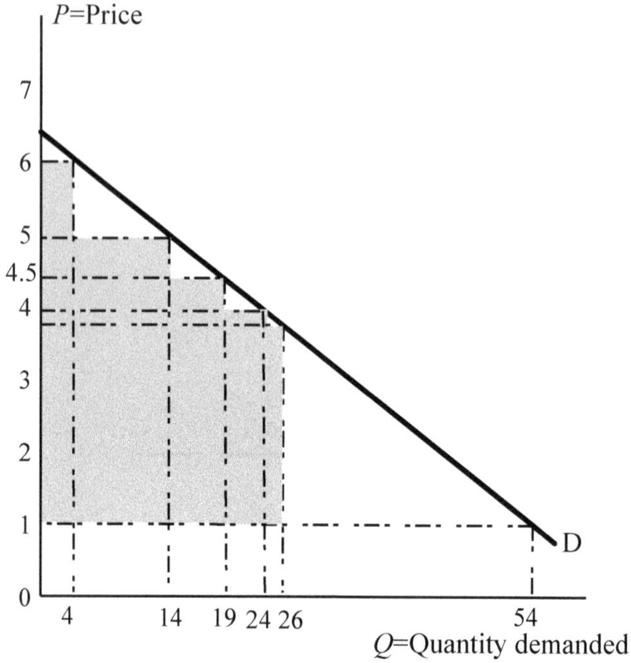

Figure 2.3: Consumer's Surplus.

ducer's surplus. Similar to consumer's surplus, producer's surplus is approximately equal to the area above the supply curve but below price. Thus, each of the non-shaded triangles in Figure 2.4 would become part of producer's surplus for very small changes in output.

Sometimes it will be useful to represent the demand schedule by an equation describing the relation between price and quantity demanded, holding other determinants of demand constant. For instance, the demand schedule represented in Table 2.7 can be represented by the equation: $P = 6.4 - 0.1 \times Q_d$ where P represents willingness to pay and Q_d represents quantity demanded. Similarly, the supply schedule in Table 2.7 can be represented by $MC = 0.46 + 0.01 \times Q_s$, where MC is the marginal cost of production and equals the producer's willingness to accept and Q_s is the quantity supplied. The equilibrium price and quantity are found by equating the consumer's willlingness to pay with the producer's marginal cost: $P = MC$. Substituting the demand and marginal cost equations into the equilibrium condition yields $6.4 - 0.1 \times Q = 0.46 + 0.01 \times Q$ and solving for the equilibrium quantity yields $Q_E = 54$. Substituting the equilibrium quantity into either the willingness to pay function or the marginal cost function gives the equilibrium price: $P_E = 6.4 - 0.1 \times 54 = 0.46 + 0.01 \times 54 = \$1$.

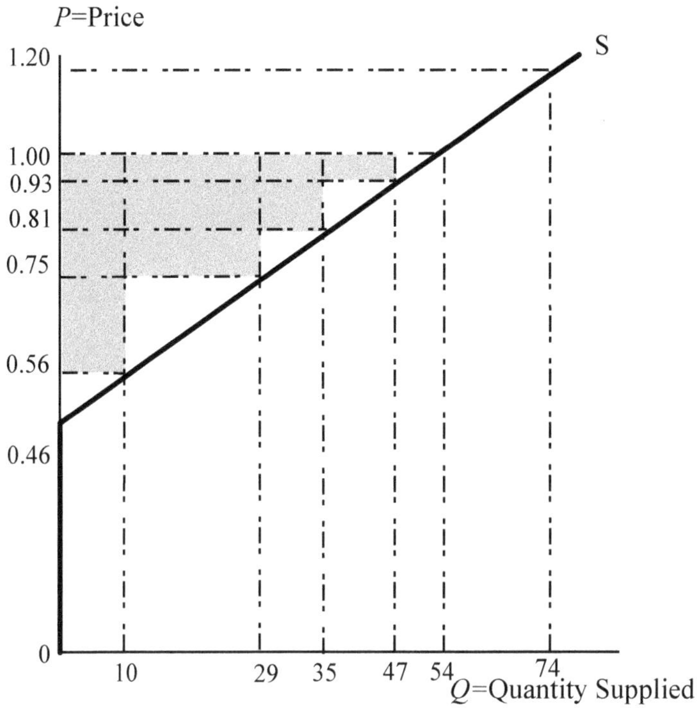

Figure 2.4: Producer's Surplus.

2.1.5 Technical Note on Consumer's Surplus

The theoretically correct way to calculate consumer's surplus is to calculate the area under what is called the compensated demand curve. The compensated demand curve holds utility constant and gives the alternative quantities demanded at different relative prices. Therefore, it looks at changes in quantity demanded that are due only to the substitution effect, not the income effect. To take an example, assume that Mark is constrained to spend no more than $50 on water and other goods. Water costs $4 per unit and other goods costs $1 per unit. Mark is indifferent between bundles A and B, where bundle A has 7 units of water and 22 units of other goods and bundle B has 9 units of water and 17 units of other goods. Since the cost of acquiring bundle B ($53) is greater than the cost of acquiring bundle A ($50), Mark chooses bundle A. Now, suppose the price of water falls to $1. Since bundle A now costs $29 and bundle B now costs $26 Mark chooses bundle B instead. Given that he was spending $50 before on bundle A, Mark saves $24 which is the true measure of consumer's surplus. With his budget of $50, Mark might now choose 13 units of water and 37 units of other goods. The total effect of the price change is for Mark's quantity demanded to increase from 7 units to 13 units as illustrated in Figure 2.5. The substitution effect causes Mark to buy 2 extra units of water, while the income effect causes Mark to buy 4 extra units of water. The area under the compensated demand curve represented by the trapezoid AEBC equals \$24 and corresponds to the true measure of the change in consumer's surplus due to the lower price of water. In contrast, the area of trapezoid AEFC equals \$30 and is an over-estimate of the increase in Mark's consumer's surplus due to the price change.

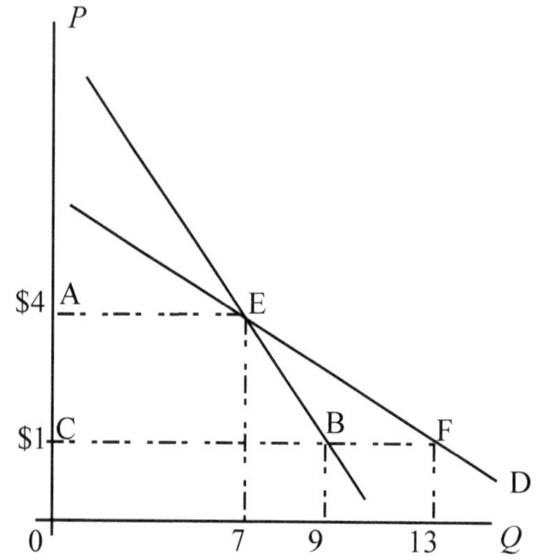

Figure 2.5: Compensated and Uncompensated Demand Curves.

2.2 Changes in Demand and Supply

Besides price, there are other determinants of demand and supply. Changes in a consumer's tastes and preferences, income, expectations of future prices, taxes and subsidies, and changes in the price of related goods—substitutes and complements—will change demand holding the price of the good constant. When demand increases consumers are willing to pay more for the same quantity or, alternatively, are willing to purchase more at the same price. On the supply side, technological progress/regress, changes in the prices of resources used to produce the good, taxes and subsidies for the producer, and producer expectations of future prices will change supply. In addition, changes in the prices of goods that are jointly produced will cause supply to change. For instance, electricity and emissions of sulfur dioxide, nitrogen oxide, and carbon dioxide are jointly produced. If emissions were taxed, electric utilities would rationally try to reduce emissions to lower their tax liability, an action that would also reduce the amount of electricity generated. When supply increases sellers are willing to sell more quantity at the same price or, alternatively, are willing to accept a lower price for the same quantity. Increases (decreases) in demand or supply cause the respective curve to shift to the right (left).

Hans-Werner Sinn [221] has formulated what he refers to as "The Green Paradox." This paradox can be easily understood with the aid of a supply/demand model of the market for fossil fuels, say oil. When oil prices are low consumers drive gas guzzling autos and spend little on insulating their house. At high oil prices consumers drive less, add insulation to the attics of their houses, and take the bus or subway to work. This behavior results in the negatively sloped demand curve, D, in Figure 2.6. Similarly, when oil prices are low oil producers offer very little oil for sale, but when prices are high, the amount offered for sale increases. The initial equilibrium is represented by P_1 and Q_1.

Now, suppose that countries become concerned about greenhouse gas emissions resulting from the burning of fossil fuels. As a consequence, legislatures pass incentive programs to reduce fossil fuel use and countries sign on to global agreements like the Kyoto Protocol. Incentives might include tax credits for purchasing fuel efficient cars, or subsidies to consumers who install solar panels or insulation, or subsidies to producers of renewable energy such as bio-fuels, solar, and wind power. The effect of these subsidies is to decrease demand to D' in Figure 2.6. If only demand changes, then the decline in demand due to efforts to reduce fossil fuel usage should reduce the equilibrium quantity of oil.

However, Sinn argues that the suppliers of oil and other fossil fuels will also be affected. Reductions in the demand for fossil fuels can be expected to reduce future fossil fuel profits. If producers expect renewable energy substitutes to become widely and cheaply available, thereby reducing their future profits, they will rationally respond by increasing supply today and hold their wealth as money assets rather than oil resources in the ground. The decrease in demand and increase in supply will both put downward pressure on the price of fossil fuels. However, the equilibrium quantity can either increase or decrease depending on the magnitude of the increase in supply relative to the decrease in demand. Figure 2.6 shows an increase in the equilibrium quantity due to a relatively larger increase in supply than the decrease in demand and illustrates the potential "Green Paradox."

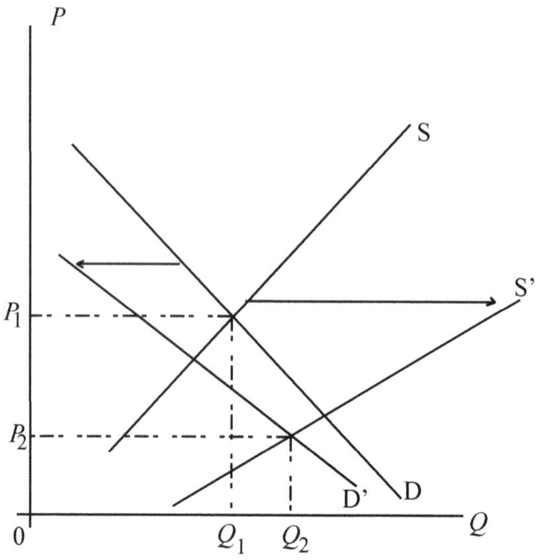

Figure 2.6: Changes in Demand and Supply.

2.3 Returns to Scale

A firm's average cost of production, ATC, equals total cost per unit of output: $ATC = \frac{TC}{Q}$. When producing goods and services the producer can generally vary output by varying the amounts of raw materials, energy, and labor, but in the short run, the producer is constrained by at least one fixed input, usually capital (K) or plant capacity. In the long run all inputs can be varied, including plant capacity. A firm produces in the range of increasing returns to scale if it increases all inputs by some proportion and output increases more than proportionally. For instance, increasing returns occur when a firm doubles all inputs and output triples. Increasing returns frequently occur because firms must be large to efficiently utilize some kinds of capital. For example, if a farmer buys a four wheel drive tractor it is more cost efficient for the farmer to be plowing and planting a crop on hundreds of acres than it is on a small plot of ground—say a quarter acre vegetable garden. When a firm experiences increasing returns to scale ATC falls as it becomes larger. As Figure 2.7 illustrates, when a firm expands capital (K) from K=1 to K=4 to K=8, each short-run ATC curve shifts to the southeast and minimum average cost falls as the firm becomes larger and produces more output.

Decreasing returns to scale occurs when all inputs are proportionally increased but output increases less than proportionally. In this case, small firms are more efficient than large ones. Decreasing returns to scale might occur because of the inefficiencies associated with operating large organizations. When

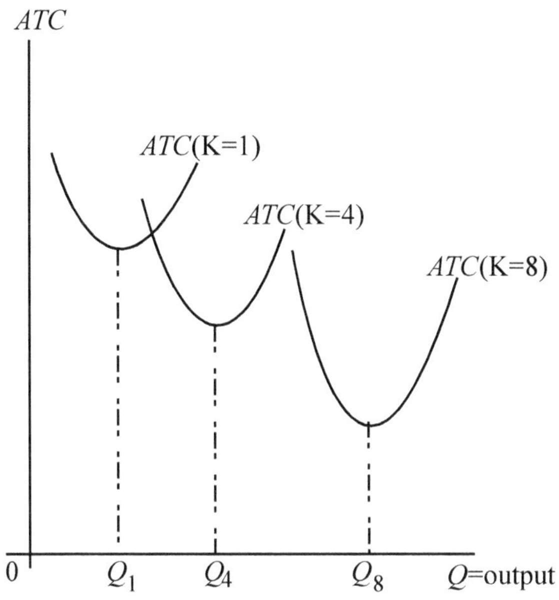

Figure 2.7: Increasing Returns to Scale.

decreasing returns to scale occur, *ATC* increases as Q increases. As Figure 2.8 shows, a consequence of decreasing returns to scale is that the short-run *ATC* curves shift to the northeast as the firm expands capital and produces more output. Here, large firms have higher per unit costs than small firms.

Finally, a firm operates in the range of constant returns to scale when a proportional expansion of all inputs causes a proportional expansion in output. As depicted in Figure 2.9 when constant returns to scale occur there are no size advantages or disadvantages. Large and small firms are equally efficient.

The competitive market process and the firm's goal of maximizing profits provides an incentive for it to choose a plant size that minimizes the long-run average cost of production. If the firm chooses a plant size that does not minimize average costs then there is the potential to increase profits by building a plant size that does. The optimal plant size is one which can produce a given output at minimum average cost. The firm has an incentive to build a plant size that minimizes ATC because if it built any other plant size and produced any other level of output it would lose money. At the equilibrium price, P_E, the firm earns a normal profit, or, a zero economic profit. The price, P_E, just covers the marginal cost of production, and the equilibrium quantity corresponds to an efficient allocation of resources: it is Pareto efficient. In addition, the firm operates at the minimum point on the long-run average cost curve and is consistent with constant returns to scale. If the firm were instead operating in the range of increasing returns to scale it could increase profits

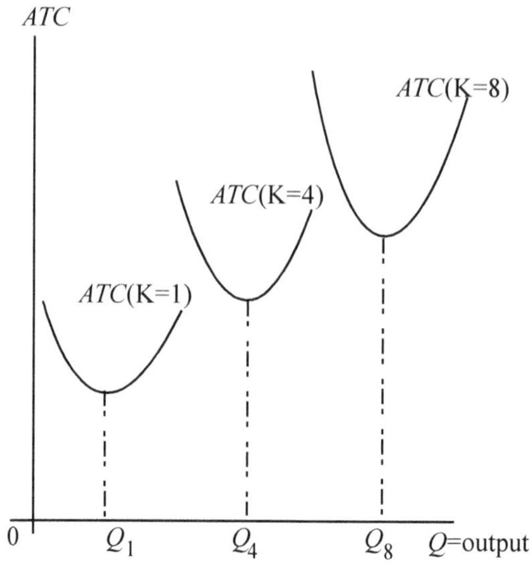

Figure 2.8: Decreasing Returns to Scale.

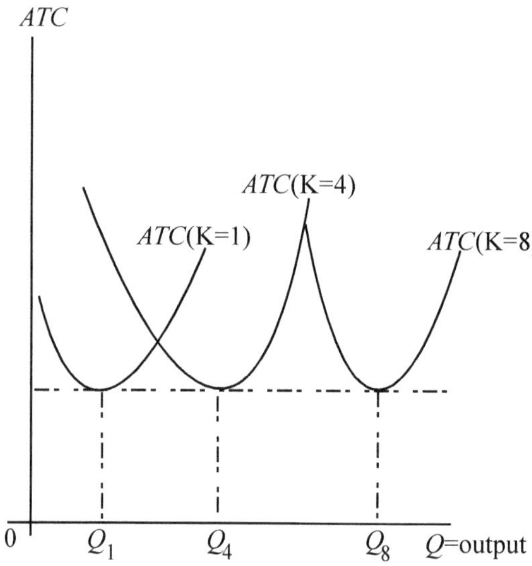

Figure 2.9: Constant Returns to Scale.

by expanding. If instead, the firm were operating in the range of decreasing returns to scale, it could increase profits by contracting (downsizing).

2.4 Equal Marginal Cost Principle

The marginal and average values of a variable are related such that if the marginal change in the variable is less than its average, the average will fall and if the marginal change in the variable is greater than its average, the average will rise. For instance, suppose an environmental organization like The Nature Conservancy owns 1000 sites that have an average of 100 different plant species per site. If the organization were to add another (marginal) site that had 125 different plant species the new average for the 1001 sites would increase. However, if the marginal site had only 90 different plant species the new average for the 1001 sites would decrease.

A similar line of reasoning holds true for the relation between MC and ATC. If MC<ATC, the ATC decreases. If MC>ATC, the ATC increases. When scale economies exist the MC of producing another unit of output is less than the ATC, and an expansion in output causes average costs to fall.

When production takes place by two or more firms and a given amount of total output is to be produced, each increment in output should be produced by the firm that has the lowest marginal cost of production. Suppose 100 units of water are to be delivered to the citizens of Dry Gulch. Two firms—Acme Water Company and Beta Water Company—can produce all or part of the 100 units of water. Acme has fixed costs of $10 and Beta has fixed costs of $4. The cost functions for the two firms are given in Table 2.8.

Table 2.8: Equal Marginal Cost Principle

Acme				Beta			
Q	TC	ATC	MC	Q	TC	ATC	MC
0	10	x	x	0	4	x	x
10	12	1.20	0.2	10	10	1.00	0.6
20	15	0.75	0.3	20	17	0.85	0.7
30	19	0.63	0.4	30	25	0.83	0.8
40	24	0.60	0.5	40	34	0.85	0.9
50	30	0.60	0.6	50	44	0.88	1.0
60	37	0.62	0.7	60	55	0.92	1.1
70	45	0.64	0.8	70	67	0.96	1.2
80	54	0.68	0.9	80	80	1.00	1.3
90	64	0.71	1.0	90	94	1.04	1.4
100	75	0.75	1.1	100	109	1.09	1.5

Since Acme can produce 100 units of water at a total cost of $75 and Beta can only produce 100 units of water at a total cost of $109 it is tempting to think that Acme should produce all 100 units of water. However, to minimize costs, each 10 unit increment in water should be produced and supplied by the firm that has the lowest marginal cost. In this case, Acme should produce the first 40 units of water. Then, Acme and Beta should both produce another 10 units since the MC to Acme from producing the 50th unit is \$0.60 and the MC to Beta from producing the 10th unit is also \$0.60. Then, both Acme and Beta should produce another 10 units each since the MC to Acme from producing the 60th unit is \$0.70 and the MC to Beta from producing the 20th unit is \$0.70. A total of 80 units of water are now produced by the two companies: 60 by Acme and 20 by Beta. To bring the total water produced to 100, both companies should produce another 10 units of water since the MC to Acme from producing the 70th unit is \$0.80 and the MC to Beta from producing the 30th unit is also \$0.80. If Acme had produced all 100 units of water the Acme's cost would have been \$75, but Beta would have incurred costs of \$4 since it would still have to pay its fixed costs bringing the total cost to the two companies to \$79. But if Acme produces 70 units of water at a cost of \$45 and Beta produces 30 units of water at a cost of \$25, the total cost to the two companies is only \$70. Thus, to minimize costs of producing a given level of output, the marginal cost of production should be the same for all producers.

As a final look at the water production problem note that Acme has higher fixed costs associated with operating a larger plant than Beta, but it has a lower marginal cost of production when each firm produces the same output. Looking at the average cost of production schedule, we can see that Acme's ATC falls and reaches a minimum of \$0.60 at 50 units of output. In contrast, Beta's ATC falls and reaches a minimum of \$0.83 at the 30 units of output. In this example the technology is characterized by increasing returns to scale. In the long run, when firms can adjust their plant size, Beta would have the incentive to build a larger plant. If Beta increases its plant size to be consistent with that of Acme, the 100 units of water could be produced at an even lower total cost, but the equal marginal cost principle would still hold. Acme would produce 50 and Beta would produce 50 and the total cost of producing the 100 units of water would be only \$60 = \$30(Acme)+\$30(Beta).

2.5 Pareto Efficiency and Equity

Recall from Chapter 1 that an allocation is Pareto efficient if it is impossible to make one person better off without making another person worse off. Under what conditions does the market process satisfy the condition of Pareto efficiency? The first fundamental welfare theorem states that when property rights are well defined over all resources and markets are competitive with price-taking buyers and sellers who have perfect information, the outcome generated by the market process will be Pareto efficient.

Although the first welfare theorem guarantees Pareto efficiency the distribution of utility amongst the population might be very unequal with a few people receiving a lot and many people receiving little. Can a more equitable distribution of utility be generated? The second fundamental welfare theorem states that given any lump-sum redistribution of resources the market process will generate a Pareto efficient outcome.

Although the two welfare theorems provide a foundation of support for the market process it is important to note that the conditions necessary for the two welfare theorems to hold are seldom met. A lack of private property rights in environmental resources such as wildlife and climate means that voluntary trade between two parties can affect the welfare of third parties, a situation referred to as externalities taken up in Chapter 8. When individuals demand goods and services that are collectively consumed (public goods) the market process is subject to a free-rider problem resulting in Pareto inefficiency. The conditions necessary for efficiency when citizens demand public goods are addressed in Chapter 7. In addition, markets are often not competitive and buyers and sellers often lack perfect information about prices and quality and might not trade often enough to generate sufficient trust with the counterparty. These kinds of transactions costs are assumed away in the first fundamental welfare theorem. Finally, lump-sum redistributions of resources are taxes and transfers that do not affect incentives. Almost all tax and transfer schemes affect incentives and can distort the market system's ability to generate a Pareto efficient outcome.

When markets fail to generate an equitable distribution of resources or a Pareto efficient outcome, government often attempts to correct the market failure. However, government prescriptions to correct market failures can generate their own kinds of inefficiency, a topic taken up in Chapter 13.

2.6 Problems

1. The foreword to Aldo Leopold's *A Sand County Almanac* is provided below.

> There are some who can live without wild things, and some who cannot. These essays are the delights and dilemmas of one who cannot.
>
> Like winds and sunsets, wild things were taken for granted until progress began to do away with them. Now we face the question whether a still higher 'standard of living' is worth its cost in things natural, wild, and free.
>
> For us of the minority, the opportunity to see geese is

> more important than television, and the chance to find a pasque-flower is a right as inalienable as free speech.
>
> These wild things, I admit, had little human value until mechanization assured us of a good breakfast, and until science disclosed the drama of where they come from and how they live. The whole conflict thus boils down to a question of degree. We of the minority see a law of diminishing returns in progress; our opponents do not.
>
> Aldo Leopold, *A Sand County Almanac*, 1949 [142, p. vii].

(a) Does Leopold see the environment (and wild things) as a luxury or necessity good? (Note: as income grows people spend an increasing proportion of income on luxuries and a decreasing proportion of income on necessities.)

(b) As the US economy continues to grow will people be willing to allocate more or fewer resources to environmental protection? (When did these wild things acquire value?)

(c) What does Leopold mean by the statement "We of the minority see a law of diminishing returns in progress; our opponents do not."

(d) Do you think that countries that prohibit or restrict free speech will be as successful in protecting environmental amenities like pasque-flowers as countries that protect free speech as an inalienable right?

2. The table below gives the market demand schedule for a vehicle entrance into Yellowstone Park. The current entry price is $20 per vehicle. Approximate the dollar value of consumer's surplus at that price.

Yellowstone National Park		
WTP	Vehicles=Q	WTP−P
$120	0	
$35	850,000	
$30	900,000	
$25	950,000	
$20	1,000,000	
$15	1,050,000	
$10	1,100,000	
$5	1,150,000	

3. The table below gives the supply schedule of open-space land in Megacity, USA. The current equilibrium price of land is $3,000 per

acre. Approximate the dollar value of producer's surplus at that price.

MC/acre	Megacity Q_s = acres	P–MC
\$500	0	
\$750	1,000	
\$1,000	2,000	
\$1,250	3,000	
\$1,500	4,000	
\$1,750	5,000	
\$2,000	6,000	
\$2,250	7,000	
\$2,500	8,000	
\$2,750	9,000	
\$3,000	10,000	
\$3,250	11,000	
\$3,500	12,000	
\$3,750	13,000	

4. The estimated demand schedule for Alaskan fishing trips is $Q_d = 2500 - P$, or alternatively, $P = 2500 - Q_d$.

   (a) If the price of a fishing trip is \$1,500, calculate consumer's surplus. (Hint: graph the line $P = 2500 - Q_d$ and then use the graph to calculate consumer's surplus.)

   (b) If the price of a fishing trip is \$1,750, calculate consumer's surplus.

   (c) As the equilibrium price increases consumer's surplus (increases or decreases).

5. Private business firms have a marginal cost of providing fish to the market given as $MC = \$750 + \$1500 \times Q_s$, where Q_s is tons.

   (a) What is the dollar value of producer's surplus if the current equilibrium price of fish is \$3,750 per ton? (Hint: graph the supply function: $MC = \$750 + \$1500 \times Q_s$ and then use the graph to calculate producer's surplus.)

   (b) What is the dollar value of producer's surplus if the current equilibrium price of fish is \$5,250 per ton?

   (c) As the equilibrium price increases, producer's surplus (increases or decreases).

6. Explain what will happen to consumer's surplus due to each of the following changes:

(a) Consumers experience an increase in income and the good is a normal good.
(b) Tastes and preferences turn in favor of the good.
(c) The price of a close substitute for the good in question increases.
(d) A per unit tax is imposed on the good and is to be paid by the demander.

7. Explain what will happen to producer's surplus due to each of the following changes:

(a) There is an advance in technology associated with producing the good.
(b) There is a decline in the price of energy, which is an input used to produce the good.
(c) There is a per unit tax imposed on the good to be paid by the supplier.
(d) Prices fall due to a decline in consumer demand.

8. In each of the following cases a firm operating in a competitive market is in short-run equilibrium (economic profits can be either positive or negative), but at least one input (capital) is fixed. Explain the adjustment process the typical firm will undertake in moving toward long-run equilibrium.

(a) The typical firm is earning positive economic profits and is operating in the range of decreasing returns to scale.
(b) The typical firm is earning negative economic profits and is operating in the range of increasing returns to scale.

9. The table below gives the total benefits (TB) and the total cost (TC) schedules for society for clean air (larger Q means cleaner air).

(a) Complete the marginal benefits (MB=$\frac{\triangle TB}{\triangle Q}$) schedule and the marginal cost (MC=$\frac{\triangle TC}{\triangle Q}$) schedules.
(b) Now, use the idea that more should be produced and consumed as long as MB>MC, while less should be produced and consumed if MB<MC to determine the efficient level of clean air to be provided.
(c) The net surplus to society equals TB−TC. At what Q is the net surplus maximized?
(d) Suppose a regulatory agency decrees that $Q = 9$ units of clean air be provided. What happens to the net surplus to society? Why?

10. Is buying locally grown food good for the environment? Suppose the cost function for food production is given as: $C = 2 \times Q^{0.9}$ where Q

TB	MB	Q of Clean Air	MC	TC	Net Surplus =TB−TC
0	xx	0	xx	0	0
15		1		1	
29		2		3	
42		3		7	
54		4		13	
65		5		21	
75		6		30.5	
84		7		43	
92		8		57	
99		9		73	
105		10		91	

is the amount of food grown. Transport costs (farm to market) are approximately 5% of total costs. There are two alternative methods of producing food. With Method A, eight farms produce $Q = 625$ units of food each. These farms sell locally. With Method B, one farm produces $Q = 5000$ units of food. Under method B, the one large farm must transport its food a longer distance to the market, but transport costs are still 5% of total costs.

(a) Calculate the transport costs per farm for Method A and then sum these costs over the eight farms to get the total transport costs.

(b) Calculate the transport costs for the one large farm under Method B.

(c) Compare transport costs under the two methods.

(d) Does the technology exhibit increasing, decreasing, or constant returns to scale? (Hint: calculate ATC per farmer for both methods).

11. The table below depicts the populations and calorie intake per person of developed countries (DCs) and less developed countries (LDCs).

Year	DC Pop	DC Calories	LDC Pop.	LDC Calories	World Calories	Average Calories	Marginal Calories
1980	1B	3,000	3B	1,000			xx
2000	1B	3,000	5B	1,100			

(a) Fill in the table for total calories consumed by the world population, average world calories, and the marginal calories.

(b) Look at average world calories from 1980 to 2000. Given this information, is the world population better or worse off? Does your answer make sense?

(c) What relation can we infer from average world calories and marginal calories?

3

Indicators of Environmental and Economic Well-Being

> The greatest value of a picture is when it forces us to notice what we never expected to see.
>
> J. W. Tukey [239] op. cit. [125]

In 2002, Peter Kennedy presented "Ten Commandments" that applied researchers should follow when testing theoretical hypotheses and presenting empirical results in economics. Commandment 4 [125, p. 582] is "Thou shalt inspect the data." In this chapter we examine some trends in various environmental and economic indicators. The graphs we present are meant to complement, not substitute, for other modes of data analysis and theory and are only a first step in learning about various environmental economic theories and methods of analysis. We start by examining some major environmental policies that helped enable researchers to begin to gather data and monitor environmental outcomes regarding air, water, and land pollution.

3.1 Policies and Instruments

Federal, state, and local governments have passed much environmental legislation and Table 3.1 lists a few of the major policy acts [94]. Quite often these acts were passed in the wake of some environmental disaster. For example, The Superfund Act of 1980 passed after the health hazards of buried toxic chemicals at Love Canal, New York became known. Similarly, The Oil Pollution Act of 1990 passed after the 1989 Exxon Valdez oil spill in Prince William Sound of Alaska.

In the absence of specific government policy the interaction of unfettered markets frequently results in damage to a party that is adjudicated in courts. **Common law**, where judges and jurors decide individual cases, has long played a role in helping to reduce pollution. In 1864, a St. Louis man sued his neighbor over smoke from a shed that was making his house "untenantable" and was awarded $50 from the Missouri Supreme Court. A general statute requiring minimum heights for chimneys came after the common law precedent

[87]. Common law helps the market process become more efficient by defining property rights to environmental resources such as clean air and water.

The federal government circumvented common law and assumed control over mining resources with the General Mining Law of 1872. Today, one of the most unpalatable aspects of this law allows miners to pay a $250 patent application plus $50 per claim within each application, which, if approved, allows the claimant to purchase the surface and mineral rights for $2.50 per acre for placer claims which are located in sand or gravel, or $5 per acre for lode claims. These fees tend to be well below the prices that would exist in private markets, thus providing a subsidy to mining claimants.

Various policy instruments are available to protect citizens from environmental damage. With **technology standards** regulators mandate the use of specific pollution abatement methods, such as requirements that forced coal-fired electricity generating plants to install scrubbers to reduce SO_2 emissions. Technology standards are a kind of "command and control" and tend to have costs that are higher than necessary to achieve a given level of abatement. In addition, by requiring all newly constructed generating plants to install scrubbers—including those plants that burned low-sulfur coal—the technology standard might have contributed to higher emissions than might have occurred without the standard [119]. A report commissioned by The Environmental Protection Agency [8] which reviewed sixteen studies that estimated the costs of reducing air pollution found that command and control was between 7% and 2100% more expensive than market-based approaches, with a mean of 363%.

A less costly approach to abatement is a **performance standard** where regulators limit emissions above a certain threshhold, but allow firms to choose the least costly method of meeting the standard. The 1970 Clean Air Act Amendments imposed performance standards on SO_2 (sulfur dioxide) emissions from new power plants. An unintended consequence of these standards is that by exempting existing plants from the standard, utilities delayed construction of new plants thereby extending the life of older, more polluting plants.

As an alternative to technology and performance standards, **emissions' taxes** can be imposed on polluting firms. When carefully designed, these taxes cause firms to internalize the costs of pollution leading to higher product prices and a simultaneous reduction in production and consumption of the polluting product. Some states, such as Louisiana and California, impose wastewater discharge fees into publicly owned treatment facilities. As part of the 1990 Clean Air Act Amendments states can impose air emissions fees to recover the costs of administering permit programs. The South Coast Air Quality Management District in southern California imposes some of the highest air emissions fees. Other economic incentive programs include fees for used tire disposal and household waste.

In 1960, Congress passed the Multiple-Use Sustained-Yield Act so that national forests would be administered for recreation, timber, livestock range, and wildlife/fishery purposes. Today, when one encounters a national forest boundary sign the name of the forest appears and is sometimes appended with the slogan "Land of many uses." An economic problem that arises with the policy of multiple-use sustained-yield is that the uses are often in conflict. Furthermore, without market prices, national forest superintendents lack information about the highest valued uses of those forests.

Table 3.1: Natural Resource and Environmental Legislation

Policy	Year	Summary
The General Mining Law	1872	Established federal control over mining resources; legislation in 1988 imposed $50 claim fee and $2.50/acre mining rights
Federal Insecticide, Fungicide, and Rodenticide Act	1947	Regulates pesticide use; Amended in 1972 with the Federal Environmental Pesticide Control Act
Air Pollution Control Act	1955	Provided funds for federal research in air pollution
Multiple-Use Sustained-Yield Act	1960	Natl. forests managed for fish recreation, range, timber, watershed, and wildlife
Clean Air Act	1963	Federal monitoring/control of air pollution begins; Amended in 1970, 1977, and 1990
Wilderness Act	1964	Allows Congress to designate wilderness areas
National Environmental Policy Act	1969	Established President's Council on Environmental Quality; federal agencies to prepare environmental impact statements
EPA Order 1110.2	1970	Executive order creating the EPA
Clean Water Act	1972	Issued guidelines for industrial water pollution; required firms to use best available technology; issued effluent permits; major waterways to be fishable and swimmable by 1981
Safe Drinking Water Act	1974	EPA sets safe drinking water standards

Table 3.1: (continued)

Policy	Year	Summary
Magnuson-Stevens Fishery Conservation and Management Act	1976	Establishes a 200 mile coastal zone; mandates catch limits
Comprehensive Environmental Response, Compensation, and Liability Act (CERCLA)	1980	EPA to develop a hazard ranking system and create a national list of sites eligible for Superfund; amended in 1986 to improve community info about nearby chemical hazards
Montreal Protocol	1989	Phased out ozone depleting agents (CFCs)
Oil Pollution Act	1990	Companies must have a plan to prevent spills and detailed containment and cleanup plans

In 1969, the National Environmental Policy Act established the President's Council on Environmental Quality. In addition, federal agencies had to prepare environmental impact statements for any proposed agency action. On December 4, 1970 President Nixon established the Environmental Protection Agency (EPA) via executive order 1110.2. The EPA's mission was to conduct research, write and enforce environmental protection regulations, and help assist in preventing and cleaning up pollution. In the 1970s Congress also passed the Clean Air Act, Clean Water Act, Safe Drinking Water Act, Magnuson-Stevens Fishery Act, and amendments to the 1947 Federal Insecticide, Pesticide, and Rodenticide Act.

Another way to improve environmental performance is to provide the public greater information. In 1980 Congress passed the Comprehensive Environmental Response, Compensation, and Liability Act (CERCLA) and established Superfund to clean up toxic chemical waste dumps that had been placed on a national priorities list. Amendments to CERCLA in 1986 in the form of the Emergency Planning and Community Right to Know Act required chemical companies to list the volume of chemical releases and established a Toxics Release Inventory administered by the EPA. Greater information helps channel the self-interest of corporations to maximize profits into simultaneously serving the public good. Corporations wishing to avoid costly litigation and fines from their toxic releases have the incentive to reduce those releases. In addition, corporate stock returns appear to incorporate information about known toxic releases as the firms that had the largest stock price declines on the day that toxic release information became available also had larger subsequent decreases in emissions than other firms in the same industry [130]. Thus, corporations seeking to maximize shareholder equity have the incentive to reduce emissions to the extent that the benefits of doing so outweigh the costs.

Although information can enhance decision making, the costs and benefits of acquiring the information should also be accounted for. In 1986, Californians passed Proposition 65—the Safe Drinking Water and Toxic Enforcement Act. The Act requires the state to publically list chemicals known to cause cancer and businesses to apply a label notifying consumers of any product carrying a cancer risk of 1 in 100,000. Have these warning labels reduced the incidence of cancer? In a study covering the years 1979 to 2009, Michael Marlow [164] compared cancer incidence rates in San Francisco, California and in three cities (Atlanta, Detroit, and Seattle) that are in states that did not require the warning labels. His findings indicate little to no evidence that cancer incidence rates fell more in San Francisco than the other three cities. Rather than provide useful information some businesses might use warning labels to avoid potential lawsuits or support warning labels as a means of raising competitors' costs. Furthermore, consumers faced with too many warnings on everyday products might choose to ignore the warnings. Research in behavioral economics also suggests that people tend to overrespond to low-risk events and underrespond to larger-risk events. Given this finding, the costs to businesses, consumers, workers, and taxpayers might have been better spent on some alternative public health program rather than the inefficient labeling law.

3.2 Cancer

Throughout this course we will assume that environmental amenities and disamenities enter as arguments in an individual's utility function. Environmental amenities, like national parks, scenic views, plants, and most wildlife, enhance an individual's well-being whereas disamenities, such as pollution and plant and wildlife extinctions, make individuals worse off.

In *Silent Spring*, Rachel Carson warned that man's use of pesticides such as DDT, parathion, chlordane, and other pesticides in his war on mosquitoes, fire ants, and gypsy moths was leaving collateral damage in the form of casualities to birds, worms, and fish populations. Carson also warned that "the 20th century was to create countless new cancer-causing chemicals and to bring the general population into intimate contact with them" [36, p.221]. She wrote "Today we find our world filled with cancer-producing agents. An attack on cancer that is concentrated wholly or even largely on therapeutic measures ... will fail because it leaves untouched the great reservoirs of carcinogenic agents that which would continue to claim new victims faster than the as yet elusive "cure" could allay the disease" [36, p.241].

Table 3.2 reports the peak year for age-adjusted cancer death rates per 100,000 population in the United States for males and females. Except for lung and bronchus cancers, there had been no explosion in cancer death rates in the years before or after 1962 when Carson wrote her book. In fact, most cancers are due to individuals living longer lives, not environmental pollution. Refrigeration helped reduce stomach cancers as smoked and pickled foods declined in people's diets. Mortality rates from cancers of the breast, colon, liver, stomach, and uterus all peaked before 1950. Mortality rates from cancers of the ovary, leukemia, and pancreas (males) peaked during 1967–1970. Lung cancer mortality rates peaked in 1990 for males and 1998 for females. Mortality rates from pancreatic cancer among females peaked in 2006. At all cancer sites mortality rates for females peaked in 1936 and for males in 1990. Subtracting lung and bronchus cancers, cancer mortality rates peaked among women in 1936 and among males in 1948.

Table 3.2: Peak Years for Age Adjusted Cancer Mortality Rates per 100,000 Population, 1930–2006 [256]

	Males Peak Year	Females Peak Year
All Sites	1990	1936
Breast	...	1948
Ovarian	...	1969–70
Colon	1948	1947
Leukemia	1967	1968
Liver	1930	1930
Lung	1990	1998
Pancreas	1968	2006
Prostate	1993	...
Stomach	1930	1930
Uterus	...	1932
All Sites w/o Lung	1948	1936

The point we should take away from Table 3.2 is not that pesticide and chemical use do no harm, in fact, pesticide use is thought to contribute somewhere between 1% and 10% of cancer deaths, but that the use of these chemical agents also bring benefits in terms of an agricultural surplus and in the fight against malaria. In the United States, the National Malaria Eradication Program began in 1947 and involved spraying DDT inside houses in 13 southeastern states. By 1952, malaria was considered eradicated. According to the World Health Organization [260], in 2010 there were about 219 million malaria cases with an estimated 660,000 malaria deaths worldwide. Although the United States banned the use of DDT in 1972, today the WHO considers indoor spraying of DDT an effective and environmentally safe weapon in the fight against malaria. Figure 3.1 shows indexes of insecticide, herbicide, and fungicide use from 1979 to 2007. The indexes take values of 1.0 in 1979, when 1487 million pounds of total active ingredient from insecticides (255 million pounds), herbicides (610 million pounds), fungicides (124 million pounds), and other chemicals (498 million pounds) were applied in the United States. By 2007 total active ingredient had declined to 1133 million pounds.

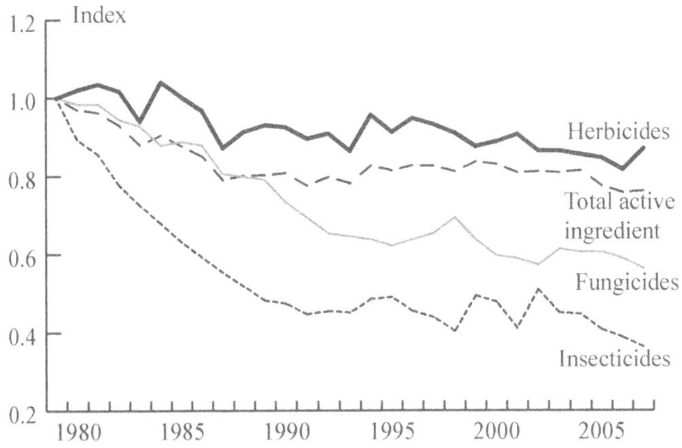

Figure 3.1: Indexes of Annual Amounts of Pesticide Active Ingredient Used in the United States (Source: Environmental Protection Agency: Pesticide Market Estimates).

3.3 Life Expectancy

Life expectancy at birth is an important indicator of societal well-being. Figure 3.2 depicts the increase in life expectancy in the United States from 1850–2008. The detrimental effects of the 1918 Spanish influenza are clearly seen as life expectancy dropped from 54.5 years in 1915 to 39.1 years in 1918 before rebounding back to 54.7 years in 1919. Worldwide, more than 50 million people died in the pandemic. The gains in life expectancy in the early years were primarily due to declines in childhood mortality and infectious diseases as knowledge of germ theory advanced. Further declines have been attributed to public health programs, better personal hygiene, and rising incomes.

Most economists would not agree with a policy goal of maximizing per capita income or maximizing the growth rate of per capita income. Other factors, such as life expectancy, access to clean air and water, leisure time, and literacy are also important. However, increases in per capita income tend to be positively correlated with these other factors. Figure 3.3 shows the positive correlation between per capita real GDP and life expectancy at birth for 182 countries. As clearly seen, life expectancy increases, albeit at a decreasing rate, as per capita income increases.

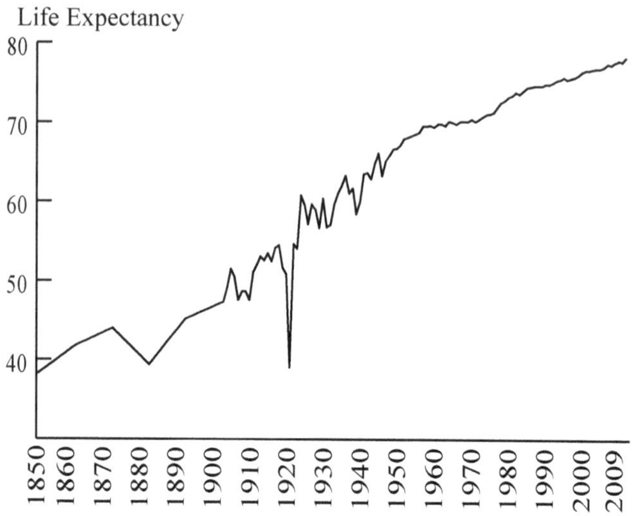

Figure 3.2: Life Expectancy in the United States, 1850–2008 (Source: Table Ab644–655 *Historical Statistics of the United States Millenial Edition* for 1850–1998, *Statistical Abstract of the United States* from 1999 to 2008).

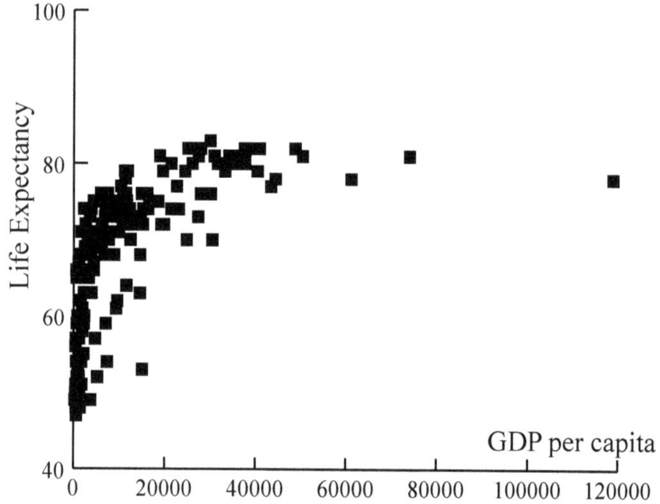

Figure 3.3: Life Expectancy and GDP Per Capita in 182 Countries in 2009 (Source: Real GDP Per Capita from the Penn World Tables, https://pwt.sas.upenn.edu/; Life Expectancy from World Health Organization, http://apps.who.int/gho/data/node.main.688?lang=en).

3.4 Grain Yields

Increased grain yields have helped feed a growing world population. Figures 3.4, 3.5, and 3.6 show increasing rice, wheat, and corn yields worldwide. From 1960 to 2010 the world's population grew by 133% from 3 billion to 7 billion people. However, during the same period rice yields grew by 139%, wheat yields grew by 76%, and corn yields grew by 179%. In addition, the amount of land under cultivation grew for rice and corn and declined slightly for wheat, so total grain production grew by 55% for wheat, 215% for rice, and 382% for corn. Of course, an increasing number of acres in cultivation also leads to land fragmentation and reduces the amount of land available for wildlife habitat.

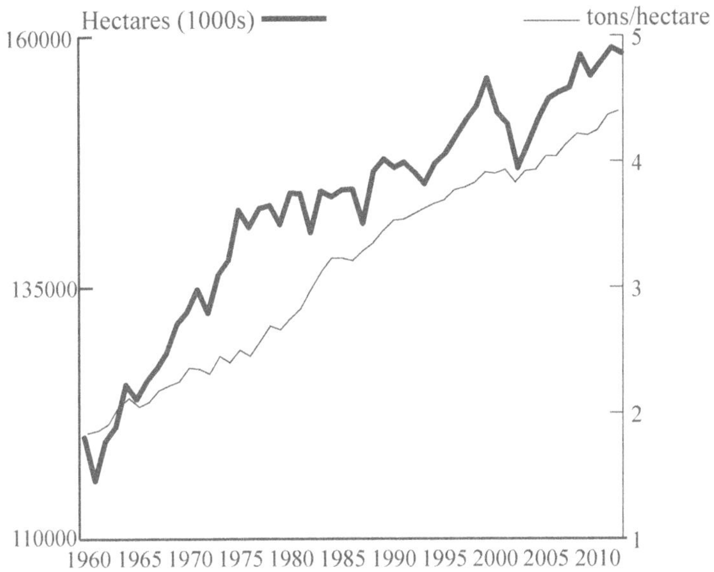

Figure 3.4: World Rice Yields, 1960–2012 (Source: International Rice Research Institutes, http://ricestat.irri.org:8080/wrs2/entrypoint.htm).

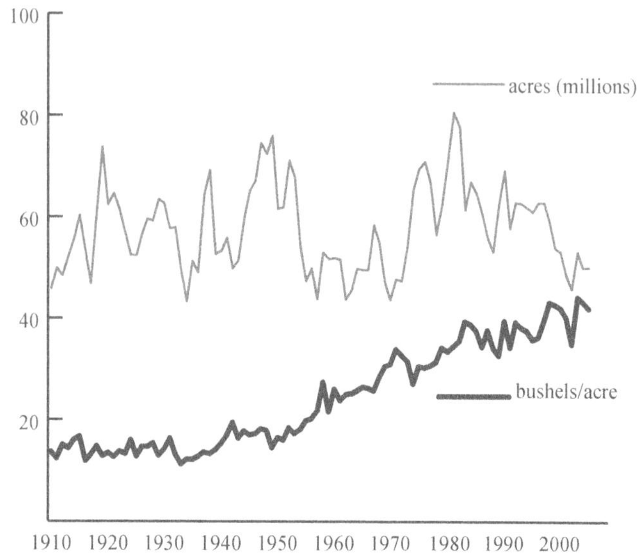

Figure 3.5: World Wheat Yields, 1910–2005 (Source: USDA Foreign Agricultural Services, http://www.ers.usda.gov/data-products/wheat-data.aspx#25377).

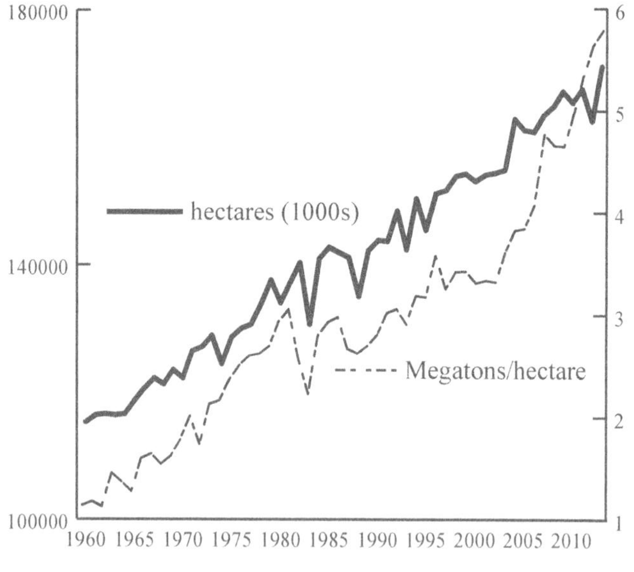

Figure 3.6: World Corn Yields, 1960–2013 (Source: USDA-ERS Wheat Outlook, http://www.fas.usda.gov/psdonline/psdQuery.aspx).

3.5 Solid Waste

The fact that matter can be neither created nor destroyed means that production and consumption will generate waste. However, the amount of materials needed to support a given level of production and consumption will tend to decrease as the production process becomes more technically efficient and as materials are recycled and reused [128]. Table 3.3 shows the amounts of materials generated, recovered, and discarded into the U.S. municipal solid waste stream from 1960 to 2010. In 1960 only 10.2% of the waste generated was recovered. Although the recovery rate increased to 36.5% by 2010, the total amount of materials discarded more than doubled from 1960 to 2010.

Table 3.3: Municipal Solid Waste, 1960–2010 (1000s of tons)

	Materials Generated	Materials Recovered	Materials Discarded
1960	54,620	5,610	49,010
1970	83,280	8,020	75,260
1980	108,890	14,520	94,370
1990	146,510	29,040	117,470
2000	178,700	53,010	125,690
2005	184,910	59,300	125,610
2007	186,390	63,100	123,290
2008	181,340	61,720	119,620
2009	172,340	61,530	110,810
2010	177,860	64,970	112,890

Source: Environmental Protection Agency [188].

In 2010, paper and paperboard made up the biggest part of the municipal waste stream followed by metals, wood, textiles, glass, and plastic. In 1960, approximately 30 million tons of paper and paperboard waste was generated with only 5 million tons recovered from the waste stream. By 2010, 71 million tons of paper and paperboard waste was generated, but 45 million tons were recovered making for a net gain of only 1 million tons. The recovery of metals in the waste stream also increased from around 50 thousand tons in 1960 to about 8 million tons in 2010 for a recovery rate of 35%. Since 1960, plastics, textiles, and wood have grown in absolute amount and as a proportion of the materials discarded.

Although greater recycling can help reduce solid waste, the recycling process also uses resources as it takes time and energy to recycle. However, some recycling might be efficient to the extent that landfill space is priced below cost. Low cost pricing of landfills helps act as an incentive to prevent illegal dumping.

3.6 Oil Spills and Water Quality

Oil spills caused by the Exxon Valdez in 1989 in Alaska's Prince William Sound and the 2010 Deepwater Horizon platform explosion and subsequent leakage caused widespread concern about the fate of ocean and coastal wildlife and vegetation. Figure 3.7 plots the number of oil spill incidents and the thousands of gallons of oil spilled in US coastal waters from 1970 to 2009. Missing from the plot are the values for the Deepwater Horizon accident, the largest oil spill in history which is estimated to have caused 4.9 million barrels of oil to flow into the Gulf of Mexico over a period of 87 days before the well was plugged.

An Oil Spill Liability Trust Fund was established in 1986. The fund imposed a $0.08/barrel charge with the funds used to clean up oil spills. Furthermore, parties responsible for the spill must repay the fund for monies expended to clean up a spill. The Oil Pollution Act of 1990 required firms to have a plan to prevent oil spills and, in their event, a plan to contain the spill. As shown in Figure 3.7 the number of oil spill incidents has sharply declined from 1970 to 2009. Furthermore, with the exception of the Deepwater Horizon accident, the annual amount of oil spilled has declined since 1990.

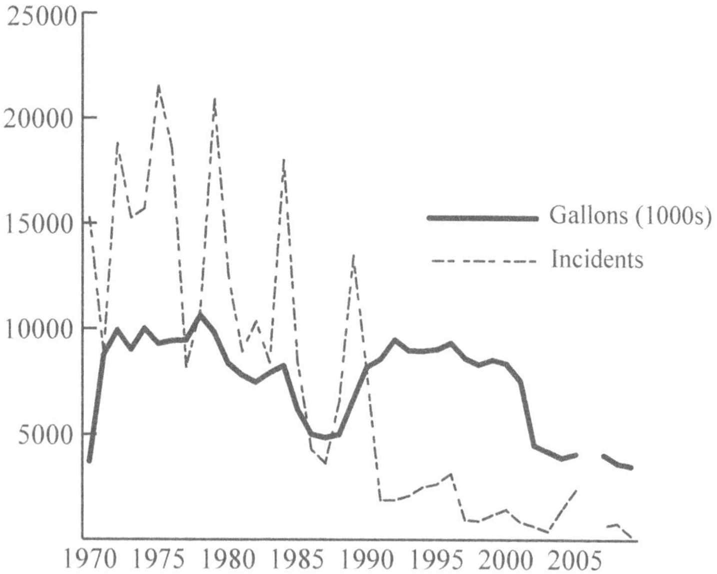

Figure 3.7: US Oil Spill Incidents and Amounts, 1970–2009 (1,000s of gallons) (Source: US Census Bureau, Statistical Abstract of the United States).

Indicators of water quality tend to be specific to a body of water or basin and recent trends show water quality improving in some basins and declining in others. Figure 3.8 depicts biological oxygen demand (BOD) in kilograms per day for the nine countries with the greatest BOD among reporting countries in 1997. Biological oxygen demand refers to the amount of oxygen that bacteria will consume in breaking down waste. As more oxygen is consumed, fishes and invertebrates are stressed and sometimes die. Six out of the nine countries experienced declines in BOD with the United States leading the way. In fact, the decline in US BOD more than offset the sum total increase in BOD in Germany, Indonesia, and Spain. However, data on BOD were not available for China until 2003 when it reported BOD of 7,066,070 kg/day. By 2006, China's BOD had risen to 9,428,874 kg/day, more than 4 times US levels.

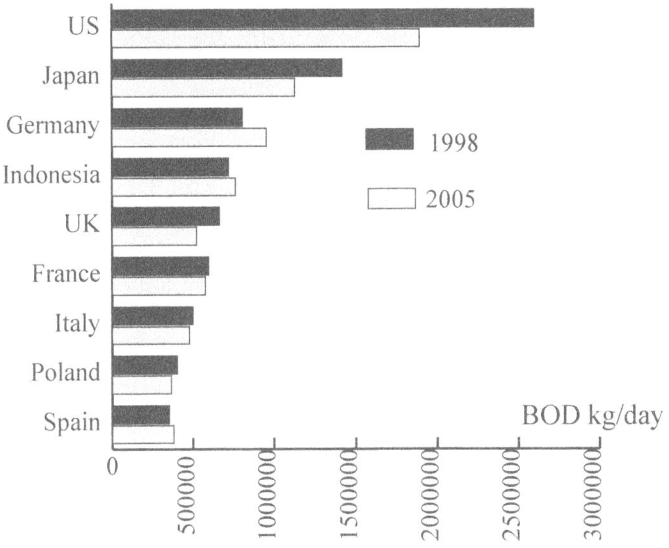

Figure 3.8: Biological Oxygen Demand, 1997 and 2005, kg/day (Source: The World Bank, http://data.worldbank.org/indicator.EE.BOD.TOTL.KG/).

3.7 Air Pollution

From 1963 to 2010 US real per capita disposable income grew from \$2,245 to \$36,090 and the US population aged 16 and over grew from 122 milllion to 238 million. What effects did income and population growth have on various air pollutants?

Sulfur dioxide is an undesirable by-product from burning fossil fuels and is a precursor to acid rain which causes acidification of forest and water ecosystems and causes damage to crops. In addition, fine sulfur particulates less than 2.5 micrometers in diameter (PM2.5) derived from sulfur dioxide emissions cause harmful human health effects [24, 209]. Figure 3.9 shows U.S. sulfur dioxide emissions from 1900–2010 in 1000s of tons. Emissions peaked in 1973 at 31.7 million tons and by 2010 were lower in the aggregate than they were in 1900. Amendments to the Clean Air Act in 1970 set the stage for federal regulation of sulfur dioxide. Further amendments to the Clear Air Act in 1990 helped reduce emissions by establishing a permit trading program in sulfur dioxide allowances. However, in per capita terms, sulfur dioxide emissions peaked in 1950 and sulfur emissions per unit of GNP peaked in 1921 [87].

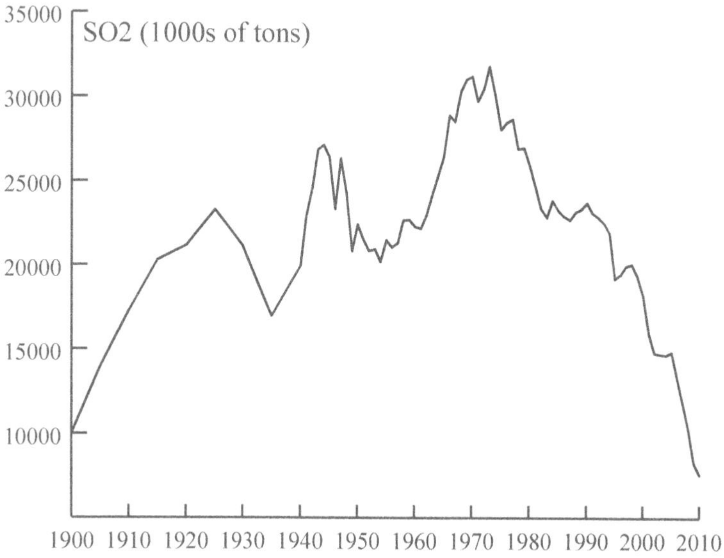

Figure 3.9: US Sulfur Dioxide Emissions, 1900–2010, 1000s of tons (Source: US Census Bureau, Statistical Abstract of the United States).

Nitrogen oxides are also an undesirable by-product of burning fossil fuels and fertilizer use and contribute to smog (haze), ozone, and greenhouse gas emissions. Ozone causes respiratory problems in levels above those that are naturally occurring. Figure 3.10 graphs emissions of nitrogen oxides from 1900–2012. U.S. nitrogen oxide emissions peaked in 1978 at 25 million tons, and by 2012 were lower in the aggregate than in any year since 1954. Nitrogen oxide emissions per person peaked around 1960 and emissions per unit of GNP peaked in 1933.

Emissions of volatile organic compounds (VOCs) contribute to climate change and are formed in the process of extracting and refining oil and natural gas and through burning of biomass [24]. Figure 3.11 graphs US emissions of

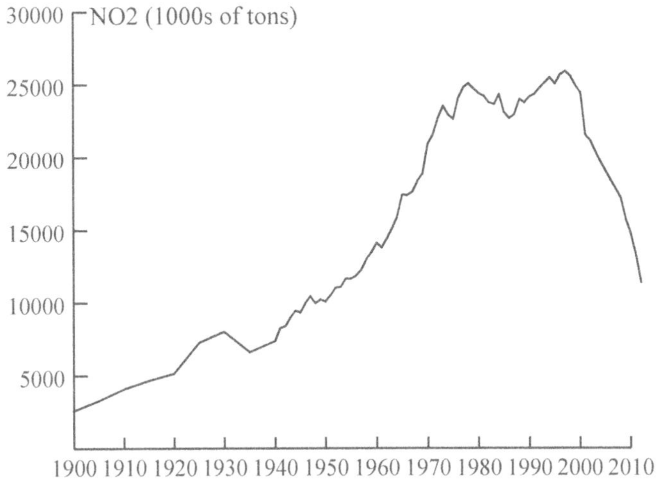

Figure 3.10: US Nitrogen Dioxide Emissions, 1900–2012, 1000s of tons (Source: US Census Bureau, Statistical Abstract of the United States).

VOCs from 1900 to 2012. Aggregate emissions peaked during 1970–1971 at 129 million tons.

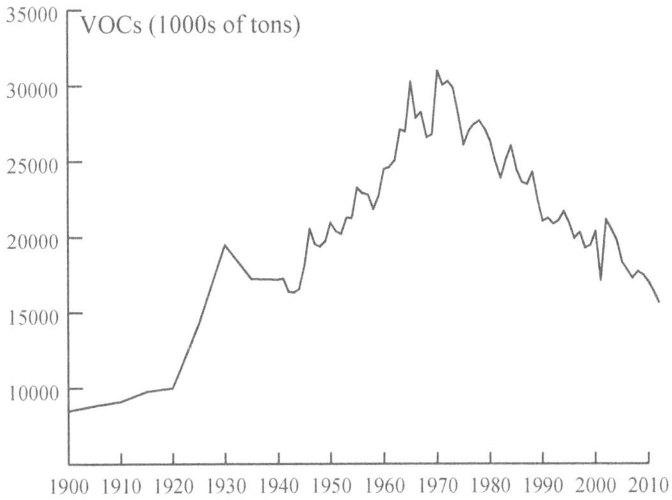

Figure 3.11: US Volatile Organic Compound Emissions, 1900–2012, 1000s of tons (Source: US Census Bureau, Statistical Abstract of the United States).

Carbon monoxide emissions occur from the partial combustion of fossil

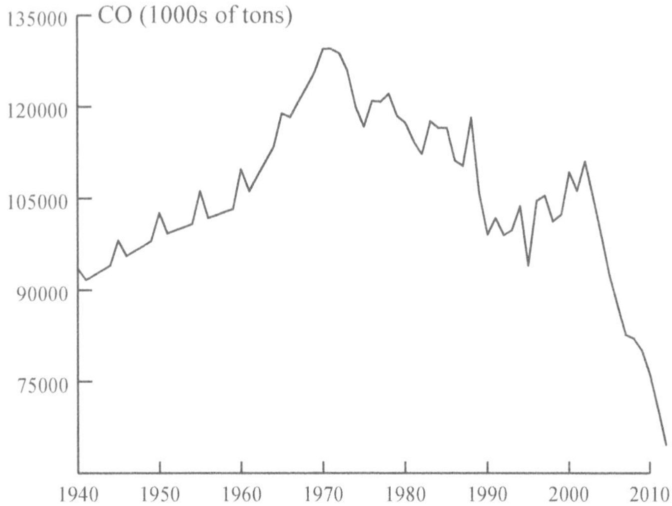

Figure 3.12: US Carbon Monoxide Emissions, 1940–2012, 1000s of tons (Source: US Census Bureau, Statistical Abstract of the United States).

fuels and are a cause of ozone. In urban areas carbon monoxide levels arise from automobile and truck transportation, lawnmowers, and the use of other internal combustion engines. Indoor carbon monoxide poisoning occurs from improperly vented space heaters, furnaces, and generators. As seen in Figure 3.12 carbon monoxide emissions peaked in 1971 and then were cut in half by 2011.

3.8 Carbon and Chlorofluorocarbon Emissions

In the 1980s scientists discovered a hole in the ozone layer in Antarctica. The hole was due to emissions of chlorofluorocarbons (CFCs) that were used as propellants in aerosols and as refrigerants. CFCs are a known greenhouse gas and the destruction of the ozone layer caused harmful ultraviolet radiation to reach earth and increase the risk of skin cancer. The Montreal Protocol of 1989 was an international agreement to phase out use of ozone depleting agents. Figure 3.13 graphs the sum of emissions of CFC11 and CFC12. While rising rapidly from 1950 to 1970, CFC production peaked in 1989 and fell to near zero by 2004. The widespread agreement among countries to cut CFC emissions can be attributed to the high costs of continued CFC use and due to developed countries helping compensate less developed countries for cuts in CFC use.

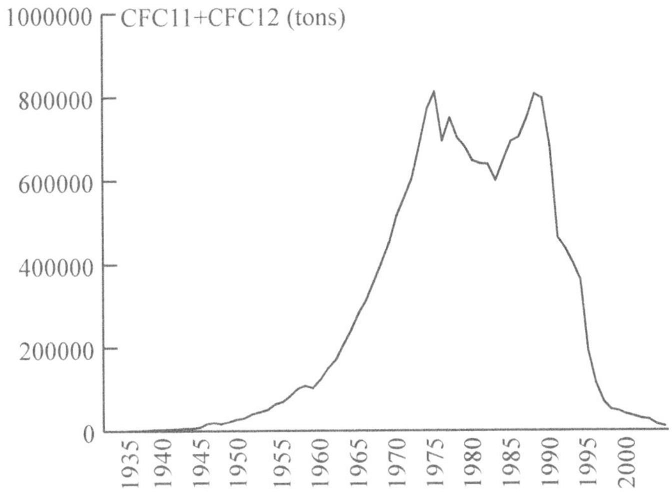

Figure 3.13: Chlorofluorocarbon (CFC11 + CFC12) Emissions, 1931–2004, metric tons (Source: AFEAS production and sales data, from http://www.afeas.org/data.php?page=prod_table).

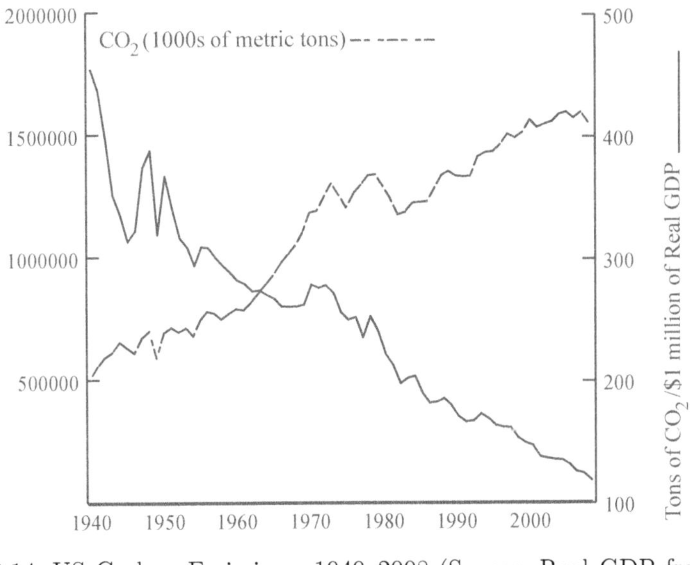

Figure 3.14: US Carbon Emissions, 1940–2008 (Source: Real GDP from Table 10.1 http://www.whitehouse.gov/omb/budget/historicals; Carbon from http:// cdiac.ordnl.gov/trends/emis/tre_usa.html [23]).

Today, there is great concern about the potential global warming effects of growing greenhouse gas emissions. According to measurements taken at

Mauna Loa in Hawaii, atmospheric concentration of CO_2 surpassed 400 ppm in May 2013. The Kyoto Protocol of 1997 committed developed countries to reduce greenhouse gas emissions to 95% of the amounts emitted in 1990 in an effort to hold atmospheric concentrations of CO_2 to 450 ppm. Although the United States signed the treaty, it was not ratified by Congress. Figure 3.14 graphs US carbon emissions for the period 1940–2008. Recent emissions have stabilized and fallen slightly during the 2008–2012 period, although the decline might be due to the Great Recession that occurred in the aftermath of the financial crisis. However, carbon intensity in real GDP has fallen from 456 tons per $1 million in real GDP in 1940 to only 118 tons per $1 million in real GDP in 2008.

3.9 Questions to Ponder

1. Does the evidence presented in this chapter suggest optimism or pessimism regarding the standard of living for present and future generations?
2. A wealth of data on various environmental indicators is available. Interesting internet sources include Yale University's Environmental Performance Index available at http://epi.yale.edu, The SEA Around Us available at http://www.seaaroundus.org, The Environmental Protection Agency http://www.eia.gov, the US Energy Information Administration at http://www.eia.gov, and The World Bank at http://data.worldbank.org/indicator. Visit one of these websites or one of your own choosing. For an environmental issue of your interest download data and present a graph or chart that depicts time series or cross-section data of an environmental indicator. Does the trend indicate environmental degradation or improvement? If the environmental indicator is for a cross-section of states or countries or regions do the data indicate a problem that will increase in severity if people in society earn more money income, or will the problem become less acute?

4

Economic Growth and the Environment

Assuming then my postulata as granted, I say, that the power of population is indefinitely greater than the power in the earth to produce subsistence for man.

Population, when unchecked, increases in a geometrical ratio. Subsistence increases only in an arithmetical ratio. A slight acquaintance with numbers will shew the immensity of the first power in comparison of the second.

Thomas Robert Malthus, *An Essay on the Principle of Population*, Chapter 1 [161]

4.1 Malthusian Trap

In this chapter we will focus on the causes of economic growth, the link between income and population, and the inter-twining links between population, income, and the health of the environment. Trying to link overpopulation and food shortages is not a recent endeavor, but dates at least from the Reverend Thomas Malthus's warning in *An Essay on the Principle of Population.*

Gregory Clark [40] points out that women over the course of their reproductive lives can give birth to twelve or more children, although prior to 1800, the number of children surviving to adulthood was always only a little over two, say 2.005. The world's population growth rate before 1820 was small, averaging 0.01% between AD 1 and 1000, 0.1% between 1000 to 1500, and 0.27% between 1500 to 1820. Even given these small rates of population growth, the world's population grew from around 100,000 in 130,000 BC to approximately 230 million in AD 1, to 1 billion in 1820 [132]. What mechanism kept the equilibrium growth rate at this low level?

The Malthusian model of population makes three assumptions. First, there is a positive relation between real income per person (material living standards such as food, shelter, and clothing) and the birth rate. Second, the death rate falls as material living standards increase. Third, there is an inverse relation between the population and real income per person. Figure 4.1 depicts these relations. Let y represent real income per person, N the size of the population,

D the death rate, and B the birth rate. The dotted horizontal line at M in the top diagram corresponds to the biological maximum birth rate, around 12 to 13 children per adult female. In the bottom diagram, the inverse relation between per capita income and population is given by the line Ny.

When the birth rate equals the death rate, income per person equals y^* and the size of the population is constant at N^*. To visualize the equilibrium process toward (y^*, N^*) suppose the population is initially N_0. In this case, income per capita is y_0 and the death rate (D_0) exceeds the birth rate (B_0) causing the population to decline and income per capita to rise. Instead, if the population is initially N_1, income per capita is y_1 and the birth rate (B_1) exceeds the death rate (D_1) causing the population to increase and income per capita to fall back to the equilibrium y^*.

Suppose an economy is in Malthusian equilibrium but experiences some technological advance, such as better tools. Technological progress shifts the population-income line to the right to Ny' and initially results in a disequilibrium per capita income of y'. The new higher per capita income causes birth rates to rise, death rates to fall, and the population to increase to a new higher population (N'), returning per capita income to its initial level at y^*. Because technological progress cannot permanently raise per capita income in the Malthusian model, population has to be checked, either through preventive checks, such as birth control, abstinence, delayed marriage, or nonprocreative sex, or by what Malthus termed "positive checks" such as premature death, infanticide, abortion, famine, or disease. While the upper classes tended to rely more on preventive checks, Malthus thought the lower classes would be more susceptible to positive checks. Any technological progress that increases the food supply removes one of the checks on population, resulting in an expansion of population to a new level where it must be checked again.

The positive relation between the birth rate and per capita income is due to preventive checks in the Malthusian model where people defer marriage and having children when income is low. Without preventive checks to birth rates only positive checks are present. In this case, women bear the biological maximum number of children and the equilibrium income occurs at the intersection of the death rate line and the horizontal line emanating from the vertical intercept at M. Clearly, the presence of preventive checks results in a higher level of equilibrium income and lower population than when only positive checks are present.

One kind of preventive check arose in Western Europe beginning in the fourteenth century: women married later and a relatively large share of women never married. This preventive check is known as the European Marriage Pattern (EMP). Why did it arise? One theory [243] is that the Black Death of 1348–1350 changed patterns in agriculture in favor of women. Prior to the Black Death the ratio of land to labor was low resulting in high land rents and low wages. The Black Death took the lives of around one-third to one-half of the western European population. This loss of life increased the ratio of land to labor and caused a switch from grain production toward livestock

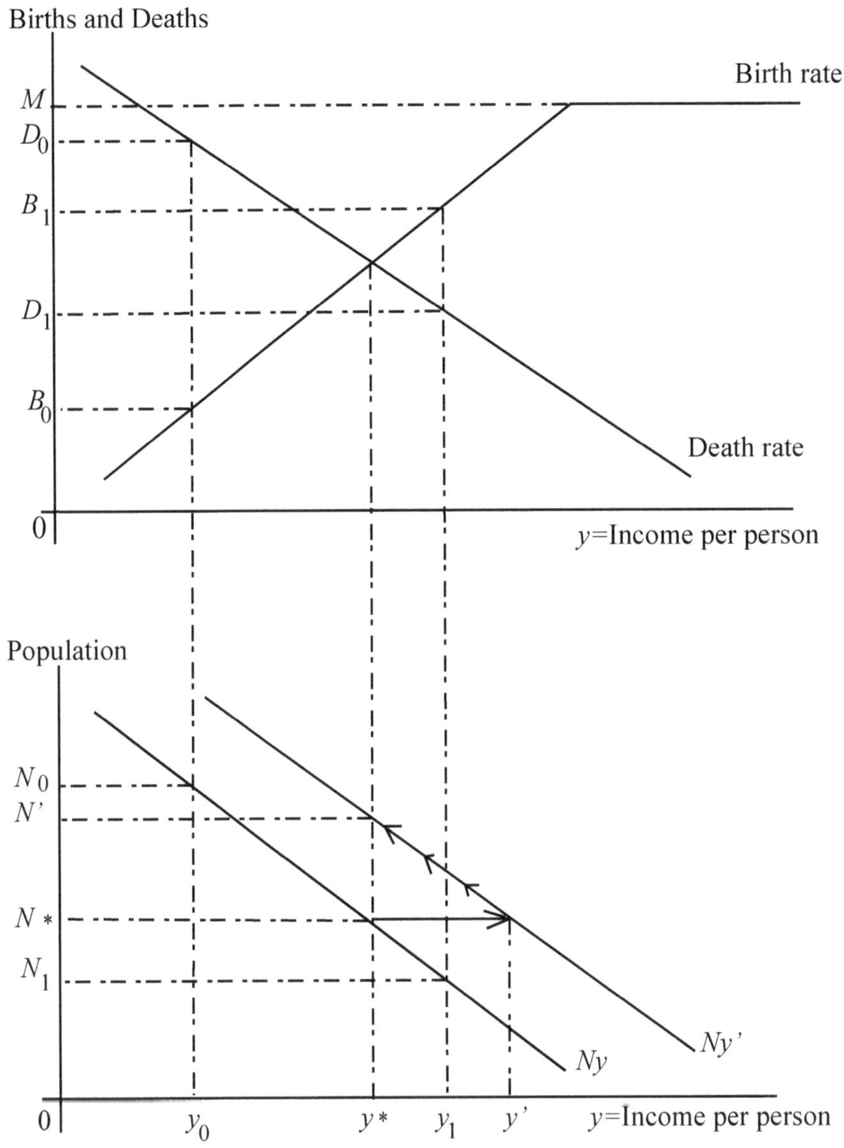

Figure 4.1: Birth Rates, Death Rates, and Income Per Capita in a Malthusian World.

production. With a low land to labor ratio prior to the Black Death, agricultural production in grain was labor intensive with men having a comparative advantage because of the strength needed for plowing. After the Black Death, the land to labor ratio increased and agricultural production shifted toward livestock and animal husbandry and away from grain. The lower land to labor ratio increased female labor opportunities outside of the peasant household. Women gained a comparative advantage in animal husbandry milking cows and herding sheep. Contracts provided women with room, board, and money, but as a condition of employment women were required to remain celibate, with pregnancy terminating the contract. The celibacy requirement was a consequence of animal husbandry tending to have year round labor requirements, unlike grain production. As women's wages rose, more women worked in animal husbandry, marriage was delayed, and fertility fell. In turn, lower rates of fertility reduced population growth and caused per capita incomes to grow. With respect to Figure 4.1 one can think of the EMP as shifting the birth function down or to the right.

A positive check that also helped reduce population growth in Europe is that after the Black Death, highly fragmented European countries were often at war. Such wars contributed to higher death rates, primarily due to hunger and disease among soldiers. According to Nico Voigtländer and Hans-Joachim Voth [242] death rates for the Soviet Union during World War II were 15%, in Poland 17%, and in the United States and United Kingdom the death rate was less than 1%. "In contrast, the Religious Wars in late 16th-century France and the Thirty Years War in Germany claimed approximately 20 and 33 percent of the population, respectively" [242, p. 172]. Higher death rates raised the land to labor ratio and contributed to higher per capita incomes. As income grew, governments found it easier to raise tax revenues which were then used to finance war and the process continued. In the context of the Malthusian model presented in Figure 4.1, wars shifted the death function up or to the right.

4.2 Escape from the Malthusian Equilibrium

For much of human history advances in technology, such as better tools in hunting, increased total income but tended to cause population growth so that per capita well-being remained constant. For a given stock of animals and plants, population growth increased the likelihood of a tragedy of the commons. Harold Demsetz [63] argues that inter-tribal violence was a natural response; by reducing hunting pressure, a tribe could increase their own harvest. Shortages of foods preferred by the hunter-gatherer, along with potential climate change, were likely precursors to the rise of agriculture. Early farmers were nomadic and practiced swidden agriculture where trees would be slashed in an area during one year and in the following year the area would be burned releasing nutrients. Since fields would quickly lose productivity these early farmers would move on to different areas after a period of farming. Gradually, a slow accumulation of knowledge regarding the benefits of crop rotation, greater use of animal fertilizer including human night waste, and irrigation allowed farmers to remain in one place for longer periods and to store grain. These stationary farmers slowly began to acquire private property rights to their possessions and collective farms gave way to private ownership. In England, private ownership accelerated in the seventeenth and eighteenth centuries with the enclosure movement which converted commonly owned property to private ownership.

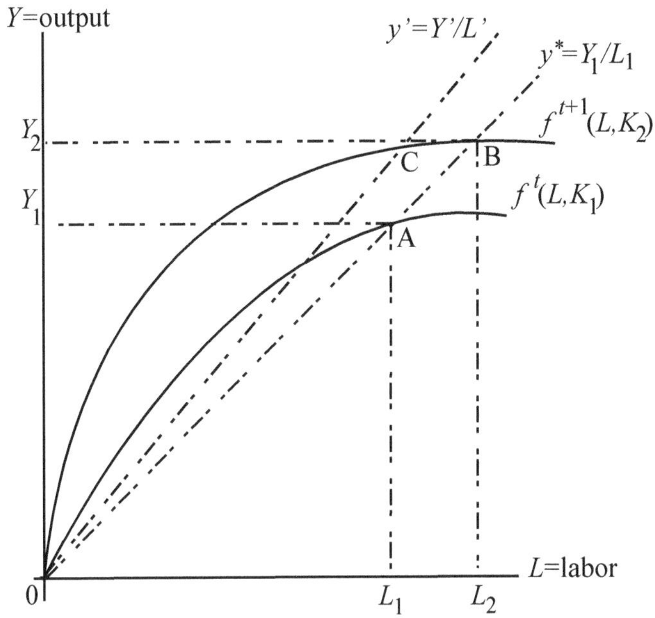

Figure 4.2: Simple Production Function.

Figure 4.2 illustrates a production function exhibiting diminishing returns to labor (L) for a given level of capital (K) used to produce output (Y). In period t the stock of capital is K_1 and the production function is given by $Y = f^t(L, K_1)$. To keep things simple we assume that the labor force is proportional to the population, so labor productivity (output per worker) defined as $y = Y/L$ is also proportional to population. In period t production takes place at point A. Labor productivity, $y_1 = \frac{Y_1}{L_1}$, corresponds to the Malthusian equilibrium level of per capita income, y^*, given in Figure 4.1. From period t to $t+1$ the capital stock increases from K_1 to K_2 and technological progress occurs so that the production function shifts upward to $Y = f^{t+1}(L, K_2)$. In a Malthusian world, the initial increase in output subsequently causes the population to increase along with the labor force to L_2. Although aggregate output grows to Y_2, the accompanying increase in the labor force causes labor productivity to remain constant as production takes place at point B.

To escape the Malthusian trap output has to grow faster than population and the labor force. Advances in technology and/or capital accumulation shift the production function upward and yield higher labor productivity and a rising standard of living if production in period $t + 1$ takes place at point C instead of point B. What factors contributed to this at first slow and then more rapid rise in labor productivity?

The slow rise of private property rights increased the expected probability that wealth could be accumulated and retained. Nuclear families could acquire income and wealth via work done by parents and work by children. According to Demsetz, parents chose family size by balancing the tradeoff from income earned by children with the cost of feeding, clothing, and housing them. As the Industrial Revolution began and offered opportunities for young adults to work in factories, parental control of children declined, as did family size. In rural areas, the decline in family size provided the incentive for farmers to substitute capital for labor which increased the agricultural surplus and caused further rural to urban migration. Greater freedom of young adults offered by factory work in urban areas caused per capita income to rise. Describing conditions in eighteenth century London, Dorothy George writes that "Drink, ignorance, and dirt were positive checks on the population of London whose effects progressively diminished" [83, p.27].

Several important innovations preceded the Industrial Revolution and helped set the stage for more rapid future growth. Large populations have a higher probability of producing innovative thinkers [132] and doers and China, with its large population, was responsible for many early important innovations [136]. The water wheel, invented in China, allowed circular motion, but in Europe, technical adjustments brought forth reciprocating motion which was used to pound pulp into paper. Paper, also invented in China, allowed a broader dissemination of ideas. Gunpowder, which was also invented in China but was primarily used for fireworks, spread to Europe, where knowledge of bell foundries allowed it to become more explosive. Centuries later, Eli Whitney set the stage for mass production with his assembly line process to make guns. Another early innovation was the mechanical clock, which allowed sailors to fix longitude at sea. Before invention of the mechanical clock, the Roman Catholic Church had tried to fix an equal number of daylight and nighttime hours. The mechanical clock changed that and, with it, the notion of circular time gave way to linear time. With linear time the early capitalists began to think of the amount of output produced per unit of time and the search for improvements in labor productivity was on. Finally, spectacles allowed craftsmen, whose skills were beginning to peak just when their eyes could no longer focus, to extend their years of useful employment. Such master craftsmen helped design the machine tools necessary for mass production.

Given China's large population and early technological lead, why did the Industrial Revolution begin in England rather than China and why did it not start earlier than around 1780 [40]? One hypothesis [145] argues that China was imperially and ideologically unified, unlike Europe, where there was competition between church and state, church and church, and state and state. While such unification allowed a semblance of stability, it was also indifferent, if not hostile, to new ideas and methods. Moreover, China's vast irrigation system in agriculture relied on bureaucratic oversight. Government service was the most honorable and worthwhile occupation, with entry into the bureaucracy the final goal of upward mobility. However, to enter the bureaucracy the applicant had to pass the civil service exam which required them to memorize the Confucian classics. With 431,286 characters, an applicant who memorized 200 characters a day could complete the task in six years. So, China's large population and early technological lead was lost as their best and brightest prepared for life in the bureacracy. Meanwhile, a merchant class began to arise in Europe and the Enlightenment and Scientific Revolution spread novel ideas throughout the population.

Gregory Clark [40] argues that there were four main drivers of the Industrial Revolution that occurred over centuries. First, rich and middle class families had a surplus of children and the poor did not. As marriage between richer and poorer families occurred, the desirable values of the rich and middle class flowed down the social ladder, reaching those at the bottom of the socioeconomic hierarchy. Interest rates declined because individuals became more patient and willing to defer gratification. Clark gives evidence that forager and hunter societies have very high rates of time preference; they strongly prefer current consumption to future consumption. For example, the Yanomamo of Brazil have such a high rate of time preference that they cut the branches off berry bushes to make picking easier, although it reduces future harvests and can even kill the bush. The decline in interest rates allowed greater accumulation of capital which helped fuel growth. Second, economic growth occurred because of people's willingness to work more. People in forager societies lived at the subsistence level but worked little; on average the male Yanomamo worked between three and six hours a day, while male laborers in England worked between eight and nine hours per day, but also, according to Clark, existed at the subsistence level. However, the longer work hours became a cultural norm and helped enhance efficiency as England transitioned to lower population growth. Third, "literacy and numcracy went from a rarity to the norm" [40, p. 166].' In contrast to barter, numeric skills were needed to facilitate the use of money as a medium of exchange and ultimately those skills allowed technological innovation to occur. Finally, there was a decline in interpersonal violence which enhanced trust between individuals. Without trust, voluntary trade between strangers is more difficult preventing the potential gains from specialization afforded by comparative advantage.

Clark provides evidence on a wide range of economic and social indicators between various hunter/gatherer societies and more industrial societies to build his case. He looks at height, fertility rates, calorie consumption, work hours, transportation costs, population, profits, interest rates, land rents, and wages. Clark's overall theme suggests that the Industrial Revolution occurred in England and Europe because of cultural changes, rather than institutional changes, such as greater reliance on private property rights, markets, and limited government.

Robert Allen [6] provides several criticisms of Clark's interpretation of the escape from the Malthusian trap. First, Clark uses height to make his case that no increase in the standard of living occurred before 1800. However, he uses male height, when female height is actually a better indicator of deprivation. Furthermore, height in England in the mid-eighteenth century was 172 cm, a height that was considered tall in nineteenth century societies where laborers earned four times the subsistence wage. Second, the move to agriculture from a forager society increased fertility, since mothers in sedentary societies did not have to carry their children (thus a lower cost of children). However, there is not much evidence that those agricultural societies were richer than foragers and some evidence that the standard of living fell. Third, Clark argues that institutions—private property rights, rule of law, and democracy—did not cause growth because those institutions were already features of the English economy well before the start of the Industrial Revolution. However, Allen argues that medieval serfdom was not an efficient institution, since it did not have secure property rights or free labor markets for serfs. Lords could impose arbitrary tax assessments and restrict the movement of serfs. Allen argues that efficient institutions which came about in Britain, Europe, North America, and Japan helped promote economic growth.

Fourth, Clark offers two behavioral explanations for the onset of growth: a fall in interest rates due to greater patience and a rise in literacy due to a greater preference for education among the rich, whose children moved down the social ladder to find work and mates. The downward social mobility helped spread the aristocratic virtues of patience and education. Clark also argues that citizens became more thrifty and prudent which helped spur growth by encouraging savings and investment. However, Allen argues that the invention of the printing press reduced the cost of books and promoted literacy and numeracy. In addition, workers in urban areas were more literate than their rural counterparts and rural to urban migration proceeded rapidly. In regard to time preference, Allen argues that the decline in interest rates occurred on the European continent before it occurred in Britain, possibly because of institutional changes. William and Mary assumed the throne in 1689 and William, who previously was Prince of Orange and Stadtholder in the Dutch republics, brought with him modern continental financial institutions. Attitude changes have also been attributed to the Protestant Ethic, greater rationality due to the Enlightenment, and the Scientific Revolution of the seventeenth century. Here, we might think in terms of correlation vs. causation. Did literacy and numeracy give rise to commercial life as Clark would argue, or did commercial life give rise to literacy and numeracy? Fifth, Allen offers evidence that the Industrial Revolution began in England due to a high wage to capital price ratio and a high wage to energy price ratio that made it profitable to substitute capital and energy for labor. This substitution effect took the form of famous inventions such as the spinning jenny and the steam engine.

Consistent with Allen [6], Daron Acemoglu and James Robinson [1] argue that a major cause of economic growth arises from institutional choices. Countries that have inclusive governments where there is a rule of law, private property rights, and where political power is distributed widely across the population are more likely to experience growth than countries which have extractive institutions where political power is narrowly distributed, taxes are high, property rights are tenuous, and slavery or serfdom limit much of the population's ability to seek better opportunities elsewhere. Such extractive institutions include tsarist Russia, dictators in various African and Central American countries, and the Premier and Politburo in the former Soviet Union, all of which transferred wealth from the masses to a small elite group. The Glorious Revolution of 1688 in England helped limit the power of the king, established a rule of law, and helped the growing merchant class gain power in parliament at the expense of the landed aristocracy. In addition, property rights for ideas—patents—were enforced which helped promote innovation. Acemoglu and Robinson [1] argue that once England had moved toward inclusive institutions a virtuous circle helped to reinforce property rights and the rule of law. Those rights also helped foster incentives for innovation and promoted greater trust between citizens which helped promote trade.

Many hypotheses have been put forth for the rise in the standard of living brought forth by the Industrial Revolution; the Protestant ethic toward hard work and greater savings and investment, a rise in foreign trade, an increase in human capital, greater rationality and enlightenment, scale economies, discovery of natural resources, geographical advantages, and better institutions of government and property rights. However, Deirdre McCloskey argues that none of those reasons can fully explain the sixteenfold rise in the standard of living. Instead, she argues that innovation occurred because there was a change in the way that people talked and thought about the "previously scorned" bourgeoisie that had positive spillover benefits to society as a whole. "The external effects thus revealed were a new dignity for the bourgeoisie in its dealings and a new liberty for the bourgeoisie to innovate in economic affairs" [166, p. 394].

According to McCloskey, liberty was necessary for innovation and economic growth because at the frontiers of knowledge, no single person can predict the uses of new inventions and so society must allow many individual experiments to flourish. Dignity was also necessary for innovation and economic growth because innovations tend to cause what Joseph Schumpeter [212] called "creative destruction." Creative destruction occurs when new innovations destroy the value of old capital, such as when a company like Netflix and then streaming of videos destroys the value of the bricks-and-mortar video rental store. Such innovations are inherently disruptive as jobs are lost and people search for someone to blame and sometimes jail as was done in the former Soviet Union. Rather than respect for the warrior or knight or landed aristocracy, dignity and liberty for the bourgeoisie promotes trade and innovation which makes everyone richer.

Finally, for our purposes it is important to recognize the correlation, if not causation, of changes in energy sources that occurred. Table 4.1 shows that between the sixteenth and nineteenth centuries there was a large substitution away from renewable energy sources toward nonrenewable energy, primarily coal. During 1561–70, annual energy consumption was 19,167 megajoules per capita with 89% of that energy derived from renewable sources. By 1850–59, annual energy consumption had increased almost five times to 96,462 megajoules per capita, but only 8% of that amount was derived from renewable sources [262]. For comparison 1 megajoule equals 3.6 kilowatt hours. Reiner Kümmel defines an "energy slave" as the equivalent of 2500 kilocalories required by a very heavy human workload. Each person had approximately one energy slave (their self) 100,000 years ago, rising to ten in medieval times, to between 40 and 100 in the developed world today. Using Kümmel's definition of an energy slave, the average person in England and Wales during 1850–59 had approximately 25 energy slaves.

Table 4.1: Annual Energy Consumption in England and Wales per Head of Population (megajoules)

Year	Human	Draught animals	Firewood	Wind	Water	Coal	Total
1561–70	4,373	6,210	6,324	59	162	2,039	19,167
1600–09	4,161	4,647	4,729	85	152	3,153	16,925
1650–59	4,521	4,802	3,849	153	156	6,772	20,253
1700–09	4,789	5,744	3,939	238	173	14,719	29,602
1750–59	4,519	5,113	3,429	427	198	21,403	35,089
1800–09	4,233	3,471	1,877	1,282	111	41,373	52,347
1850–59	3,564	2,633	118	1,280	89	88,779	96,462

Source: P. Warde [244] and E.A. Wrigley [262].

When the primary source of energy is renewable, outputs such as iron are produced locally since it is too costly to haul wood and ore long distances and deforestation proceeds rapidly [135]. The substitution to nonrenewable energy helped reduce deforestation but the laws of thermodynamics could not be overturned, with rising waste and entropy the consequence. Since 2000, economic growth in the developed world has slowed. Robert J. Gordon [91] argues that one of the six headwinds that future economic growth must face is paying the costs of global warming that has arisen from past energy use. The role of the economist will be to work with scientists, engineers, and policy makers in assessing the benefits and costs of alternative energy sources.

4.3 Demographic Transition

During the Malthusian era, death rates and birth rates were in close equilibrium and population grew slowly. Innovations that produced short-term income gains were quickly offset by a higher population counteracting those gains. But as societies escaped the Malthusian trap, population growth accelerated.

Figure 4.3 depicts the demographic transition [140]. We begin with a small population that experiences high death rates and birth rates resulting in a relatively stable population. During this phase of development, most people live short lives and die young because of disease, famine, and injury. Such a phase is consistent with hunter/gatherer societies that slowly transition into swidden agriculture eventually giving way to sedentary agriculture. With sedentary agriculture birth rates may rise as women no longer have to carry children. Sedentary agriculture brings humans and domestic animals into close proximity. As a consequence, zoonose diseases spread from animals to humans causing death rates to temporarily increase until people gain immunity [168]. During a third phase of the transition, populations gained immunity to those zoonose diseases, the germ theory of disease became known, and improvements in medical care and sewage and sanitation helped to reduce mortality rates, with the greatest impact on the young. To the extent that families try and choose the number of surviving children, not the number of births, lower mortality rates among children causes a decline in the birth rate. Since the decline in the birth rate lags the decline in mortality rates, population grows relatively rapidly. In later phases of industrialization falling birth rates gradually catch up to lower mortality rates resulting in slower population growth. During these phases of the transition families might be substituting quality for quantity in the number of offspring. In addition, women who gain education and greater freedom and choose to work in the market economy experience a higher opportunity cost of bearing children, further decreasing birth rates. In some industrialized countries such as Japan, Germany, and France the birth rate has fallen below replacement, which, if in effect for a long enough time, will cause population to actually begin to fall.

Average fertility rates for high income countries, middle income countries, and low income countries are illustrated in Figure 4.4 for the period 1963–2008. High income countries have completed the demographic transition and currently have fertility rates below the replacement rate. Fertility rates in middle income countries have rapidly declined, and, in 2008, averaged only 2.5 per female. Although still high, fertility rates are also falling in low income countries and the World Bank predicts that by 2100 those countries will have also completed the transition to a stable population.

Figure 4.5 graphs average birth and mortality rates at various stages of development measured by log real GDP per capita. Birth rates are higher than

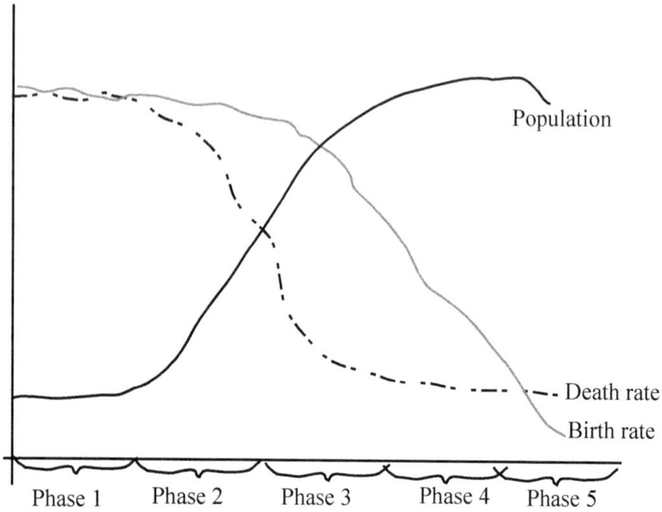

Figure 4.3: Demographic Transition.

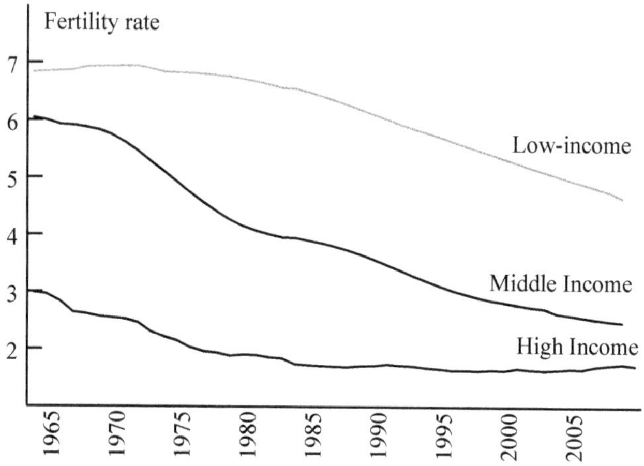

Figure 4.4: Fertility Rates in Low, Middle, and High Income Countries, 1963–2008 (Source: Extended Penn World Tables [165] and author's calculations).

mortality rates in high, middle, and low income countries, but declining birth rates are rapidly catching up with mortality rates in high income countries. Starting in 1987–88, middle income countries began to have lower mortality rates than high income countries. Such a phenomenon occurred because higher income countries achieved higher life expectancies before middle income countries did so. The higher life expectancies increased the ratio of older to younger citizens in the high income countries relative to the middle income countries. With an older population, mortality rates eventually increase. As the middle income countries age, their mortality rates will gradually rise.

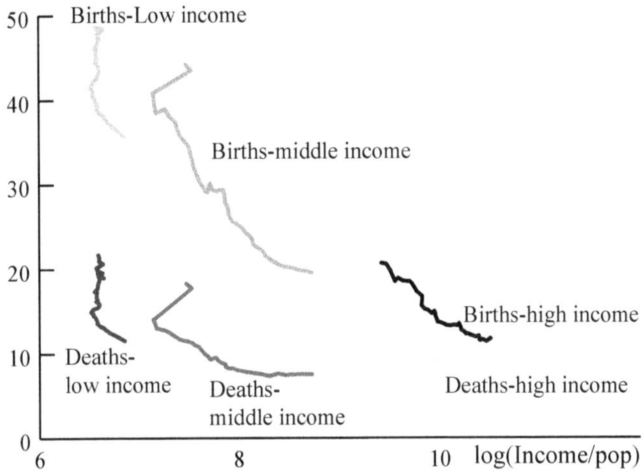

Figure 4.5: Birth and Mortality Rates in Low, Middle, and High Income Countries per 1000 population, 1963–2008 (Source: Extended Penn World Tables [165] and author's calculations).

4.4 Stages of Economic Development

Economic growth and development are ongoing and evolutionary. First, people have to feed themselves and early bands of people were hunter/gatherers. Although hunter/gatherers had low rates of population growth, they ran into common property resource problems as game became more scarce. Gradually, nomadic slash and burn (swidden) agriculture evolved as tribal bands would slash the bark on trees and then return a year or so later to burn the area. Burning released nutrients that could sustain limited agriculture for several years before the people would move on to a different area. Gradually, swidden agriculture gave rise to sedentary agriculture. In early modern stages of development labor intensive agriculture predominates. During this period mining activities and timber harvesting also account for a relatively large share of economic activity, but then tend to decline over time as seen in Figure 4.6. Later, capital intensive technologies are introduced into agriculture allowing labor to be released to other sectors of the economy. The surplus agricultural labor migrated to cities where it was employed in manufacturing which rose in relative importance. The next stage we have observed is a declining manufacturing sector accompanied by a relatively larger service sector as depicted in Figure 4.7. What tomorrow brings is a question still open to debate.

Each stage of development has environmental consequences. Hunters and gatherers and bands of people practicing swidden agriculture could move on to greener pastures, leaving behind their waste. As sedentary agriculture came to predominate, the stockpile of waste increased. Modern agricultural methods allowed easier removal of trees, the construction of irrigation ditches, and drainage canals.

Missouri, like other states, has experienced the various stages of development. Much of the prime agricultural land in Southeast Missouri in what is known as the Mississippi Delta Region was once a rich timber/wetland ecosystem. The cutting of timber, draining of swamps, and building of levees had dramatic effects on the ecosystem. Mechanized agriculture brought with it higher fertilizer and chemical requirements with their own deleterious effects, ranging from contaminated aquifers, oxygen deprived deadzones in the Gulf of Mexico, and a loss of wildlife due to habitat destruction. Mining and timber activities have had some of the most severe environmental consequences in Missouri. Clear-cutting forests causes erosion which deposits silt in the beds of rivers and streams smothering fish eggs. Timber sawmills leave large heaps of highly acidic sawdust as waste. Cutting trees to make charcoal in south-central Missouri resulted in severe air pollution problems in areas adjacent to charcoal kilns. Mining activities left behind huge waste piles and degraded ecosystem services. In the former Lead Belt area of Missouri, abandoned tiff and lead mines left giant contaminated sand piles that were part of the waste of mining. The Big River in the Missouri counties of Jefferson and St. Francois suffered fish kills from settling ponds' dams breaking and toxic waste overflowing into the watershed. Today, the elimination of lead in gasoline and in paints has left behind the toxic waste with declining populations surrounding the mining towns with place names such as Potosi, Ironton, Viburnum, and Hematite. Toxic particles from a lead smelter left behind a Superfund site at Herculaneum.

Early manufacturing resulted in heightened air pollution and other toxic land and water releases. Even service sector activities have detrimental environmental effects. Today, hospitals must dispose of large amounts of radioactive waste. As ski resorts in the west expanded they have been charged with threatening endangered species such as the lynx in Colorado and encroaching upon wilderness areas. The computer and Internet have indirectly contributed to environmental degradation by allowing people to live in once remote areas leading to sprawling subdivisions of "40 acre ranchettes" and increased concentrations of septic waste leaching into mountain streams.

There are signs, though, that the evolution of the economy from agriculture/mining/timber to manufacturing to services has helped to turn the tide of environmental degradation. Two effects have been noticed. First, although

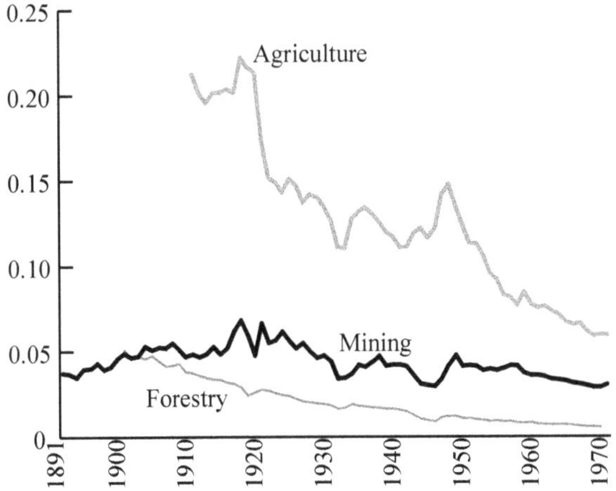

Figure 4.6: Agriculture, Mining, and Timber Shares of GNP, 1891–1969 (Source: *Historical Statistics of the United States: From Colonial Times to 1970*).

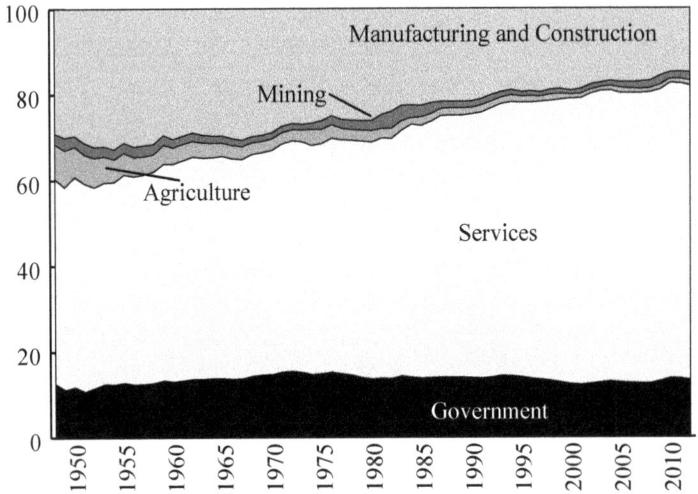

Figure 4.7: Industrial Shares of GDP, 1947–2011 (Source: US Department of Commerce, Bureau of Economic Analysis, http://www.bea.gov/industry/gdpbyind_data.htm).

service sector activities have harmful environmental effects, the damage from services tends to be less than the damage caused by mining and manufacturing. Second, if the demands for a healthy ecosystem and environmental amenities are income elastic, economic growth will lead society to demand more environmental services.

4.5 Environmental Kuznets Curve Hypothesis

How do economic growth and development affect the quality of our environmental resources? Does pollution increase as an economy grows? How will the demand for parks, clean air, clean water, and a healthy ecosystem change as aggregate income changes? What level of economic growth is sustainable and consistent with a healthy ecosystem? Are people willing to trade material wealth for a cleaner environment? Do low income people bear a disproportionately high cost of the environmental consequences of economic growth? Researchers have expended great effort seeking answers to these questions, and some consensus has started to emerge. In this section we examine some of the evidence about how income growth affects various environmental indicators and who bears the burden of environmental degradation.

The environmental Kuznets hypothesis suggests that as per capita income increases, per capita measures of pollution increase at a decreasing rate resulting in an inverted-U relation between pollution and income. Figure 4.8 illustrates this hypothesis.

At first, as income increases, pollution increases up to income y^*. After the turning point, further increases in income lead to less pollution. The rationale behind the turning point is that as economies grow, activities such as mining, timber harvesting, and certain manufacturing activities decline as the service sector expands. This changing economic structure means more economic activity takes place in less polluting industries. In addition, economists think that the demands for environmental amenities, like clean air and water, are income elastic. When demand is income elastic, increases in income cause a more than proportionate increase in the demand for the environmental amenity; after individuals have gained access to basic needs such as food, housing, minimal health care and educational services, they have a greater willingness to pay for environmental services. It is hard to imagine countries where infant mortality rates are high or where food is hard to come by caring much about the effects of global warming. However, potable water is likely to be a basic need, so that there is likely to be some demand for it even at low levels of income.

Copeland and Taylor (2004) suggest four main explanations for the Environmental Kuznets Curve hypothesis (EKC). These explanations are (1) sources of growth; (2) income effects; (3) threshold effects; and (4) increasing returns to abatement. A simple model is used to derive the sources of growth theory. Suppose that two factors of production, labor and capital, are used to produce a clean good and a dirty good. The dirty industry is assumed to

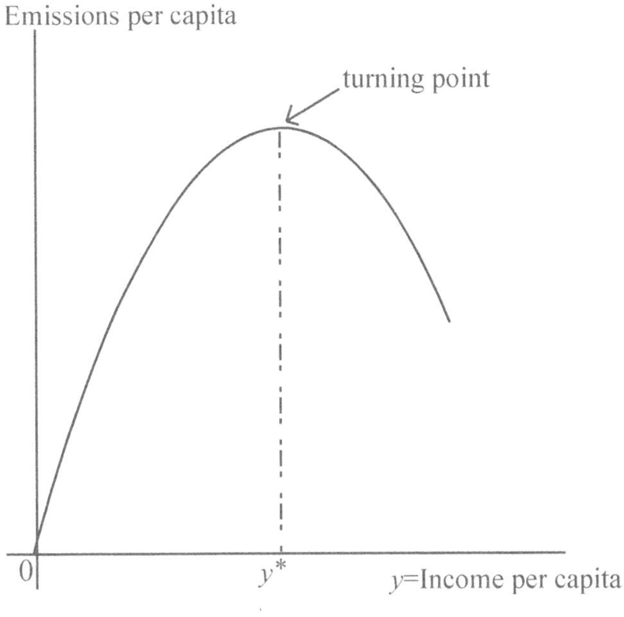

Figure 4.8: Environmental Kuznets Curve.

be capital intensive and the clean industry is labor (human capital) intensive. If countries grow primarily via physical capital accumulation in the early stages of development and via human capital accumulation in the later stages, then pollution will first rise and then fall as income growth occurs. The income effects' explanation of the EKC assumes that environmental quality is a normal good, but at low incomes pollution increases with growth because citizens place a higher relative value on increased consumption of material goods. However, as income continues to grow, the income elasticity of demand for environmental quality increases and, eventually, citizens are willing to make larger and larger sacrifices of consumption to have greater environmental quality. The third explanation of the EKC is based on threshold effects. At low levels of economic activity pollution might be unregulated, or regulation might have little impact on the cost of pollution abatement activities by firms. Therefore, pollution rises as income grows. However, after some threshold has been reached, pollution policies are either implemented or start to have an impact on costs, leading to declines in pollution as income continues to grow. If the threshold effect theory of the EKC holds then there should be a period of little or no regulatory policy followed by increasingly tougher policies over time. The fourth explanation for the EKC is increasing returns to abatement. As the scale of abatement rises its efficiency might increase. As best practice abatement technologies become more widely adapted, firms become more efficient at reducing pollution, even if pollution taxes do not change.

London Fog

Because of its geography London has always been subject to fog events. Before the onset of the Industrial Revolution, London averaged about 12 fog events per year. After 1800, coal use increased. The particulate matter from burning coal combined with water droplets in the fog and fouled the air causing adverse health effects. From 1850 to 1880 the number of foggy days increased from 25 to more than 70 annually.

In 1903 British climatologist Frederick J. Brodie presented a paper to the Royal Meteorological Society that examined the annual number of foggy days in London between 1871 and 1903. The data indicated that the number of foggy days increased throughout the 1870s and 1880s and then reversed trend and declined through the remainder of the period. Brodie attributed the pattern to variation in the production of coal smoke. During the first part of the period households and business firms burned coal with "reckless abandon." However, after 1890, London fog events and their adverse health effects declined.

Karen Clay and Werner Troesken expand Brodie's data set and find that fog related events are associated with extreme spikes in the weekly crude death rate in London. Several events transpired after 1890 to reduce fog related events. "We tend to think of Victorian Britain as a place where the expedience of cholera and child labor trumped the expense of clean water and a decent wage, as a place where environmental and social degradation served as the handmaidens of avarice and economic development" [41, p. 282]. However, the evidence from London is consistent with the EKC hypothesis. As London fog events increased the death toll a threshold was passed and citizens lobbied government to pass the Public Health Act in 1891. Police enforcement of smoke abatement policies via fines caused coal users to generate less coal smoke. Although hard anthracite coal was more expensive than soft bituminous, it also generated less smoke and users switched to hard coal to avoid paying fines. Furthermore, hard coal was more difficult to light than soft coal so producers installed coal grates to enhance efficiency. Greater attention to proper stoking methods also enhanced efficiency as did the removal of soot from chimneys. Moreover, households switched from using coal to gas in their stoves, helping to clear the air. Finally, population density declined from over 400 persons per hectare in 1861, to only 103.6 persons per hectare by 1911. During the period coal use increased, but as the population dispersed the coal smoke was diluted and resulted in less smog.

Kenneth Arrow et al. [12] offer four reasons for caution in interpreting the EKC. First, while the relation appears valid for pollutants involving local short-term costs, the evidence on an EKC has not been shown for the accumulation of stocks of waste or for pollutants involving long-term and more dispersed costs. Specifically, the stock and cost of CO_2 in the atmosphere might be an increasing function of income. Second, while the EKC has been shown to hold for the flow of pollutants, a different relation might hold for resource stocks: "the relation is less likely to hold wherever the feedback effects of resource stocks are significant, such as those involving soil and its cover, forests, and other ecosystems" [12, p. 520]. Third, as countries grow, emissions might be exported to other countries, or firms might reduce one pollutant that is regulated, but increase another, unregulated pollutant. Fourth, while emissions' reductions have been found to occur for some pollutants as income grows, the emissions' reductions have often been due to local institutional reforms. When pollutants flow across political borders and across generations, local institutional reforms might be lacking or insufficient to reduce pollution. Furthermore, when the costs of environmental degradation are borne by the poor, or by future generations, or by citizens of other countries, the incentives to address the problems even at the national level are likely to be weak.

To test the EKC hypothesis researchers collect data on various kinds of pollutants (or other environmental indicators), per capita income, and other variables which might affect the level of pollution. Two kinds of data sets can be employed to test the EKC hypothesis: cross-section data and time-series data. Cross-section data hold time constant and makes observations for different economic entities, usually countries or states. For example, a cross-section data set might collect information on SO_2 emissions and income for the fifty states in 1998. Time-series data examine a single economic entity at various points in time. For example, we might collect information on CO_2 emissions and per capita income for Canada from 1963–2008. A pooled data set combines cross-section and time-series data. For instance, pooled data for the United States, Canada, and Mexico for per capita incomes and per capita CO_2 emissions can be represented by a scatter diagram like the one in Figure 4.9 which shows carbon dioxide emissions (in kilograms) for the period 1963 to 2008.

One way to test the EKC hypothesis is to estimate a quadratic form for the relation between pollution, per capita income, and other variables (z). Such a function takes the form

$$E = \frac{Emissions}{Population} = b_0 + b_1 \times \frac{income}{Population} + b_2 \times \frac{income}{Population}^2 + b_3 \times z. \quad (4.1)$$

Regression analysis chooses values for b_0, b_1, and b_2 such that the fitted line goes through the mean level of emissions and income and minimizes the variance between actual emissions and predicted emissions. The coefficient b_0 is the intercept and corresponds to the level of emissions that do not depend on income. The coefficients b_1 and b_2 are the slope coefficients and tell us how emissions will change as income changes. The estimated sign of the slope coefficients reveals to us whether or not we should reject the inverted-U hy-

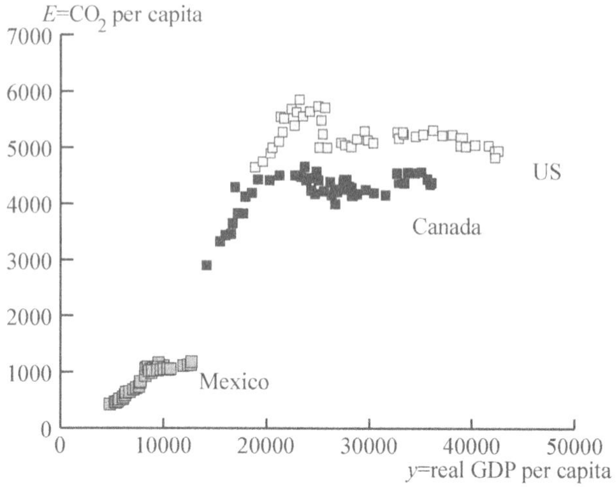

Figure 4.9: Carbon Dioxide Emissions and Per Capita Income, 1963–2008 (Source: Extended Penn World Tables [165]).

pothesis. If the estimated curve through the data reveals that b_1 is positive and b_2 is negative then we would be unable to reject the inverted-U hypothesis. Alternatively, if b_1 is positive and b_2 is also positive then the evidence would lead us to reject the inverted-U hypothesis. In this case, emissions are increasing at an increasing rate.

Other variables, z, might also affect emissions and include energy prices, the percent of population in a country that lives in an urban area, income inequality, literacy, and economic and political freedoms. For instance, if energy prices rise the law of demand indicates that people will consume less energy, causing the EKC to shift downward. Greater urbanization and higher population density can reduce transportation expenses per person as mass transportation becomes more scale efficient.

How emissions change as per capita income changes is found by taking the derivative of the emissions' function with respect to per capita income. The derivative of the equation gives its slope and at the turning point, the slope of the emissions' function is zero. Let $y = \frac{income}{population}$ represent income per capita. Taking the derivative we get

$$\frac{dE}{dy} = b_1 + 2 \times b_2 \times y. \tag{4.2}$$

To find the turning point set the derivative equal to zero and solve for income. Performing this calculation the turning point of income, y^*, is

$$y^* = -\frac{b_1}{2 \times b_2}. \tag{4.3}$$

If $b_1 > 0$ and $b_2 < 0$ then for income levels below y^*, increases in income will cause emissions to increase, while for income levels greater than y^*, increases in income will cause emissions to fall.

The scatter diagram in Figure 4.9 graphs kilograms of carbon dioxide emissions (E) per person and real per capita GDP (y) for Canada, the United States, and Mexico during the period 1963–2008. Two indicator variables, US and CAN, are used to identify the country corresponding to each observation. The indicator variable US takes a value of 1 if the observed data are from the United States and takes a value of 0 if the observed data are from Canada or Mexico. Similarly, the variable $CAN = 1$ if the observed data are from Canada and $CAN = 0$ indicates the data are from the United States or Mexico. Note that if $US = 0$ and $CAN = 0$ then the observed data come from Mexico. Depending on their estimates these two indicator variables shift the vertical intercept for the United States and Canada up or down relative to Mexico. The estimated quadratic regression line takes the form:

$$E = -472.994 + 0.18855 \times y - 3.12418\text{E} - 6 \times y^2 + 3159.9 \times US + 2147.61 \times CAN. \qquad (4.4)$$

These estimates are consistent with an inverted U-shaped EKC. The estimated coefficient for the variable US indicates that per capita emissions in the US are 3159.9 kilograms higher than in Mexico. The coefficient for the variable CAN indicates that emissions are 2147.6 kilograms per person higher in Canada than in Mexico. The turning point of income is estimated as

$$y^* = -\frac{0.18855}{2 \times 3.12418\text{E} - 6} = \$30,176. \qquad (4.5)$$

In 1988, the US real GDP per capita increased to $30,701 surpassing the turning point. In 1998, Canadian real GDP per capita increased to $30,627 surpassing the turning point. Thus, the EKC model predicts falling CO_2 emissions per capita in both countries. However, in Mexico, real GDP per capita in 2008 was only $12,752 indicating that further growth in Mexico will increase per capita CO_2 emissions.[1]

Table 4.2 lists a few of the many studies that have examined the relation between per capita income and various indicators of environmental quality.

One factor that the studies cited in Table 4.2 do not account for is the effect of economic freedom on emissions. Economic freedom consists of a variety of elements including the existence of private property rights, lack of

[1] As we will learn in Chapter 5 desirable and undesirable outputs are jointly produced. Joint production means that emissions and per capita income are simultaneously determined. Simultaneity introduces bias in the simple regression model reported in (4.4). A method called two-stage least squares can help correct for that bias. This correction entailed using labor per capita and the real value of physical capital per capita as instruments with these instruments taken from the Extended Penn World Tables [165]. The new estimated turning point was $30,460.

Table 4.2: Summary of EKC Studies

Study	Indicator	Summary
Berrens, Bohara, Gawande and Wang [22]	Toxic waste sites per manu. firm	$y^* = \$20,253$
	Hazardous waste per waste generator	$y^* = \$17,679$
deBruyn, van den Bergh and Opschoor [61]	CO_2, SO_2, NO_x	1960–93, sustainable growth rate of income for $CO_2 = 0.3\%$, for $SO_2 = 2.6\%$, for $NO_X = 3.8\%$
Harbaugh, Levinson, and Wilson [101]	SO_2, TSP, smoke	No relation between emissions and growth
Hilton and Levinson[108]	Gasoline lead emissions	$y^* \approx$ \$4,000–\$11,000
Selden and Song [213]	SO_2	1973–85, Inverted U
	NO_x	1979–81, Inverted U
	CO	1982–84, Inverted U
Lipford and Yandle [147] review of Cole et al. [47]	CO_2	$y^* \approx \$35,100 - \$54,500$
	NO_x	$y^* \approx \$23,300 - \$24,000$
Grossman and Krueger [93]	Smoke (PM10)	$y^* \approx \$9,800 - \$15,000$
Khanna and Plassman [126]	BOD	$y^* = \$12,000$
	Dissolved Oxygen	$y^* = \$4,290$
Hettige, Mani, and Wheeler [107]	BOD	1989–1995, Positive relation to \$7,000, then constant
Azomahou, Laisney, and Van [14]	CO_2	1990–1996, Positive relation between CO_2 and income
Kahn [121]	Hydrocarbons per mile	2001, $y^* = \$44,000$
Shafik [215]	Deforestation	Inverted U
	Municipal waste	Positively related to income
	Lack of safe water	Inversely related to income
	Lack of urban sanitation	Inversely related to income
	Dissolved oxygen in rivers	Inversely related to income
	Fecal coliform in rivers	Inverted U at low incomes, positive relation at high incomes
	Carbon	Positively related to income
Hervieux and Darné [106]	Ecological footprint	No relation between income and footprint in 29/30 countries

tariff barriers, low government consumption and taxes, stable monetary policies, the rule of law, absence of a military draft, no interest rate controls, an open banking system, and few state-owned enterprises [98]. Interestingly, countries with a high degree of economic freedom have relatively high levels of per capita GDP and experience faster economic growth, lower unemployment rates, higher life expectancies, higher adult literacy, and lower poverty levels than countries with low levels of economic freedom [95]. An even more profound finding is that countries with low levels of political freedom but high levels of economic freedom tend to gain political freedom over time, but countries with high levels of political freedom and low levels of economic freedom tend to lose political rights over time. Therefore, an important topic of research is to investigate the effects that economic freedom exerts on emissions and other environmental indicators. Is there a turning point for emissions with respect to economic freedom? How does the EKC shift as countries gain or lose economic freedom? As China, the former Soviet Union, and other eastern bloc countries adopt market reforms will emissions tend to go up or down holding income constant? These questions strike at the heart of much of the environmental debate today. Do we need more government or greater reliance on property rights and markets to foster a clean environment?

4.6 Dynamic EKC Models

The **sustainable growth rate** can be defined as the growth rate of income that causes no change in emissions. Over time, ecosystems have the ability to cleanse themselves through absorption and regeneration. The length of time is dependent upon the type of emission and the half-life of the toxic release. The process of technological change may also enable income to grow without any increase in emissions, although some observers argue that technological change is the cause of the environmental problem.

Suppose we hypothesize that the growth rate of emissions depends upon the growth rate of income and last period's level of income. Here, last period's income helps determine whether a country is to the left or the right of the turning point of income on the EKC. Let $\%\triangle E$ equal the growth rate of emissions per capita, let y_{t-1} equal last period's income, and let $\%\triangle y$ equal the growth rate of per capita income. The emissions' growth function might then be represented as

$$\%\triangle E = c_0 + c_1 \times y_{t-1} + c_2 \times \%\triangle y. \tag{4.6}$$

The sustainable growth rate of income is found by setting $\%\triangle E = 0$ in Equation(4.6) and solving for $\%\triangle y$:

$$\%\triangle y = -\frac{c_0 + c_1 \times y_{t-1}}{c_2}. \tag{4.7}$$

After estimating c_0, c_1, and c_2 for CO_2, NO_x, and SO_2 the economists de Bruyn, van den Bergh, and Opschoor [61] found that the sustainable growth rate of income in the U.S. was 0.3% for CO_2, 2.6% for NO_x, and 3.8% for SO_2. For comparison, the long-run average growth rate of income in the United States has been around 2–2.5% during the twentieth century. In fact, when growth rates are below 1% many politicians from both political parties claim growth to be "anemic" indicating the need for some kind of fiscal stimulus.

Returning to the data depicted in Figure 4.9, the estimates of the growth rate of CO_2 emissions per capita by Equation (4.6) for the United States, Canada, and Mexico are

$$\%\triangle E = 0.02018 - 1.026\text{E} - 6 \times y_{t-1} + 0.6070 \times \%\triangle y. \tag{4.8}$$

There were no significant differences in the relation between $\%\triangle E$ and $\%\triangle y$ between the three countries so the two indicator variables were dropped from the estimating equation. In 2008, real per capita GDP in the United States, Canada, and Mexico were \$43,326, \$37,515, and \$12,752. Using Equation (4.6) the sustainable growth rate of real GDP in the US is

$$\%\triangle y = -\frac{0.02018 - 1.026\text{E} - 6 \times 43326}{0.6070} = 0.040 = 4\%. \tag{4.9}$$

A similar calculation for Canada, using $y_{t-1} = \$37,515$, yields a sustainable growth rate of 3% and for Mexico, using $y_{t-1} = \$12,752$, the sustainable growth rate of per capita income is -1.17%.

4.7 Environmental Justice

Who bears the burden of environmental degradation? While some research has found a turning point for emissions, policy makers and other social activists are also concerned about the distribution of emissions among different income classes and racial groups. It likely makes little difference to low-income families that overall emissions will eventually fall as aggregate income rises if those same families are the ones exposed to the emissions.

Consider the following items found in *Council of Environmental Quality: 25th Anniversary Report, 1994–95* [55, p. 109]:

- African Americans and Hispanic-origin populations are more likely than whites to live in areas with reduced air quality.
- Low-income residents living in older, poorly maintained buildings are more likely to be exposed to dangerous levels of lead.

- Migrant farm workers are more likely to be exposed to hazardous levels of pesticides and less likely to have access to adequate protective clothing.

Although concerns about environmental justice were first mentioned more than twenty-five years ago in the first annual report of the Council on Environmental Quality, it was not until 1987 when the United Church of Christ report *Toxic Wastes and Race in the United States* [49] brought the issue of environmental justice to the forefront of environmental policy issues. In 1994 President Clinton issued an executive order directing federal agencies to incorporate environmental justice principles into their day-to-day operations. Recently, the EPA has delayed or denied approval of new manufacturing plants in predominantly low-income and African American locales to help move the principle of environmental justice forward.

The empirical evidence on environmental justice is mixed, especially when it comes to the relation between race and proximity to hazardous waste dumps and new plants that emit toxic substances. Brooks and Sethi [31] found that emissions (measured in pounds of sulfuric acid equivalent) are greater in areas where the population has lower educational attainment, higher poverty levels, higher minority concentrations, and higher renter occupied housing. Controlling for the effects of wages, the percent of the population voting in the last election, population density, housing values, education, and the percent of the population in manufacturing, other researchers [134] found no relation between race and toxic releases. Still, poverty was a good predictor of exposure to toxic emissions. On the other hand, Pollock and Vittas [198] found that in Florida, low-income African Americans were more likely to live closer to facilities with toxic releases than whites, Hispanics, or Native Americans, even after controlling for the effects of urbanization, population density, manufacturing employment, and housing values. In another study, Torras and Boyce [237] found that people in countries and cities with low literacy rates and low levels of political rights and civil liberties were exposed to higher levels of sulfur dioxide and smoke and had a higher percent of the population exposed to unsafe drinking water.

A major criticism of the studies that have found evidence of environmental injustice deals with the issue of correlation versus causation. Consider the following example. Suppose that a new hazardous waste dump is opened out in the middle of nowhere. As a result land values fall within some radius of the dump and the lower land prices attract low-income residents. A researcher then goes out, observes that only low-income people live around the dump, and infers the existence of environmental injustice.

Individuals often undertake a risky endeavor after carefully weighing its costs and benefits. In the absence of jobs, many individuals might make a rational decision to live and work in areas with high levels of pollution, rather than not work in an area with low levels of pollution. Bernard Cohen [45, 46] reports that the loss of life expectancy is between 7.2 and 9 years for individuals living in poverty. Living in poverty results in the second largest loss of life expectancy following alcoholism at number one, but preceding smoking for males which is number three. In contrast, the loss of life expectancy due to exposure to hazardous waste is 2.5 days, air pollution is 40 days, and radioactive releases is 0.06 days.

4.8 Race to the Bottom?

When markets are competitive firms and capital are free to move to countries or regions where they expect to have the greatest comparative advantage. In today's global economy capital is increasingly mobile and one worry about the EKC hypothesis is that countries might begin to compete on the basis of lax environmental regulations. Here the concern is that if one country imposes pollution regulations on a particular industry, firms in that industry might just pick up and move to countries where regulations are lax to avoid the regulatory costs. As poor countries seek to attract jobs and income by either relaxing regulations or not regulating pollution, migration of polluting firms to the country might create a haven for polluters. If the poor country then exports the polluting products back to the rich country an EKC pattern would be observed with rising pollution and incomes in the poor country along with decling pollution in the rich country. However, if job and income losses are great enough in the rich country pressure might be brought to deregulate in an effort to stop the capital flight resulting in higher pollution in the rich country.

What evidence is there of capital flight to avoid the regulatory costs of pollution? Numerous studies have found that the costs of pollution control are relatively low and that some low income countries which have low labor and material costs can actually have lower costs of abatement than high income countries [58]. In addition, David Wheeler [253] examined the relation between foreign direct investment and pollution in China, Mexico, and Brazil, three countries that accounted for 60% of foreign direct investment in developing countries in 1998. Wheeler found that suspended particulate matter tended to decline in the three countries as foreign direct investment increased. In addition, air pollution levels in major US cities also declined during the same period, so there is little evidence of a race to the bottom.

4.9 Problems

1. Besides income, other variables might affect pollution levels. For each of the changes given below explain the expected relation be-

tween the variable and emissions. That is, will increases (or decreases) in the variable cause emissions to increase or decrease?

(a) An increase in the price of energy.
(b) An increase in population density.
(c) A decrease in the literacy rate of the population.
(d) An increase in political rights and civil liberties. For example, going from a regime of government censorship to one that allows free speech and freedom of the press.
(e) Suppose that the amount of pollution depends on the concentration of power in society. Also assume that income inequality is a good proxy for power: increases in income inequality cause increases in the concentration of power. What will happen to total emissions and the geographic distribution of those emissions if income inequality decreases?

2. Let y = per capita income and let y^* be the turning point for emissions. In each of the following cases calculate y^* by taking the derivative of each quadratic emissions' function with respect to y, setting the derivative equal to zero and solving for y.

(a) $SO_2 = 1000 + 17,500 \times y - 0.50 \times y^2$
(b) $CO_2 = 1250 + 25,000 \times y - 0.25 \times y^2$
(c) $NO_x = 1500 + 35,000 \times y - 0.75 \times y^2$

3. For each of the following chemicals and its relation to income growth, calculate the sustainable growth rate of income. Assume that last period's income was $y_{t-1} = \$15,000$.

(a) $\%\triangle SO_2 = -0.155 + 0.00001 \times y_{t-1} + 1 \times \%\triangle y$
(b) $\%\triangle CO_2 = -0.10 + 0.00001 \times y_{t-1} + 1 \times \%\triangle y$
(c) $\%\triangle NO_x = -0.05 + 0.00001 \times y_{t-1} + 1.25 \times \%\triangle y$

4. Consider the equation for the growth rate of emissions: $\%\triangle E = c_0 + c_1 \times y_{t-1} + c_2 \times \%\triangle y$.

(a) How would you interpret a negative sign for the coefficient c_0?
(b) What effect does a positive sign for the coefficient c_1 have on the sustainable growth rate?
(c) As the coefficient c_2 becomes smaller, but remains positive, what happens to the sustainable growth rate?
(d) Suppose that $c_1 = 0.00001$. What is the effect of an increase in last period's income, y_{t-1}, on the sustainable growth rate of income?

5. Return to Equation (4.6). Suppose the United States would like to cut CO_2 emissions and that $y_{t-1} = \$43,326$.

(a) What is the sustainable growth rate of real GDP per capita if an annual reduction in emissions equal to $\%\triangle E = -1\% = -0.01$ is desired?

(b) What is the sustainable growth rate of real GDP per capita if an annual reduction in emissions equal to $\%\triangle E = -2\% = -0.02$ is desired?

6. Suppose that the proponents of environmental justice are correct in their argument that low-income and minority citizens are more likely to be exposed to higher levels of toxic emissions. Is this exposure necessarily bad?

7. In a controversial memo, Lawrence Summers who was Chief Economist of the World Bank argued that "I think the economic logic behind dumping a load of toxic waste in the lowest-wage country is impeccable and we should face up to that." Summers went on to say that "I've always thought that under-populated countries in Africa are vastly UNDER polluted." Rightly or wrongly, Summers has been skewered by numerous commentators. Regardless of whether you agree or disagree with Summers' statements, what arguments can you make in support of his proposal to dump toxic waste in low wage countries? Why might the residents of low wage countries agree with Summers?

5

Joint Production of Desirable and Undesirable Outputs

Frequently unnoticed and too often disregarded, undesirable by-products (as well as certain valuable, but unpaid-for natural inputs) are linked directly to the network of physical relationships that govern the day-to-day operations of our economic system.

Wassily Leontief [141, p. 262]

The production process for goods and services often generates undesirable "bad" outputs as well as desirable "goods." For example, when electric utilities burn coal to generate electricity (a desirable output) they also generate pollution (an undesirable output). This pollution might include the discharge of sulfur dioxide, nitrogen oxide, or mercury into the air, or it might entail nuclear waste. Even hydroelectric dams generate a kind of undesirable output when the dam halts upstream migration of spawning salmon and other fishes. When undesirable outputs are jointly produced by-products of desirable output production there is an opportunity cost of reducing the undesirable output: less of the desirable output must be produced.

5.1 Production Possibilities with Two Desirable Outputs

The production possibility set consists of all possible output combinations that can be produced with a given amount of inputs. The traditional production possibilities frontier (PPF) is a subset of the production possibility set and consists of the maximum combinations of two goods that can be produced using society's scarce resources. An important assumption of the production possibility set is that the desirable outputs are freely or strongly disposable. A production technology satisfies free disposability if given some amount of two goods that can be feasibly produced with a given amount of input, smaller amounts of either or both goods can also be produced from the same amount of input.

Consider farmers who use land, labor, and capital inputs to produce coveys of quail (y_1) and bushels of corn (y_2). The production of corn reduces quail habitat and, alternatively, farmers who want better quail habitat have to forego some corn production. Suppose we observe seven different farmers who use identical amounts of land, labor, and capital, say 1,000 acres, 250 labor hours, and $150,000 of capital equipment to produce alternative combinations of corn and quail. Free disposability means that if the farmer can produce 10 units of corn and 10 coveys of quail from the given inputs, it is also possible to produce 5 units of corn and 10 coveys of quail. In this case, 5 units of corn are freely disposed of by simply not growing it (or just throwing it away). The hypothetical amounts of the two outputs produced by the seven farmers are listed in Table 5.1.

Table 5.1: Production Possibilities for Corn and Quail

Farmer	Coveys of Quail (y_1)	Bushels of Corn (y_2)	$MRT = -\frac{\triangle y_2}{\triangle y_1}$	MC per Covey= $5×$MRT$	Revenue when $p_1 = \$2,600$ and $p_2 = \$5$
A	0	7,200	x	x	$36,000
B	2	7,000	100	$500	$40,200
C	4	6,400	300	$1,500	$42,400
D	6	5,400	500	$2,500	$42,600
E	8	4,000	700	$3,500	$40,800
F	10	2,200	900	$4,500	$37,000
G	12	0	1,100	$5,500	$31,200

The production possibility set consists of all the possible combinations of corn and quail that are feasible given inputs. In Figure 5.1 the production possiblity set includes all the points bounded by line segments 0ABCDEFG0. Points inside the frontier, such as I, are feasible, but inefficient. Points outside the production possibility set, such as U, are unobtainable because of scarcity. The production possibility frontier (PPF) corresponds to the set of outputs for which it is impossible to expand both outputs simultaneously given inputs and consists of the points along the line segments ABCDEFG. Outputs are freely disposable if it is possible to produce less of at least one output holding inputs constant. Strong disposability means that if it is feasible to produce 8 coveys and 4000 bushels (point E), one can also produce 6 coveys and 4000 bushels or 0 coveys and 4000 bushels. Similarly, it is possible to produce 8 coveys and 3000 bushels, or 8 coveys and 2000 bushels, or 8 coveys and 0 bushels.

The marginal rate of transformation (MRT) equals the absolute slope of the production possibilities frontier and shows how much of one good (corn) must be foregone to produce one more unit of the other good (quail): $MRT = -\triangle y_2/\triangle y_1$. As such, the MRT measures the physical opportunity cost of producing one additional covey in terms of foregone corn production. For instance, to increase the number of coveys from 0 to 2 the MRT is 200 bushels of corn or 100 bushels of corn per covey. To increase quail from 2 to 4 coveys

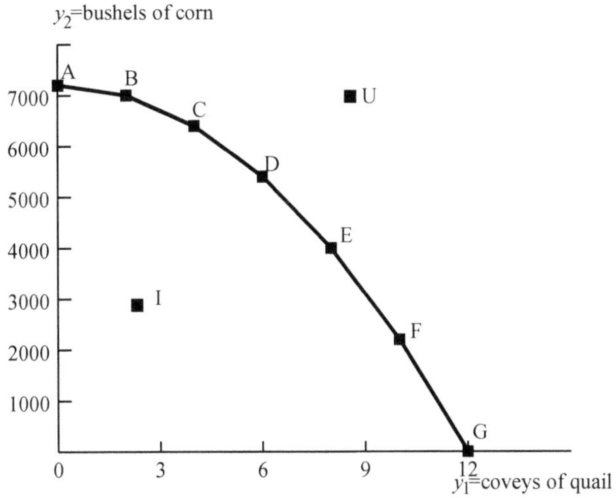

Figure 5.1: PPF for Quail and Corn.

requires 600 fewer bushels of corn be produced or 300 bushels per covey. As quail production increases, the opportunity cost (MRT) increases because resources tend to be specialized with some resources better suited to produce one thing than another. For instance, the best quail habitat is generally an edge between a field and open woods (as opposed to a closed canopy woods) with land that has undergone minimum tilling. After that habitat has been put in place, farmers looking to expand quail production must use land that is less well suited to producing coveys of quail; this is the idea of diminishing returns or increasing opportunity costs.

Figure 5.1 depicts the idea of increasing opportunity costs since the PPF is bowed outward or concave to the origin. As we move from A to B to C, etc., the slope of the frontier becomes more negative, indicating an increasing opportunity cost to producing more quail. Similarly, if we were to start at point G (12 coveys and 0 bushels) and move to F and then E and then D, we would give up 2 coveys each time, but receive smaller and smaller increments to corn production. This pattern reflects diminishing returns, which is just the flip side of increasing opportunity costs.

If more inputs are available, say 2000 acres instead of 1000 acres, the production possibilities frontier will shift outward toward the northeast expanding the production possibilities of the two outputs and possibly allowing the output represented by point U in Figure 5.1 to be produced. Technological progress in production via better seeds or bioengineering would also shift the frontier to the northeast. If the land has some biological maximum carrying capacity, the PPF will intersect the horizontal axis at the same point, such as 12 coveys of quail, but the point where the PPF intersects the vertical axis will increase.

5.2 Production with Desirable and Undesirable Outputs

5.2.1 Theory

When undesirable outputs like pollution are jointly produced by-products of desirable outputs the assumption of free disposability no longer holds for the undesirable outputs. In this case, a milder assumption called weak disposability is employed in modeling the production process. For a given amount of input, weak disposability of outputs implies that only proportional contractions in outputs are feasible. Consider an electric utility producing megawatt hours (mWh) of electricity and simultaneously emitting sulfur dioxide (SO_2) by burning coal. The utility is observed to produce 100 mWh of electricity and 50 tons of SO_2. Weak disposability means that the utility can produce any proportionally smaller amount of both outputs. For instance, given inputs, it could produce 90% of both: 90 mWh of electricity and 45 units of SO_2. Or, given inputs it could produce 40% of both outputs: 40 mWh of electricity and 20 units of SO_2. However, it is not possible for the utility to produce 100 mWh of electricity and 0 units of SO_2. Weak disposability implies an opportunity cost of reducing the undesirable output: if SO_2 is to be cut, the desirable output of electricity must also be cut.

Suppose we observe seven firms that use the same inputs, say 100,000 Btus of energy, 8 units of labor, and \$1.5 million in capital to produce different quantities of electricity (y) and sulfur dioxide (b). The observed outputs of the seven firms are presented in Table 5.2 and depicted graphically in Figure 5.2.

Table 5.2: Production Possibilities of Desirable and Undesirable Outputs

Firm	Electricity mWh (y)	Sulfur dioxide tons(b)	MRT $=\triangle y/\triangle b$	Revenue= $R = p \times y - q \times b$ $p = \$1$, $q = \$4$
origin	0	0	x	\$0
A	15	1	15	\$11
B	28	2	13	\$20
C	38	3	10	\$26
D	43	4	5	\$27
E	46	5	3	\$26
F	42	6	−4	\$18
G	27	5	inefficient	\$7

The small amounts of y and b produced by a firm like A are a consequence of the firm choosing to not burn all of the coal it has available. The assumption of weak disposability means that firms are able to also produce any proportional contraction in both outputs. Such a contraction can be depicted by drawing a line back to the origin from each point as illustrated in Figure 5.3.

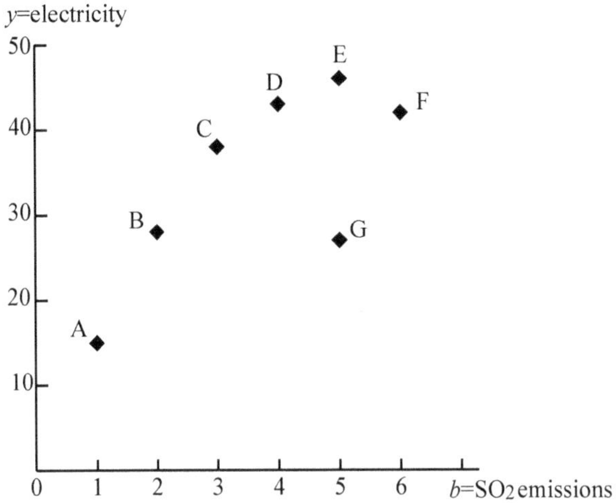

Figure 5.2: Plot of Desirable and Undesirable Outputs.

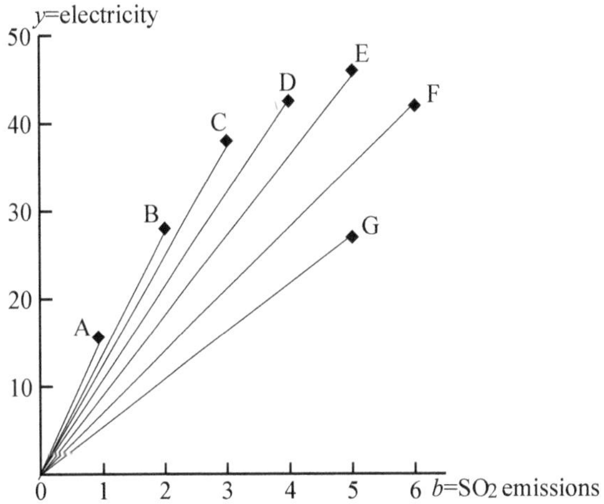

Figure 5.3: Illustration of Weak Disposability of Undesirable Outputs.

We will continue to allow the desirable output to be freely disposable. This means that a firm can produce the same amount of SO_2, but smaller amounts of electricity. The assumption of free or strong disposability is illustrated in Figure 5.4 by the vertical lines drawn from each observation down to the horizontal axis. The production possibility set is bounded by the points 0ABCDEFH0 as shown in Figure 5.5. Firm G produces inside the production

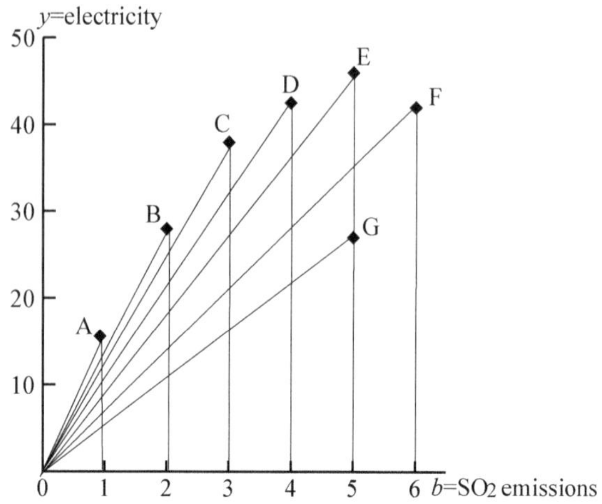

Figure 5.4: Illustration of Strong Disposability of Desirable Outputs.

possibility frontier and is inefficient. If it were to operate efficiently it could reduce SO_2 to 2 tons as firm B has done, or it could increase electricity to 46 mWh holding SO_2 constant as firm E has done. Firm F produces along a negatively sloped portion of the production possibility frontier. This firm could reduce pollution and simultaneously increase electricity up to point E. Sometimes a situation like this occurs when a firm produces so much of an undesirable output that it interferes with the production of desirable outputs. One might imagine a situation where the air is so dirty that workers are unable to breathe and could actually be more productive if the firm produced less pollution.

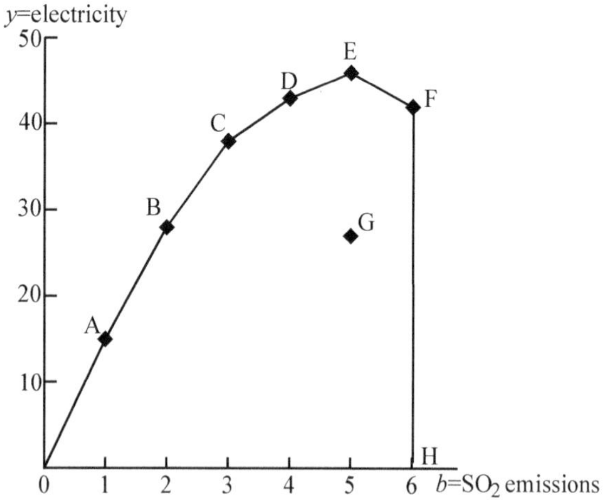

Figure 5.5: PPF with a Desirable and Undesirable Output.

5.2.2 View of the Two Koreas at Night

The production possibility frontier model provides a powerful theoretical method of accounting for opportunity costs, value maximization, technical efficiency, and productivity change. A real life visualization of the PPF is provided in Figure 5.6 which views North Korea and South Korea from space at night. Many commentators have correctly remarked that the darkness of North Korea represents the very high opportunity cost to the country of building a nuclear bomb. But let's think of it another way: taking the desirable output to be electricity and the undesirable output to be carbon emissions, North Korea has a very low carbon footprint compared to South Korea, Japan, and China. In this case, the low carbon footprint means that North Korea is very near the origin of their PPF that accounts for the production of desirable and undesirable outputs.

Why the profound difference between North and South Korea given that they share a common people and geographic area? One possible reason is that the two countries have chosen different institutions to allocate resources and those institutions impact the size of the production possibility set and where the country produces in that set. South Korea uses a form of democratic capitalism where both markets and government play a role in production and consumption decisions. In contrast, North Korea is a communist dictatorship. In 1980, North Korea generated 6.5 tons of carbon per person and its per capita real GDP (in 1990 US dollars) was $2,841. Its counterpart, South Korea, produced $4,114 in per capita GDP, but generated only 3.5 tons of carbon per person. By 2010, South Korea's real GDP had grown to more than 16 times per capita GDP in the North: $30,400 in South Korea vs. $1,800 (estimated) in North Korea. Per capita emissions of carbon also moved in opposite directions; by 2010, North Korea emitted only 2.8 tons per person while in South Korea, emissions averaged 11.9 tons per person [39, 70, 154].

Most people would probably argue that the cost of cutting carbon experienced by North Korea was too high relative to its benefits. We take up the issue of where on the PPF a country should produce if it wants to maximize value in Section 5.4.

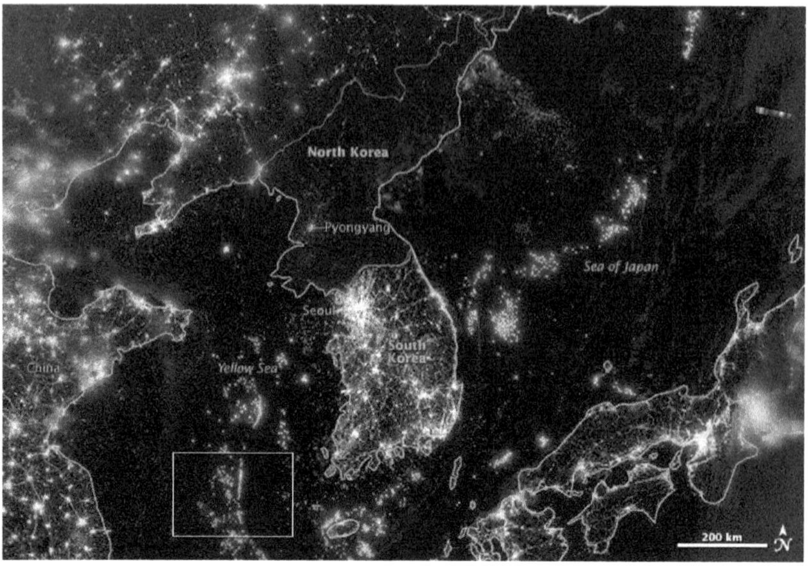

Figure 5.6: North Korea's Carbon Footprint (Source: NASA Earth Observatory, http://earthobservatory.nasa.gov/NaturalHazards/view.php?id=79796).

5.3 Technical Efficiency

5.3.1 Technical Efficiency with Two Desirable Outputs

When a firm or country produces an output combination on the PPF they are technically efficient. When a producer is technically efficient the only way it can produce more of a desirable output is by producing less of another desirable output, or by increasing an undesirable output. Similarly, if a technically efficient producer wants to reduce an undesirable output it must forego the production of some amount of desirable output.

There are a variety of ways to measure the degree of technical efficiency. One such way uses the output distance function to calculate the proportional expansion of all (desirable) outputs that would be feasible given inputs. Figure 5.7 illustrates the calculation of the output distance function using two of the efficient producers (farmers C and D) of corn and quail from Table 5.1 and an inefficient producer, farmer H, who produces 4 coveys of quail and 4720 bushels of corn. If farmer H operated efficiently he could produce either 6,400 bushels of corn and 4 coveys of quail, or 4,720 bushels of corn and between 6 and 8 coveys of quail, or the farmer could expand both corn and quail production proportionally to the frontier at point J.

Let x represent the inputs the firm employs and y represent the desirable

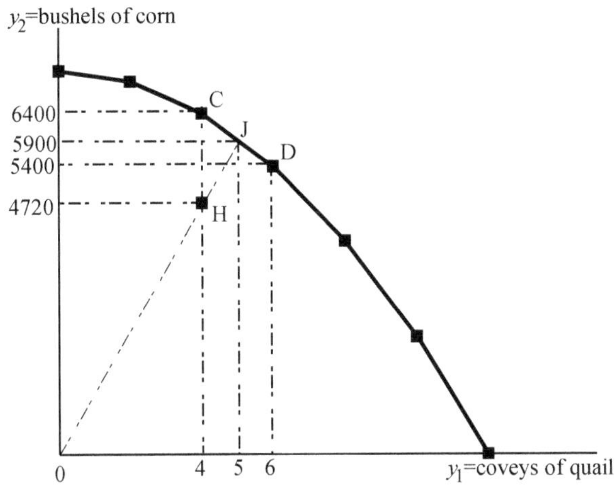

Figure 5.7: Estimating Technical Efficiency.

outputs it produces. As our example indicates, sometimes a producer, like farmer H, is inefficient. The output distance function, $D_o(x,y)$, equals the ratio of actual outputs to maximum potential output: $D_o(x,y) = \frac{\text{actual } y}{\text{maximum } y}$. The value of $D_o(x,y)$ serves as a measure of technical efficiency with higher values indicating greater efficiency. When $D_o(x,y) = 1$ a firm is technically efficient and produces on the production possibilities frontier. When $D_o(x,y) < 1$, a firm is inefficient. The reciprocal of the output distance function, $\frac{1}{D_o(x,y)}$, gives the maximum proportional expansion of two (or more) outputs that can be produced if resources were efficiently employed in the production of output.

To calculate the output distance function draw a ray from the origin through point H to the production possibility frontier at point J and calculate the ratio of the distances 0H to 0J. To perform this operation we first have to calculate the coordinates of point J. These coordinates are found by finding the intersection of the ray through point H and the equation of the line through points C and D. The equation of the ray through point H is $Corn = \frac{4,720}{4} \times Quail = 1,180 \times Quail$. The equation of the line segment CD is $Corn = 8,400 - 500 \times Quail$. To find their intersection at point J we set the two equations equal to each other:

$$\begin{aligned} 1180 \times Quail &= 8400 - 500 \times Quail \\ 1680 \times Quail &= 8400 \\ Quail &= \frac{8400}{1680} = 5. \end{aligned} \tag{5.1}$$

Substituting $Quail = 5$ into either $Corn$ equation we can derive point J's quantity of corn as $Corn = 1,180 \times 5 = 5,900$ or $Corn = 8,400 - 500 \times 5 = 5,900$. Since the output distance function equals the ratio of actual output to maximum potential output we have $D_o(x,y) = \frac{4,720 \text{ bushels}}{5,900 \text{ bushels}} = \frac{4 \text{ coveys}}{5 \text{ coveys}} = 0.80$.

Therefore, farmer H is only 80% efficient relative to the hypothetical farmer J. Also, since $\frac{1}{D_o(x,y)} = \frac{1}{0.80} = 1.25$, farmer H could proportionally expand both corn and quail production by a factor of 1.25.

The output distance function can also be calculated by measuring the distance of the two line segments 0H and 0J and taking their ratio: $D_o(x,y) = \frac{0H}{0J}$.[1] The output distance function depends on inputs, x (such as land, labor, fertilizer, seed, energy, etc.), and outputs y (in this case quail and corn). Firms C and D operate on the frontier and are technically efficient so that $D_o(x,y) = 1$. In addition, the farmers represented by A, B, E, F, and G are also efficient.

A property of the output distance function is that $D_o(x,y) \leq 1$ if and only if the outputs are part of the production possibility set. When $D_o(x,y) > 1$ the outputs are outside the production possibility set. These properties mean that we can use $D_o(x,y)$ to evaluate policies that seek to expand various outputs. For instance, suppose the federal government wants to pursue a biofuels policy that expands the production of corn ethanol but simultaneously maintains the existing quail population. Let y represent the existing corn and quail outputs and let y' represent the new corn and quail outputs, where $y \leq y'$. A feasible policy would imply that $D_o(x,y') \leq 1$. Similarly, the policy is infeasible if $D_o(x,y) > 1$, but could possibly become feasible if the level of inputs, x, are somehow increased, or if sufficient technological progress occurs.

5.3.2 Technical Efficiency with Desirable and Undesirable Outputs

Now, let's return to the case where firms produce a desirable (good) output like electricity and an undesirable (bad) output such as sulfur dioxide. Using the data from Table 5.2 we note that firm G is technically inefficient since it produces inside the production possibility frontier. In the case of desirable outputs, y, and undesirable outputs, b, we consider two alternative measures of technical efficiency. The first measure of technical efficiency equals the ratio of actual desirable output to maximum potential desirable output, holding undesirable outputs constant. If firm G continued to produce 5 tons of SO_2, it could produce up to 46 mwh of electricity as firm E has done. Therefore, technical efficiency is $D_o(x,y,b \text{ constant}) = \frac{\text{actual } y}{\text{maximum } y} = \frac{28}{46} = 60.9\%$.

A second measure of technical efficiency would be to hold the desirable output constant and calculate the ratio of the minimum amount of undesirable output to the actual amount of undesirable output produced. If firm G were

[1] The Euclidean distance function (derived from the Pythagorean Theorem) takes the form: $d(y,x) = \sqrt{(y_2 - y_1)^2 + (x_2 - x_1)^2}$ where y is the value of the variable on the vertical axis and x is the variable on the horizontal axis. The distance OH$=\sqrt{(4,720-0)^2 + (4-0)^2} = \sqrt{22,278,416}$ and the distance OJ$=\sqrt{(5,900-0)^2+(5-0)^2} = \sqrt{34,810,025}$. Therefore, $D_o(x,y) = OH/OJ = \frac{\sqrt{22,268,416}}{\sqrt{34,810,025}} = 0.80$.

efficient in minimizing the undesirable output it could produce 46 mwh of electricity with only 2 tons of SO_2 as firm B has done. Therefore, technical efficiency is $D_o(x, y \text{ constant}, b) = \frac{\text{minimum } b}{\text{actual } b} = \frac{2}{5} = 0.40 = 40\%$.

Sometimes firms or policy makers might have multiple objectives and want to know how much the undesirable output could be reduced while the desirable output is simultaneously increased. [2] If firm G were to simultaneously increase the desirable output (y) and reduce the undesirable output (b) it could produce somewhere on the frontier between points BCDE in Figure 5.5. Finally, we note that firm F is technically efficient in producing the desirable output. It cannot produce more of y given the amount of undesirable output it produces. However, it could reduce its production of the undesirable output from 6 to 4 if it were to hold the desirable output constant and produce like firm D. Alternatively, it could reduce the undesirable output from 6 to 5 and expand the desirable output from 42 to 46 if it were to produce like firm E.

5.4 Allocative Efficiency

5.4.1 Two Desirable Outputs

The overall efficiency of a firm or decision-making unit depends on whether the firm operates on the production possibility frontier, i.e., is technically efficient, and whether the firm is producing the revenue maximizing mix of outputs. When a firm chooses the revenue maximizing mix of outputs it is allocatively efficient. When a firm is allocatively inefficient it might be operating on the production possibility frontier but it could increase revenues by producing more of one good and less of another good. Let the quantities of two goods be represented by y_1 and y_2. Recall that the marginal rate of transformation is defined as $MRT = -\frac{\triangle y_2}{\triangle y_1}$ and measures the opportunity cost of producing one more unit of y_1 in terms of the foregone y_2.

Let the prices of two desirable goods be represented by p_1 for y_1 (coveys of quail) and p_2 for y_2 (bushels of corn). The price ratio, $\frac{p_1}{p_2}$, measures how much of y_2 the consumer is willing to give up for one more unit of y_1, or, alternatively, it measures the consumer's marginal rate of substitution between two goods. For instance, if the price of a covey of quail is $p_1 = \$2,600$ and the price of a bushel of corn is $p_2 = \$5$, then the price ratio is $\frac{\$2,600}{\$5} = 520$, meaning that consumers are willing to pay 520 bushels of corn for 1 additional covey of quail.

Referring back to Table 5.1, suppose that production takes place at point A

[2] One way to measure technical inefficiency when desirable outputs are expanded and undesirable outputs are contracted is by the directional output distance function [64, 75, 77, 246]. This function seeks the maximum simultaneous expansion in desirable outputs and contraction in undesirable outputs given inputs.

on the production possibility frontier. Moving toward point B, the opportunity cost of an additional covey is 100 bushels per covey (or 200 bushels for 2 coveys). Since the price ratio indicates that consumers are willing to forego 520 bushels of corn for an additional covey of quail but the opportunity cost is only 100 bushels of corn, the producer could increase value (revenues) by reallocating some of the inputs toward the production of quail and away from the production of corn. Therefore, if the price ratio of the two goods is greater than the opportunity cost, $\frac{p_1}{p_2} > MRT$, more of good 1 and less of good 2 should be produced. Alternatively, when the price ratio is less than the opportunity cost, $\frac{p_1}{p_2} < MRT$, less of good 1 and more of good 2 should be produced. Resources are allocated efficiently when, given inputs, revenues are maximized. Such a condition corresponds to an output mix such that $\frac{p_1}{p_2} = MRT$. This condition means that any further reallocation of resources to produce more of good 1 and less of good 2 or vice versa will cause value to decrease. That is, the allocation is Pareto efficient.

The revenues, R, earned from selling quantities y at prices p are $R = p_1y_1 + p_2y_2$. In Figure 5.1 we put corn, y_2, on the vertical axis and quail, y_1, on the horizontal axis. Therefore, we solve the revenue equation for corn y_2:

$$\begin{aligned} R &= p_1y_1 + p_2y_2 \\ p_2y_2 &= R - p_1y_1 \\ y_2 &= \frac{R}{p_2} - \frac{p_1}{p_2}y_1. \end{aligned} \tag{5.2}$$

The iso-revenue line represented by (5.2) gives the alternative combinations of y_1 and y_2 that yield the same revenue: $y_2 = \frac{R}{p_2} - \frac{p_1}{p_2}y_1$. For instance, for $R = \$36,000$, $p_1 = \$2,600$, and $p_2 = \$5$, one combination would be $y_2 = 7,200$ bushels and $y_1 = 0$ coveys. Another combination that yields $R = \$36,000$ is $y_2 = 4,600$ bushels and $y_1 = 5$ coveys. Still another combination would be $y_2 = 2,000$ bushels and $y_1 = 10$ coveys. For $R = \$36,000$, one can choose any quantity for y_1, and solve for y_2. Three iso-revenue lines are superimposed on the production possibility set for corn and quail taken from Table 5.1. Notice that when the absolute value of the slope of the iso-revenue line, $\frac{p_1}{p_2}$, is greater (less) than the absolute slope of the production possibility frontier (MRT), revenues can be increased by increasing (decreasing) coveys of quail and reducing (increasing) bushels of corn. Revenues are maximized when $y_1 =$ 6 coveys and $y_2 = 5,400$ bushels of corn are produced at point D.

One can easily show that when the price of one of the good outputs increases relative to the other, value will be maximized by producing more of the higher priced good and less of the relatively lower priced good.

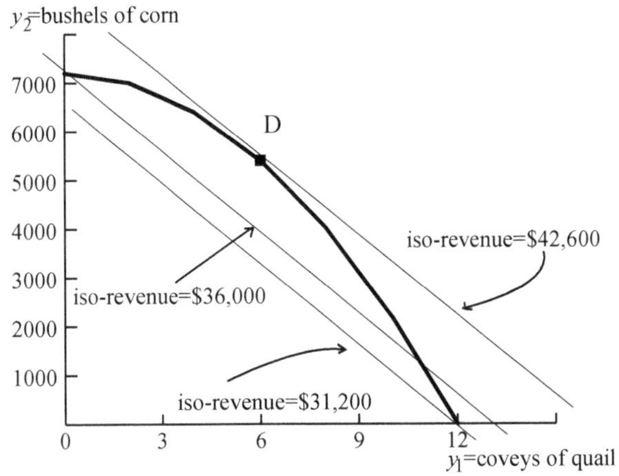

Figure 5.8: Production Possibilities and Iso-Revenue Lines.

5.4.2 Allocative Efficiency with a Desirable and Undesirable Output

When a desirable output such as electricity and an undesirable output such as SO_2 are jointly produced, revenues equal $R = py - qb$ where q is the charge or fine incurred from producing the undesirable output. For instance, suppose the price of electricity is $p = \$1$ and the charge for generating SO_2 is $q = \$4$. Then, for a firm like A, which produces $y = 15$ electricity and $b = 1$ units of SO_2, revenues equal $R = \$1 \times 15 - \$4 \times 1 = \$11$. Since we graphed electricity (y) on the vertical axis in Figure 5.2, the iso-revenue line can be derived from the revenue function by solving for y:

$$
\begin{aligned}
R &= py - qb \\
py &= R + qb \\
y &= \frac{R}{p} + \frac{q}{p}b.
\end{aligned} \tag{5.3}
$$

For instance, given $p = \$1$ and $q = \$4$, a few of the (y, b) combinations that would yield $R = \$11$ include $(y, b) = (15, 1)$, $(y, b) = (19, 2)$, $(y, b) = (23, 3)$, $(y, b) = (31, 5)$, and $(y, b) = (91, 20)$.

With a desirable and undesirable output the marginal rate of transformation gives the extra desirable output that can be generated if one more unit of undesirable output is simultaneously produced, or, alternatively, the amount of desirable output that must be foregone if one less unit of undesirable output is produced. The price ratio, $\frac{q}{p}$, represents the consumer's willingness to accept increased pollution (b) if they receive one more unit of the desirable good (y). Alternatively, the price ratio, $\frac{q}{p}$, represents the amount of desirable good a consumer is willing to forego if the undesirable good is reduced by 1 unit. For instance, given $q = \$4$ and $p = \$1$, the price ratio is $\frac{\$4}{\$1} = 4$ and consumers would accept one more unit of pollution in exchange for four more units of electricity. Alternatively, consumers would be willing to forego 4 units of electricity if SO_2 could be reduced by 1.

If a producer is at a point on the PPF where the $MRT = 8$, then the producer could increase electricity by 8 if SO_2 increases by 1. Given $\frac{q}{p} = 4$ the producer could add to revenues by simultaneously increasing electricity and SO_2. Therefore, when $\frac{q}{p} < MRT$, revenues can increase by simultaneously producing more of the desirable good and more of the undesirable good. If instead $\frac{q}{p} > MRT$, revenues can increase by producing less of the desirable good and less of the undesirable good. Here, when the firm reduces SO_2 emissions, it reduces the charges (fines or taxes) by more than any lost revenue from producing less electricity. Firms that maximize revenues have allocated resources efficiently and are Pareto efficient. Allocative efficiency also implies that $\frac{q}{p} = MRT$.

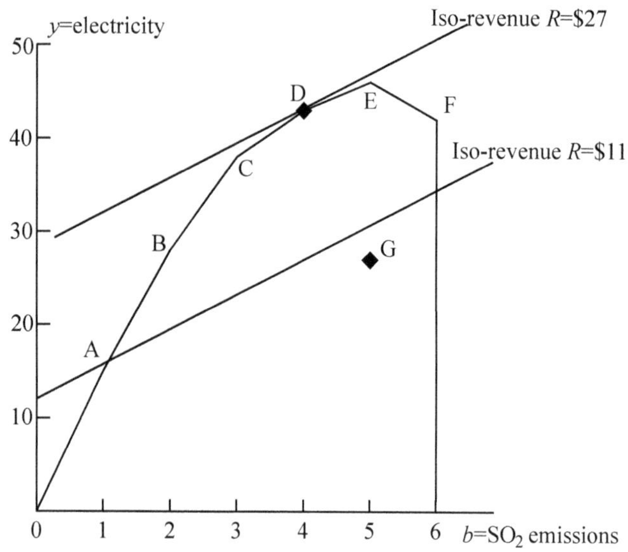

Figure 5.9: Iso-Revenues with Desirable and Undesirable Outputs.

Figure 5.9 superimposes two iso-revenue lines on the production possibilities from Table 5.2. Combination D maximizes revenue for the firms depicted and corresponds to the highest iso-revenue line that is tangent to the production possibilities frontier.

5.4.3 Input Allocative Efficiency

For input allocative efficiency, firms must not only minimize the physical amount of inputs used to produce a given level of output, they must also choose the cost minimizing mix of inputs. A firm is technically efficient with respect to inputs if it is impossible to reduce at least one input and still produce the same level of output. For example, suppose there are three firms each producing 5 mWh of electricity and 5 tons of SO_2. The three firms use different input combinations as listed in Table 5.3.

Table 5.3: Input Technical Efficiency and Allocative Efficiency

Firm	Labor L	Capital K	cost when $w = \$5, p_K = \$10$	cost when $w = \$15, p_k = \$10$
A	4	2	\$40	\$80
B	2	4	\$50	\$70
C	4	4	\$60	\$100

Firm A uses 2 units of capital and 4 units of labor. Firm B uses 4 units of capital and 2 units of labor. Firm C uses 4 units of capital and 4 units of labor. Clearly firm C is technically inefficient since it could use either the same amount of capital and 2 fewer units of labor or it could use the same amount of labor and 2 fewer units of capital and still produce the same level of output. Firms A and B are technically efficient. Now suppose the price of capital is $p_K = \$10$ and the price of labor is $w = \$5$. Total costs of production are $C = w \times L + p_K \times K$ and are reported in Table 5.3. Firm A has allocated resources efficiently since it has chosen the cost-minimizing mix of inputs. Firms B and C have failed to minimize costs and are allocatively inefficient. Now, suppose the price of labor increases to $w = \$15$ while the price of capital remains constant at $p_K = \$10$. Firm B is now allocatively efficient since it minimizes costs. Note the change in resource use. As the relative price of labor increased from $\frac{w}{p_K} = \frac{\$5}{\$10} = 0.5$ to $\frac{w}{p_K} = \frac{\$15}{\$10} = 1.5$, the cost minimizing quantity of labor decreased from 4 to 2 units. The same amount of output can still be produced, but it is efficient for the firm to use more capital and less labor.

An understanding of both allocative efficiency and technical efficiency is important for the design of efficient environmental policy. For instance, if we are concerned with cleaning up a certain amount of pollution generated by electric utilities, regulators might require that all plants install scrubbers on their smokestacks. This policy may or may not be efficient and will depend on whether burning low sulfur coal is a cost-effective alternative to scrubbers. It is also important to keep in mind the trade-off that must occur in the production process of desirable and undesirable outputs. If reducing pollution is the goal, we would like the decline in the value of the desirable good to be less than the gain in value from reducing pollution. If not, society will be worse off.

For instance, if pollution causes $100 damage and can only be reduced at a cost of $250 in lost production of the desirable output it would be allocatively inefficient to reduce pollution.

5.5 Productivity Change

5.5.1 Productivity Change with Desirable Outputs

Over time the production possibility frontier will shift due to technological progress or regression. In addition, firms might become either more or less efficient thereby moving either closer or further from the PPF. Figure 5.10 illustrates technological progress in the case of two desirable outputs and Figure 5.11 illustrates the case of technological progress in the case of a desirable and an undesirable output.

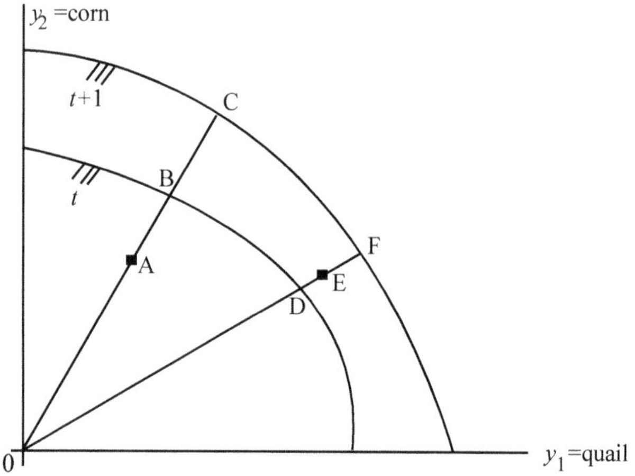

Figure 5.10: Technological Change with Two Desirable Outputs

Labor productivity is defined as the amount of output per unit of labor, y/L, where y equals the amount of output produced and L equals labor. When two or more outputs are produced using two or more inputs, measures of total factor productivity are employed as an alternative to labor productivity. Total factor productivity change can be decomposed into efficiency change and technological change. Efficiency change measures whether the firm is getting closer to or farther from the production possibility frontier from period t to $t+1$ and technological change measures the shift in the frontier holding inputs constant.

In Figure 5.10 we observe a firm producing at point A in period t and at point E in period $t+1$. We know the level of efficiency in period t is given by the output distance function: $D_o^t(x,y) = \frac{0A}{0B}$. In period $t+1$ the firm's efficiency is $D_o^{t+1}(x,y) = \frac{0E}{0F}$. We can form an index of efficiency change as

$$EFFCH = \frac{D_o^{t+1}(x,y)}{D_o^t(x,y)} = \frac{0E/0F}{0A/0B}. \tag{5.4}$$

Increases in efficiency from t to $t+1$ are indicated by $EFFCH > 1$ and decreases in efficiency from t to $t+1$ are given by $EFFCH < 1$.

Technological change measures the shift in the production possibility frontier from period t to period $t+1$. Once again, in Figure 5.10 we can see technological progress by the northeasterly shift in the production possibility frontier. If the firm represented by A held its mix of outputs constant it could expand all the way to point C in period $t+1$ if it were to operate efficiently. The ratio $0C/0B$ measures potential output growth due to technological progress. If the firm producing at E in period $t+1$ were to hold the mix of outputs constant it could have produced at D in period t, but if it were efficient in period $t+1$ it could have produced at point F. The potential growth of outputs due to technological progress for this mix of outputs would be represented by the ratio $0F/0D$. The period $t+1$ PPF might not be parallel to the period t PPF so the ratio $0C/0B$ might not equal the ratio $0F/0D$. Such a nonparallel shift would be due to biased technological progress where it becomes relatively easier to produce one of the outputs. To account for this possibility an index of technological change can be constructed as the geometric mean of the two measures of technological progress. That is,

$$TECH = \left(\frac{0C}{0B} \times \frac{0F}{0D}\right)^{0.5}. \tag{5.5}$$

Values of $TECH > 1$ indicate technological progress and values of $TECH < 1$ indicate technological regress. Multiplying the index of efficiency change and the index of technological change gives what is called the Malmquist index (for economist Sten Malmquist) of total factor productivity change:

$$MALM = EFFCH \times TECH. \tag{5.6}$$

When $MALM > 1$ total factor productivity has increased and when $MALM < 1$ total factor productivity has decreased.

5.5.2 Productivity Change with Desirable and Undesirable Outputs

When undesirable outputs (bads) and desirable outputs (goods) are jointly produced we can also measure efficiency in the two periods and construct indices of efficiency change and technological change. Figure 5.11 illustrates the PPF for period t and period $t+1$ and illustrates technological progress between the two periods. Technological progress is evident in that the PPF shifts toward the northwest so that holding inputs constant, it is possible to produce less of the undesirable output and more of the desirable output in period $t+1$ than was possible in period t. A firm is observed to produce at point A in period t with desirable output efficiency given by the ratio HA/HB. In period $t+1$ the firm produces at E with desirable output efficiency given as GE/GF. Efficiency change is $EFFCH = \frac{D_o^{t+1}(x,y,b \text{ constant})}{D_o^t(x,y,b \text{ constant})} = \frac{GE/GF}{HA/HB}$. Values of $EFFCH > 1$ indicate greater efficiency and $EFFCH < 1$ indicates less efficiency.

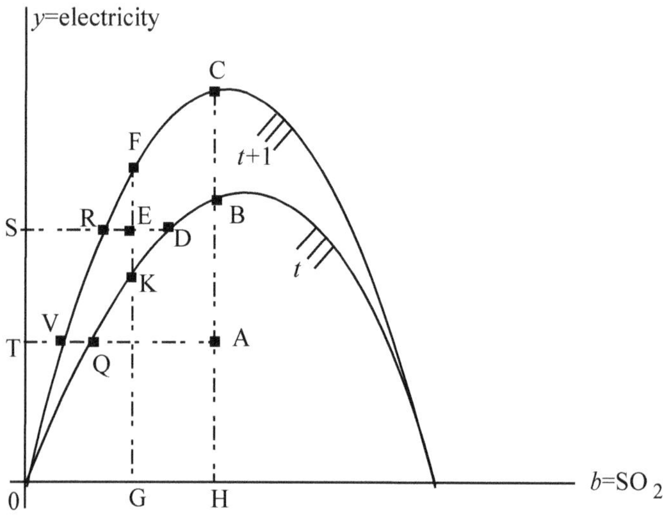

Figure 5.11: Technological Change with a Desirable and Undesirable Output.

We can also construct an index of technological progress when firms produce both a desirable and an undesirable output. Assuming we measure technological change as the vertical shift in the production possibility frontier from period t to period $t+1$, the index of technological change for the firm represented in Figure 5.11 is $TECH = \left(\frac{CH}{BH} \times \frac{FG}{KG}\right)^{0.5}$. Once again, total factor productivity equals the product of efficiency change and technological change. Although we have measured productivity change "vertically" in the case of a desirable and undesirable output by holding the undesirable output constant, we could also measure productivity change "horizontally" by holding the desirable output constant and examining the reduction in the undesirable output due to efficiency change and due to technological change. We leave this case as one of the end of chapter problems.

There are many policy questions related to productivity change that can be addressed. For instance, do firms in regulated industries experience greater or less productivity growth relative to firms in unregulated industries? Are regulators accounting for all of the desirable and undesirable outputs in their policies? How will the regulation of one pollutant affect the quantities of other pollutants? Recently, the relicensing of hydroelectric dams by the Federal Energy Regulatory Commission (FERC) has been subject to utilities maintaining in-stream flows for downstream fisheries and the installation of fish ladders so that migratory species can pass around the dam [236]. Today, the regulatory process even considers the removal of dams as a potentially cost effective means of restoring migratory fish populations such as Atlantic salmon in New England's rivers [112].

5.6 Aggregate Production Possibilities Frontier

An aggregate or industry production possibilities frontier can be constructed from the individual firm's production possibilities frontiers. The aggregate PPF will have the same properties of the individual PPFs: weak disposability of undesirable outputs, strong disposability of desirable outputs, convexity, and the desirable output will increase at a decreasing rate as more undesirable output is jointly produced. Consider two firms that use different levels and mixes of inputs to jointly produce a desirable output and an undesirable output with their production possibilities given in Table 5.4.

Table 5.4: Individual Firms' PPF

Firm 1			Firm 2		
y	b	$MRT = \frac{\triangle y}{\triangle b}$	y	b	$MRT = \frac{\triangle y}{\triangle b}$
0	0	x	0	0	x
8	1	8	10	1	10
15	2	7	19	2	9
21	3	6	27	3	8
26	4	5	34	4	7
30	5	4	40	5	6
33	6	3	45	6	5

Note that the two firms have different marginal rates of transformation at each level of output. According to the principle of comparative advantage the firm with the highest MRT should undertake the production of the marginal amount of desirable output. Suppose neither firm is producing any desirable output or undesirable output. Since Firm 2 has the higher MRT it should produce the first 10 units of output. In addition, Firm 2 should also produce the next 9 units of output. Next, since Firm 1 and Firm 2 have an $MRT = 8$, both firms should produce the next 16 units of output. Similarly, after 35 total units of y and 4 total units of b are produced, each firm should produce an additional 7 units of output since their respective MRTs equal 7. The aggregate PPF is constructed in Table 5.5.

Table 5.5: Aggregate PPF

y	b	$MRT = \frac{\triangle y}{\triangle b}$
0	0	x
10	1	10
19	2	9
35	4	8
49	6	7
61	8	6
71	10	5
75	11	4
78	12	3

Efficiency requires that the MRT for the last unit of desirable output produced should be the same for all firms in a particular industry. This efficiency condition is equivalent to the equal marginal cost principle investigated in Chapter 2. For instance, suppose that policy makers and citizen voters have somehow decided that it is optimal for only $b = 8$ units of undesirable output to be produced. The aggregate PPF shows that when $b = 8$ it is possible to produce up to $y = 61$ units of the desirable output. For the 61st unit of desirable output the $MRT = 6$. Efficiency requires that Firm 1 produce (y, b)=(21,3) and Firm 2 produce $(y, b) = (40, 5)$. As easily seen in Tables 5.4 and 5.5 the $MRT = 6$ for each individual firm and in the aggregate. To show that these levels of production are efficient let Firm 1 produce 1 more unit of b for a total of $b = 4$ and let Firm 2 produce 1 less unit of b for a total of $b = 4$. Aggregate desirable output production now equals $26 + 34 = 60$, which is 1 less than when the MRTs are equal for both firms. A similar line of reasoning shows that if Firm 1 produces only $b = 2$ and Firm 2 produces $b = 6$, aggregate output equals $15 + 40 = 55$ and is less than what could be efficiently produced given a total of $b = 8$.

Alternatively, if society has some target for the desirable output, say $y = 49$, pollution is minimized when the $MRT = 7$ and Firm 1 produces $(y, b) = (15, 2)$ and Firm 2 produces $(y, b) = (34, 4)$. You can check this result by noting that if Firm 1 produced $(y, b) = (30, 5)$ and Firm 2 produced $(y, b) = (19, 2)$ the aggregate amount of $y = 49$, but now $b = 7$ is produced which is an inefficient outcome. In this case, Firm 1's $MRT = 4$ and Firm 2's $MRT = 9$ indicating that Firm 2 should produce more of both outputs and Firm 1 should produce less of the two outputs.

5.7 Regulated Production Possibilities

When pollutants are unregulated or when there are no property rights to environmental resources producers can freely dispose of these pollutants, although some producers might choose to self-regulate and limit pollution. For instance, homeowners who live in a municipality where there are no restrictions on wood-burning stoves or fireplaces can freely dispose of wood smoke. However, if a family member suffers some respiratory ailment, the family might voluntarily forego using the fireplace at times when the afflicted family member is outside. Similarly, farmers who own access to streams and like to fish might choose to fence the stream to keep livestock out or they might provide a buffer zone between stream and field to prevent fertilizer or livestock wastewater runoff.

Regulations, pollution taxes, and/or the assignment of private property rights to scarce resources increase the cost of disposing of pollutants and will restrict the firm's production possibilities. As an example, the US EPA has restrictions on ozone emissions. Counties that fail to meet these restrictions can be assigned nonattainment status which can, in turn, curtail some economic activity within the county.

As a hypothetical example, suppose we observe four counties that each use the same amount of input (say $x = 1$ unit), but generate different levels of money income (y) and ozone (b) as listed in Table 5.6.

Table 5.6: Regulated Production Possibilities

County	b =Undesirable Output	y =Money Income
P	2	6
Q	4	10
S	6	12
T	4	8

The points from Table 5.6 are plotted in Figure 5.12. If ozone is freely disposable the production possibilities set is bounded by 0FSE0 and the PPF consists of the horizontal line FS and the vertical line SE. Counties P, Q, and T produce inside the frontier and are inefficient. If these counties were to increase efficiency through better management or incentives they could generate more income. County S is efficient as it produces on the frontier.

When ozone is regulated, there is an opportunity cost to its disposal. This opportunity cost equals the lost income that occurs because some resources that were producing income can no longer be employed. The regulated production possibility set is bounded by the lines 0PQSE0 and the PPF consists of 0PQSE. The positive slope of the frontier over the range 0PQS represents the opportunity cost of reducing ozone pollution. Counties P, Q, and S all produce on the regulated frontier and are efficient while county T produces inside the frontier. County T can either reduce ozone, moving to a point like W on the regulated frontier, or it can increase income, moving to a point like Q on the regulated frontier, or it can simultaneously reduce ozone and increase income moving to some point on the line segment WQ.

The vertical distance between the unregulated frontier and the regulated frontier represents the total cost of the regulation in terms of foregone income and the slope of the regulated PPF represents the marginal cost of reducing pollution. For instance, county P produces income of $y = 6$ but it could have produced income of $y = 12$ if it were allowed to freely dispose of pollution. The total cost of the regulation for county P is 6 (= 12 – 6). Since the slope of the PPF between the origin and point P equals 3, county P's marginal cost of reducing ozone is 3 in lost income. County Q produces $y = 10$ but could produce $y = 12$ if it were efficient. Therefore, the total cost of the regulation for County Q is 2 (= 12 – 10) in lost income. The slope of the PPF between P and Q equals 2 and represents the marginal cost of reducing ozone for county Q. For County S, there is no cost to the regulation since it already produces on both the regulated and unregulated frontiers. However, if County S is required to make cuts in the undesirable output, the marginal cost of the cut equals the slope of the PPF between S and Q which equals 1. Furthermore, as larger cuts in the undesirable output are made, the marginal cost of reducing it further increases since the slope increases as one moves from S to Q to P to the origin. Applying the principles of equal marginal cost, we can easily see that if County S were to reduce ozone by two units and County P were to increase ozone by two units, total income among the three counties would increase even though total ozone remained constant.

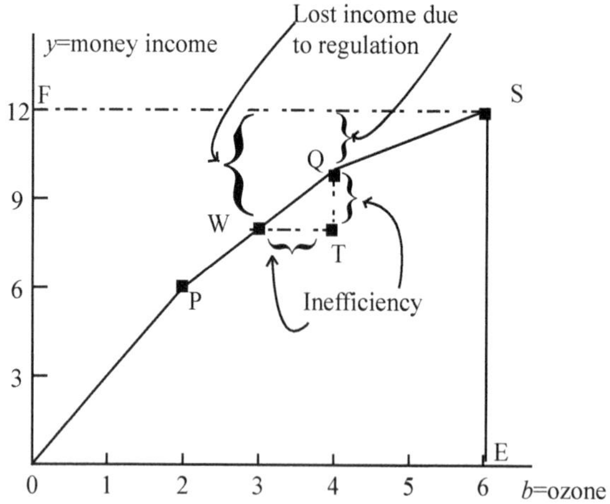

Figure 5.12: Regulated Production Possibilities Frontier

Michael Porter and Claas van der Linde [199] have suggested that the regulatory process might force firms to innovate and adapt more quickly than their nonregulated counterparts. The Porter hypothesis argues that properly designed market-based policies, such as pollution taxes or emissions' permits, can be used to induce innovation by reducing organizational inefficiencies. Innovation would shift the production possibility frontier outward and if organization inefficiencies are reduced, firms would move closer to their production possibilities frontier. If true, the regulatory process can be win-win for the environment and the economy. On the other hand, if environmental regulation can induce innovation and greater firm efficiency, then firms face an unfilled profit opportunity and will have an incentive to harness the resources to exploit this profit opportunity, even in the absence of regulation. Is evidence available to support the Porter hypothesis? Studies examining productivity change of oil and gas production in the Gulf of Mexico [162], manufacturing sectors in Quebec during 1985-1994 [137], and Swedish manufacturing [29, 30] have found very little evidence to support the Porter hypothesis.

5.8 Sustainability

Given the production model with desirable and undesirable outputs a formal definition of sustainability is possible. To illustrate the sustainable changes from period t to period $t+1$ we assume that a producer uses the same amount of input in both periods: $x = x^t = x^{t+1}$. This assumption can be dropped but is maintained so as to illustrate sustainability in a two dimensional diagram. Let the production possibility set in period t be represented by $P^t(x)$. This set gives all the feasible combinations of desirable outputs (y) and undesirable outputs (b) including inefficient combinations of y and b that can be produced by inputs (x). Similarly, the production possibility set in period $t+1$ is $P^{t+1}(x)$.

Strong sustainability: A change in production from $(y^t, b^t) \in P^t(x)$ to $(y^{t+1}, b^{t+1}) \in P^{t+1}(x)$ is strongly sustainable if $y^t \leq y^{t+1}$ and $b^{t+1} \leq b^t$.

A change in production satisfies strong sustainability if more desirable output and less undesirable output is produced in period $t+1$ than in period t. Figure 5.13 shows the PPF for a firm that is observed to produce at I= $(5, 5)$ in period t. Technological progress occurs between t and $t+1$ and is represented by the shift to the northwest in the production possibility set, $P(x)$. At I $= (y^t, b^t) = (5, 5)$ the firm is inefficient in period t, but the following changes in production are sustainable: M $= (y^{t+1}, b^{t+1}) = (6, 4)$, N $= (y^{t+1}, b^{t+1}) = (5, 2)$, and T $= (y^{t+1}, b^{t+1}) = (10, 5)$. In case M, the firm produces more y and less b in period $t+1$ than they did in period t. In case N, the firm produces the same amount of y but less b in period $t+1$ than they did in period t. In case T, the firm produces more y but the same amount of b in period $t+1$

than they did in period t. Two cases that do not satisfy the definition of strong sustainability are Q $= (y^{t+1}, b^{t+1}) = (3, 1)$ and S $= (y^{t+1}, b^{t+1}) = (11, 9)$. In case Q, the firm produces less b but it also produces less y. In case S, the firm produces more y, but it also produces more b.

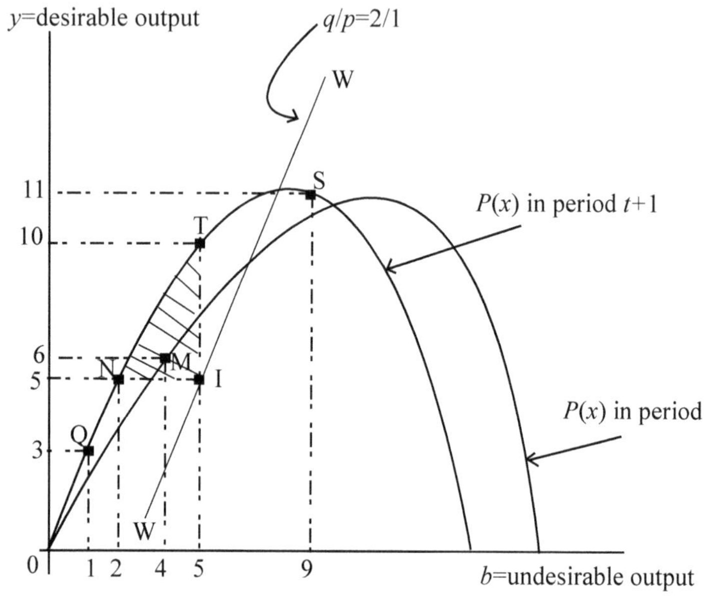

Figure 5.13: Strong Sustainability with Desirable and Undesirable Outputs

The shaded area to the northwest of point I represents the set of strongly sustainable changes in production. Part of any sustainable change can be attributed to greater efficiency and part of the change can be attributed to technological progress. For instance, if the firm moves from I to M they experience gains in efficiency relative to the period t frontier, but are still inefficient relative to the period $t+1$ frontier. Move from I to N or I to T are strongly sustainable changes where the firm experiences both greater efficiency and technological progress.

To evaluate changes from a point like I in Figure 5.13 to a point such as Q or S we need an alternative notion of sustainability that depends not just on the quantities of the desirable and undesirable outputs produced in the two periods, but also on their prices. Again, holding inputs constant $x = x^t = x^{t+1}$ the revenues a firm earns when they face a desirable output price p and a fine or charge for the undesirable output of q equal $R = py - qb$.

Weak sustainability: A change in production from $(y^t, b^t) \in P^t(x)$ to $(y^{t+1}, b^{t+1}) \in P^{t+1}(x)$ satisfies weak sustainability if $R^{t+1} = py^{t+1} - qb^{t+1} \geq R^t = py^t - qb^t$.

The above definition checks whether the value generated in period $t+1$ is greater than the value generated in period t. To check whether the changes from I to Q and I to S satisfy weak disposability let $p = \$1$ and $q = \$2$. The revenue generated from point I is $R = \$1 \times 5 - \$2 \times 5 = -\$5$. At point Q the revenue generated is $R^{t+1} = \$1 \times 3 - \$2 \times 1 = \$1$. Since more revenue is generated at Q than at I the change from I to Q satisfies weak sustainability. In contrast, revenue at S is $R^{t+1} = \$1 \times 11 - \$2 \times 9 = -\$7$ and so the change from I to S does not satisfy weak sustainability. In Figure 5.13 all of the points to the northwest of the line WW drawn through point I with slope $q/p = 2/1$ satisfy weak sustainability.

We can also evaluate changes in production when multiple undesirable outputs are produced. For instance, electric utilities often have coal-fired plants which generate greenhouse gases and nuclear-powered plants where radioactive waste is the undesirable by-product. If greenhouse gas emissions are to be reduced, electric utilities might turn to nuclear plants as an alternative, causing one undesirable output to be substituted for another.

Table 5.7 lists hypothetical production possibilities for electric utilities that all use the same inputs and produce the same amount of desirable output (electricity) in period t and Figure 5.14 graphs the alternative combinations of undesirable outputs. Recall that in the case of two desirable outputs the PPF is concave to the origin. When two undesirable outputs are produced the PPF is convex to the origin. The negative slope implies that holding the desirable output constant, if less of one undesirable output is produced more of another undesirable output will take its place. The convex shape of the PPF with two undesirable outputs represents increasing opportunity costs: as production moves from E to D to C to B to A, the opportunity cost of reducing radioactive waste increases. Alternatively, as radioactive waste is substituted for CO_2 emissions and production moves from A to B to C to D to E, the reduction in CO_2 emissions becomes less and less for each one unit increase in radioactive waste.

Table 5.7: Undesirable Output Possibilities

Firm	mwh of Electricity$= y$	Nuclear Waste$= b_1$	Carbon emissions$=b_2$	Revenue$= R = py - q_1b_1 - q_2b_2$
A	100	1	25	\$45
B	100	2	15	\$60
C	100	3	10	\$65
D	100	4	7	\$66
E	100	5	5	\$65
F	100	4	15	\$50

Firm F produces more radioactive waste than firms B or C and produces more carbon emissions than firms C or D and is inefficient. Technological progress would shift the trade-off between radioactive waste and carbon emissions to the southwest and result in fewer undesirable outputs for a given amount of desirable output. In period t, the set of strongly sustainable changes for firm F are in the shaded area where less of at least one undesirable output and no more of the other undesirable output are produced.

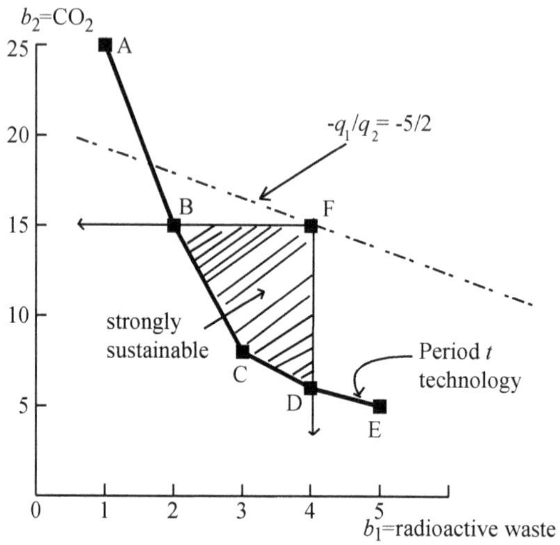

Figure 5.14: Substitution between Two Undesirable Outputs

If firm F were to move to a point like A or E along the period t undesirable output frontier, one of the undesirable outputs would be reduced but the other undesirable output increased and so the move would not satisfy strong sustainability. However, such a change might satisfy weak sustainability. Let the amount of electricity produced by all six firms be $y = 100$ and let the price of electricity be $p = \$1$. Let the charge on radioactive waste be $q_1 = \$5$ and the charge on CO_2 emissions be $q_2 = \$2$. Revenues are calculated as $R = \$1 \times 100 - \$5 \times b_1 - \$2 \times b_2$ and are reported in the last column of Table 5.7. Firm D maximizes revenues and is allocatively efficient. The dashed line running through point F has a slope equal to the price ratio of the two undesirable outputs: $-\frac{q_1}{q_2} = -\frac{5}{2}$. For firm F, the set of undesirable outputs that would satisfy weak sustainability includes all of the points to the southwest of the dashed line. Note that the shaded area corresponding to the set of outputs that are strongly sustainable also satisfy weak sustainability. In addition combination E satisfies weak sustainability since it generates more revenues than combination F. However, combination A does not satisfy weak sustainability for firm F since it generates less revenues than F and is above the dashed line.

Such trade-offs between undesirable outputs should be accounted for in any discussion of sustainability. In addition, many such changes in production will involve trading one undesirable output for another and will not satisfy the condition of strong sustainability. For instance, many utilities have substituted the use of natural gas for coal in generating electricity. Such a change reduces greenhouse gas emissions, but recently exploited sources of natural gas on the Green River Plateau in Wyoming have caused natural wildlife migration corridors to be disrupted and might reduce winter grazing lands for antelope, deer, and moose [211]. In addition, the use of hydraulic fracturing to force natural gas toward the earth's surface has been associated with contamination of wells and a reduction in stream flows in the Marcellus region of the Appalachian Plateau of Pennsylvania and West Virginia [152]. It might very well be desirable to substitute natural gas for coal-fired plants as society seeks ways to reduce greenhouse gas emission. However, those trade-offs should be evaluated at the undesirable output prices to ensure that the trade-off satisfies weak sustainability.

5.9 Problems

1. The table below depicts the outputs of six timber companies who all use the same level of input (land, labor, and capital) to produce board feet of timber and wildlife such as owls and woodpeckers.

Company	Timber	Wildlife	Revenues	Revenues'
A	25	0		
B	20	0		
C	18	4		
D	10	12		
E	5	5		
F	0	15		

(a) Draw a graph of the production possibility frontier for the timber-wildlife companies putting timber on the vertical axis and wildlife on the horizontal axis.

(b) The production possibility set corresponds to all the combinations of the two goods (both efficient and inefficient) that can be produced from the given amount of input. Shade the area of the production possibility set.

(c) Which companies are technically efficient and produce on the boundary of the production possibility frontier?

(d) Which companies represent inefficient producers of timber and wildlife? Explain.

(e) How can you see free disposability of the two desirable outputs in the production possibility set?

(f) Suppose that the price of timber is $p_{timber} = \$3$ and that timber companies can charge wildlife photographers for access to their land which results in a price of wildlife of $p_{wildlife} = \$6$. Calculate total revenues for each of the companies and determine the allocatively efficient mix of timber and wildlife.

(g) Now suppose that the price of timber is $p_{timber} = \$3$ and $p_{wildlife} = \$2$. Calculate total revenues for each of the companies and determine the allocatively efficient mix of the two outputs.

(h) Circle the correct answer in parentheses. As the price of wildlife falls relative to the price of timber the quantity of wildlife supplied (increases/decreases) and the quantity of timber supplied (increases/decreases).

2. The table below is taken from observations of seven pulp to paper firms that all use the same amounts of input (capital, labor, and chemicals) but produce different combinations of paper (y) and chemical waste (b) which is an undesirable (bad) output.

Firm	y	b	Revenues	Revenues'	Revenues"
1	0	0			
2	10	2			
3	15	15			
4	20	15			
5	30	30			
6	40	52			
7	30	65			

(a) Graph the coordinates of paper produced (y) and the jointly produced chemical waste (b) with paper on the vertical axis and chemical waste on the horizontal axis, then draw in the boundary of the production possibilities frontier.

(b) Shade the production possibility set. This set corresponds to all the possible combinations (efficient and inefficient) of paper and the jointly produced undesirable chemical waste that can be produced.

(c) Identify the efficient and inefficient producers.

(d) How does the production possibility set illustrate the idea of weak disposability of the desirable output and free disposability of the desirable output?

(e) Suppose the price of paper is $p = \$1$ and we allow paper companies to dump chemical waste into the watershed at no charge, $q = 0$. Calculate total revenues for each firm and determine the mix of paper and chemical waste that is optimal from the firm's point of view.

(f) Now, suppose the price of paper is $p = \$1$ but we no longer allow the firms to dump their chemical waste into watersheds at no cost. Instead the charge (or fine) for dumping chemical waste is $q = \$0.50$. Calculate total revenues for each of the seven firms in the column labeled Revenues'. Determine the allocatively efficient mix of paper and chemical waste that should be produced.

(g) Finally, suppose $p = \$1$ and $q = \$1$. Calculate revenues for these new prices in the column labeled Revenues". Determine the allocatively efficient mix of paper and chemical waste that should be produced.

(h) As the cost of dumping chemical waste relative to the price of paper increases, the optimal amount of chemical waste to produce (increases/decreases) and the optimal amount of paper (increase/decreases).

3. From one period to another firms might experience productivity change. Increases in productivity occur when a firm can produce more desirable output from a given amount of input. This occurs for one of two reasons: changes in the efficiency with which a firm operates, i.e. ,moving closer to the production possibilities frontier, or technological change, i.e., shifts in the production possibilities frontier. Suppose that problem 1 refers to the observed production combinations for the timber companies in period $t = 1$. The period $t = 2$ production combinations for those same companies is given below. Assume that the timber companies use the same amounts of inputs in both periods.

Company	Timber	Wildlife	Efficient (Yes/No)	Productivity (Increase/Decrease)
A	30	0		
B	20	6		
C	22	6		
D	12	14		
E	10	6		
F	0	20		

(a) Plot the points and then sketch in the production possibilities frontier (PPF) for $t = 2$ on the same diagram as the PPF for $t = 1$.

(b) Now complete the table by determining whether each company is on the frontier in $t = 2$ and whether their productivity has increased or decreased.

(c) Has the PPF shifted from $t = 1$ to $t = 2$ in a way that indicates technological progress or technological regression?

4. Let the production possibilities for the paper firms given in problem 2 be for period $t = 1$ and let the period $t = 2$ production possibilities for those same paper firms be in the following table.

Utility	y	b	Efficient? Yes/No	Productivity Increase/Decrease
1	0	0		
2	10	5		
3	16	15		
4	22	15		
5	30	28		
6	45	45		
7	30	50		

(a) Graph the production possibilities frontier for $t = 2$ on the same diagram as the PPF for $t = 1$.

(b) Has the PPF shifted in such a way to indicate technological progress or technological regression?

(c) For each firm identify whether the firm is efficient in $t = 2$ and whether productivity increased or decreased from $t = 1$ to $t = 2$.

5. In which of the following cases can we unequivocally say that there has been an increase in productivity? A "+" means that the good or bad has increased from period $t = 1$ to period $t = 2$. A "−" means that the good or bad has decreased from $t = 1$ to $t = 2$. Assume that the amount of inputs employed remains constant from $t + 1$ to $t + 2$.

6. The graph below depicts the production possibilities for a commercial salmon fishery. In the process of using driftnets to catch coho salmon on the Skeena River in British Columbia a fishing boat labeled by point A catches a desirable output, coho salmon, and an endangered species, steelhead trout, which we will consider an undesirable output. We observe the fishing boat to produce at point A in period t and at point E in period $t + 1$. In both periods, the same amount of inputs are employed.

(a) Explain whether or not the commercial fishing boat was efficient in period t and in period $t + 1$.

(b) Explain whether technological progress occurred from period t to period $t + 1$.

Case	y	b	Productivity: Increase, Decrease, Uncertain?
1	+	+	
2	+	−	
3	−	−	
4	no change	−	
5	+	no change	
6	−	+	

Case	y_1	y_2	Productivity: Increase, Decrease, Uncertain?
1	+	+	
2	−	+	
3	−	−	
4	+	no change	
5	no change	+	
6	+	−	

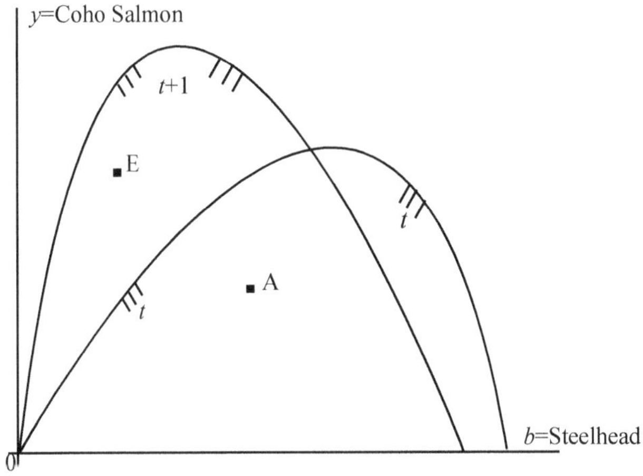

(c) Sketch in the set of points that would correspond to changes in outputs that correspond to strong sustainability for the fishing boat at point A.

(d) Identify a point Z where coho salmon production increases, but which does not satisfy strong sustainability.

(e) Identify a point X where the by-catch of steelhead trout decreases, but which does not satisfy strong sustainability.

(f) Briefly explain whether the change from A to E satisfies strong sustainability, weak sustainability, or both.

7. Return to Figure 5.11. The discussion surrounding the figure provided an illustration of how to measure efficiency change and technological change by measuring efficiency as $D_o(x, y, b \text{ constant}) = \frac{\text{maximum } y}{\text{actual } y}$, holding the undesirable output constant. Holding the desirable output constant and measuring efficiency as $D_o(x, y \text{ constant}, b) = \frac{\text{minimum } b}{\text{actual } b}$ derive the indexes of efficiency change and technological change from period t to period $t+1$ using the points on Figure 5.11 as a guide.

8. The Midwest Timber Company harvests timber (a desirable output) in the Mark Twain National Forest. Harvesting causes erosion and runoff (an undesirable output) into the Current River and the Jacks' Fork River. The price of timber is $p = \$5$ per board foot and the charge for the damage caused by runoff is $q = \$200$ per ton of silt. In period t Midwest Timber Company produces $y^t = 10,000$ and $b^t = 100$.

   (a) Calculate revenues in period t.

   (b) Determine whether the following production alternatives in period $t+1$ satisfy strong sustainability, only weak sustainability, or neither.

	Timber $= y$	Silt $= b$	Strongly Sustainable?	$Revenues^{t+1} = py^{t+1} - qb^{t+1}$	Weakly Sustainable?
A	7,000	35			
B	11,000	90			
C	10,500	110			
D	8,000	47			
E	12,000	155			

9. Big State University is considering changing its antiquated coal-fired steam generating plant to one that runs on natural gas. The proposed plant will generate an equal amount of steam (y) as the existing plant. If the proposed natural gas plant is built carbon emissions will be reduced by 10,000 tons. However, the use of hydraulic fracturing to produce the natural gas will cause 80 water wells to become contaminated at a cost of $500 per well and will reduce the stock of wildlife by 100 animals due to the loss of grazing land. Each wild animal has a value of $p = \$200$. The environmental damage caused by one ton of carbon is estimated to be $q = \$15$. Does the switch from coal to natural gas satisfy strong sustainability? Does the switch satisfy weak sustainability?

10. Three firms in the broiler chicken industry use varying amounts of land, labor, capital, and raw materials to produce tons of broiler chickens (y) and tons of waste (b) in the form of excess nitrogen and phosphorous that contaminate the local watershed. Their production possiblities are listed in the table below.

Firm A			Firm B			Firm C		
y	b	MRT	y	b	MRT	y	b	MRT
0	0	x	0	0	x	0	0	x
15	1		14	1		10	1	
28	2		26	2		19	2	
39	3		36	3		27	3	
48	4		44	4		34	4	
55	5		50	5		40	5	
60	6		54	6		45	6	
63	7		56	7		49	7	
65	8		57	8		52	8	
66	9					54	9	
						55	10	

(a) Complete the aggregate production possibilities frontier for the broiler chicken industry. In constructing the table note that the marginal amount of production of y and b should be undertaken by the firm with the highest MRT, where $MRT = \frac{\triangle y}{\triangle b}$.

(b) Suppose 175 tons of broiler chickens are to be produced. How many tons should each firm produce to satisfy efficiency?

(c) If 175 tons of broiler chickens are efficiently produced, what is the MRT of the 175th unit of output? What are the MRTs for each individual firm?

(d) If 85 tons of broiler chickens are efficiently produced, what is the MRT of the 85th unit of output? What is the MRT of the last firm's contribution to aggregate output?

Aggregate PPF		
y	b	MRT
0	0	x
	1	
	2	
	3	
	4	
	5	
	7	
103	9	
	11	
	13	
	15	
	17	
	19	
	21	
	24	
	27	

6

Estimating Environmental Performance

I often say that when you can measure what you are speaking about, and express it in numbers, you know something about it; but when you cannot measure it, when you cannot express it in numbers, your knowledge is of a meagre and unsatisfactory kind; it may be the beginning of knowledge, but you have scarcely in your thoughts advanced to the state of *Science*, whatever the matter may be.

Lord Kelvin [235, p. 73]

6.1 Data Envelopment Analysis

Chapter 5 introduced the theory for measuring the technical efficiency of a producer, where technical efficiency compares actual outputs produced with potential outputs. However, the graphical method of estimating efficiency assumes only two dimensions holding all other variables constant: for instance, when two desirable outputs are produced all other inputs constant, or when a desirable output and an undesirable ouput are produced all other inputs constant, or when two undesirable outputs are produced all other desirable outputs and inputs constant. However, real world applications of such simplified models are rare. Typically, producers generate multiple outputs—both desirable and undesirable—using multiple inputs. Substitution between two or more inputs is possible and generally researchers have little a priori knowledge of those substitution possibilities. In addition, reductions of one undesirable output might increase other undesirable outputs or reduce the amount of desirable output produced. In this chapter we investigate how a linear programming technique, known as data envelopment analysis (DEA), can be used to measure firm performance when those firms use multiple inputs to produce multiple desirable and undesirable outputs [16, 37, 73]. Estimates of performance are derived using the Solver routine in Excel [264].

6.2 DEA Reference Technology

6.2.1 DEA with Desirable Outputs and Inputs

A production technology can be defined as the set of inputs and outputs such that the inputs can produce the outputs. One representation of the technology examined in Chapter 5 was the production possibility set, $P(x)$: the set of outputs (y, b) that can be produced with given inputs (x). To develop the DEA linear programming problems we first provide some notation. Producers are represented by $k = 1, \ldots, K$ meaning that there are K producers (or firms). For instance, in Table 5.1 there are $K = 7$ producers and $k = 3$ would represent producer 3 (Farmer C). Similarly, $k = 6$ would represent producer 6 (Farmer F). Each producer uses $n = 1, \ldots, N$ inputs to produce $m = 1, \ldots, M$ desirable outputs and $j = 1, \ldots, J$ undesirable outputs. Let x_{kn} represent the quantity of input n used by firm k. Also, let y_{km} represent the quantity of desirable output m produced by firm k. Finally, let b_{kj} represent the quantity of undesirable output j produced by firm k. In the example in Table 5.1 we assumed $N = 3$ inputs: land, labor, and capital equipment. Each farmer used 1000 acres of land (x_1), 250 labor hours (x_2), and \$150,000 in capital equipment (x_3). Thus, $x_{k1} = 1000$, $x_{k2} = 250$, and $x_{k3} = 150,000$, for each of the $k = 1, \ldots, 7$ farmers. In addition we assumed that each farmer produced $M = 2$ desirable outputs of quail (y_1) and corn (y_2) and no undesirable outputs.

The quantities of inputs used and desirable outputs produced are given in Table 6.1. We include the 7 farmers from Table 5.1 and the one additional farmer (Farmer H, $k = 8$) that we calculated efficiency for in Section 5.3.1. We assume that inputs and desirable outputs are strongly disposable. That is, given inputs and some level of desirable output, it is always possible to produce less of the desirable outputs. Similarly, given some level of output, it is always possible to produce the same level of output using more input. In addition, we assume that the technology is convex. Convexity means that if we observe a group of firms producing various amounts of outputs using different (or the same) amounts of inputs, another feasible combination of outputs and inputs consists of a weighted average of all the firms' outputs and inputs. For example, if we observe two firms such as F and H in Table 6.1, any linear combination of F's outputs/inputs and H's outputs/inputs is also feasible, such as 25% of F's and 75% of H's. This feasible combination, call it Y, is such that Y uses $x_1 = 0.25 \times 250 + 0.75 \times 250 = 250$, $x_2 = 0.25 \times 1000 + 0.75 \times 1000 = 1000$, $x_3 = 0.25 \times 150,000 + 0.75 \times 150,000 = 150,000$, to produce $y_1 = 0.25 \times 10 + 0.75 \times 4 = 5.5$, and $y_2 = 0.25 \times 2200 + 0.75 \times 4720 = 4090$.

We form a piecewise linear technology such that no more output can be produced using no less input than a linear combination of the inputs and outputs of all producers. The weights for each firm are called **intensity variables** and are represented by $z_k, k = 1, \ldots, K$. In our example above we had

$z_6 = 0.25$ and $z_8 = 0.75$. Although in our example the weights summed to 1, we only require that they be greater than zero: $z_k \geq 0, k = 1, \ldots, K$. Constraining each of the z_k intensity variable to be nonnegative gives the constant returns to scale technology.

Table 6.1: DEA Production Technology

		Land	Labor	Capital	Quail	Corn
Farmer	k	x_1	x_2	x_3	y_1	y_2
A	1	1,000	250	150,000	0	7,200
B	2	1,000	250	150,000	2	7,000
C	3	1,000	250	150,000	4	6,400
D	4	1,000	250	150,000	6	5,400
E	5	1,000	250	150,000	8	4,000
F	6	1,000	250	150,000	10	2,200
G	7	1,000	250	150,000	12	0
H	8	1,000	250	150,000	4	4,720

With two desirable outputs the DEA production technology for the data in Table 6.1 is defined as:

$$
\begin{aligned}
T = (x, y) : \ x_n &\geq \sum_{k=1}^{K} z_k x_{kn}, n = 1, \ldots, N, \\
y_m &\leq \sum_{k=1}^{K} z_k y_{km}, m = 1, \ldots, M, \\
z_k &\geq 0, k = 1, \ldots, K.
\end{aligned} \tag{6.1}
$$

The inequality in the set of input equations allows for strong disposability of inputs. That is, it is always possible to use more input than the linear combination of observed inputs. Similarly, the inequality in the set of output equations allows for strong disposability of outputs; it is always possible to produce less output than a linear combination of observed outputs.

We use the data in Table 6.1 to write down the DEA technology, T, that is derived from the observations of $K = 8$ firms, $N = 3$ inputs, and $M = 2$

desirable outputs:

$$\begin{aligned}
T =&(x, y) : \\
x_1 \geq &1000z_1 + 1000z_2 + 1000z_3 + 1000z_4 + 1000z_5 + 1000z_6 + 1000z_7 \\
&+ 1000z_8, \\
x_2 \geq &250z_1 + 250z_2 + 250z_3 + 250z_4 + 250z_5 + 250z_6 + 250z_7 + 250z_8, \\
x_3 \geq &150000z_1 + 150000z_2 + 150000z_3 + 150000z_4 \\
&+ 150000z_5 + 150000z_6 + 150000z_7 + 150000z_8, \\
y_1 \leq &0z_1 + 2z_2 + 4z_3 + 6z_4 + 8z_5 + 10z_6 + 12z_7 + 4z_8, \\
y_2 \leq &7200z_1 + 7000z_2 + 6400z_3 + 5400z_4 + 4000z_5 + 2200z_6 + 0z_7 + 4720z_8, \\
z_1 \geq &0, z_2 \geq 0, z_3 \geq 0, z_4 \geq 0, z_5 \geq 0, z_6 \geq 0, z_7 \geq 0, z_8 \geq 0.
\end{aligned} \tag{6.2}$$

In (6.2) there are three input constraints, two desirable output constraints, and eight constraints for the intensity variables. The observed inputs and outputs of all eight firms enter on the right-hand side and form the reference technology. The feasible input/output combinations consist of the left-hand side variables corresponding to the three inputs and two desirable outputs. When more inputs or outputs are analyzed, the number of constraints for the inputs and outputs increases. With more firms, the number of intensity variables increases.

6.2.2 DEA with Undesirable Outputs

When undesirable outputs are produced along with desirable outputs, another set of constraints is added to the DEA technology given by (6.1). When there are J undesirable outputs, the DEA technology takes the form:

$$\begin{aligned}
T = (x, y, b) : \ x_n &\geq \sum_{k=1}^{K} z_k x_{kn}, n = 1, \ldots, N, \\
y_m &\leq \sum_{k=1}^{K} z_k y_{km}, m = 1, \ldots, M, \\
b_j &= \sum_{k=1}^{K} z_k b_{kj}, j = 1, \ldots, J \\
z_k &\geq 0, k = 1, \ldots, K.
\end{aligned} \tag{6.3}$$

Notice that the set of constraints for the J undesirable outputs are equality constraints. These J equality constraints are derived from our assumption that undesirable outputs are only weakly disposable. Therefore, for a given linear combination of the desirable outputs, it is only possible to produce linear combinations of the undesirable outputs; it is not possible to produce 0 amounts of the undesirable outputs unless 0 amounts of all the desirable outputs are also produced. The quantities of the $N = 3$ inputs, the $M = 1$ desirable output, and the $J = 1$ undesirable output for $K = 7$ electric utilities from Table 5.2 are reproduced in Table 6.2.

Table 6.2: DEA Production Technology with Undesirable Outputs

Firm	k	Energy x_1	Labor x_2	Capital x_3	Electricity y_1	SO_2 b_1
A	1	100	8	1.5	15	1
B	2	100	8	1.5	28	2
C	3	100	8	1.5	38	3
D	4	100	8	1.5	43	4
E	5	100	8	1.5	46	5
F	6	100	8	1.5	42	6
G	7	100	8	1.5	27	5

Using 6.3 the DEA technology can be written as:

$$
\begin{aligned}
T =&(x, y, b): \\
&x_1 \geq 100z_1 + 100z_2 + 100z_3 + 100z_4 + 100z_5 + 100z_6 + 100z_7 \\
&x_2 \geq 8z_1 + 8z_2 + 8z_3 + 8z_4 + 8z_5 + 8z_6 + 8z_7 \\
&x_3 \geq 1.5z_1 + 1.5z_2 + 1.5z_3 + 1.5z_4 + 1.5z_5 + 1.5z_6 + 1.5z_7 \\
&y_1 \leq 15z_1 + 28z_2 + 38z_3 + 43z_4 + 46z_5 + 42z_7 + 27z_8 \\
&b_1 = 1z_1 + 2z_2 + 3z_3 + 4z_4 + 5z_5 + 6z_6 + 5z_7 \\
&z_1 \geq 0, z_2 \geq 0, z_3 \geq 0, z_4 \geq 0, z_5 \geq 0, z_6 \geq 0, z_7 \geq 0.
\end{aligned} \tag{6.4}
$$

The observed inputs and outputs of the $K = 7$ firms enter on the right-hand side of the set of equation in (6.4) and form the reference technology. The set of feasible input/output combinations are the left-hand side variables consisting of the three inputs, one desirable output, and one undesirable output.

6.3 Measuring Efficiency Using DEA

6.3.1 Output Distance Function

Given the DEA technologies developed in the previous section our next objective is to measure the performance or efficiency of the producers. With two or more desirable outputs the output distance function, $D_o(x, y)$, serves as the measure of technical efficiency. Recall that $D_o(x, y)$ equals the ratio of actual output to maximum potential output, given the amounts of inputs the firm employs. The reciprocal of the output distance function, $\frac{1}{D_o(x,y)}$, gives the maximum proportional expansion in desirable outputs. For the DEA technology given by (6.1) the maximum proportional expansion in desirable outputs for firm k' is estimated as

$$\begin{aligned}
\frac{1}{D_o(x,y)} &= \max_{\lambda,z} \lambda : \\
x_{k'n} &\geq \sum_{k=1}^{K} z_k x_{kn}, n = 1,\ldots,N, \\
\lambda y_{k'm} &\leq \sum_{k=1}^{K} z_k y_{k'm}, m = 1,\ldots,M, \\
z_k &\geq 0, k = 1,\ldots,K.
\end{aligned} \tag{6.5}$$

The set of equations in (6.5) form a linear programming problem that chooses the intensity variables z_k, $k = 1,\ldots,K$, so as to maximize the proportional expansion in desirable outputs, λ, for firm k'. The intensity variables are chosen so as to give the largest feasible technology given the firm's own inputs and outputs. Using the inputs and outputs found in Table 6.1 and the DEA technology given by (6.2) the reciprocal of the output distance function for Farmer H ($k = 8$) is

$$\begin{aligned}
\frac{1}{D_o(x,y)} =& \max_{\lambda,z} \lambda : \\
1000 \geq& 1000z_1 + 1000z_2 + 1000z_3 + 1000z_4 + 1000z_5 + 1000z_6 \\
&+ 1000z_7 + 1000z_8, \\
250 \geq& 250z_1 + 250z_2 + 250z_3 + 250z_4 + 250z_5 + 250z_6 \\
&+ 250z_7 + 250z_8, \\
150000 \geq& 150000z_1 + 150000z_2 + 150000z_3 + 150000z_4 \\
&+ 150000z_5 + 150000z_6 + 150000z_7 + 150000z_8, \\
\lambda 4 \leq& 0z_1 + 2z_2 + 4z_3 + 6z_4 + 8z_5 + 10z_6 + 12z_7 + 4z_8, \\
\lambda 4720 \leq& 7200z_1 + 7000z_2 + 6400z_3 + 5400z_4 + 4000z_5 + 2200z_6 \\
&+ 0z_7 + 4720z_8, \\
z_1 \geq& 0, z_2 \geq 0, z_3 \geq 0, z_4 \geq 0, z_5 \geq 0, z_6 \geq 0, z_7 \geq 0, z_8 \geq 0.
\end{aligned} \tag{6.6}$$

Note that in (6.6), the left-hand side inputs/outputs correspond to those of Farmer H ($k' = 8$). Similarly, the reciprocal of the output distance function for the other seven farmers could be estimated by changing the left-hand side variables to correspond to those of the particular farmer under consideration. Recall from Chapter 5 that Farmer H is compared to a hypothetical frontier firm labeled J in Figure 5.7. The output coordinates of hypothetical firm J are a linear combination of the outputs of Farmers C ($k = 3$) and D ($k = 4$), with $z_3 = 0.5$, $z_4 = 0.5$, and $z_1 = z_2 = z_5 = z_6 = z_7 = z_8 = 0$. That is, the hypothetical outputs at point J are $(y_1, y_2) = 0.5(4, 6400) + 0.5(6, 5400) = (5, 5900)$.

6.3.2 Technical Efficiency with Undesirable Outputs

In Chapter 5 we proposed two technical efficiency measures for when undesirable outputs are produced along with desirable outputs. The first held inputs and undesirable outputs constant with technical efficiency equal to the ratio of actual desirable output to maximum potential output. The second held inputs and desirable outputs constant with technical efficiency equal to the ratio of the minimum undesirable output to actual undesirable output. These two technical efficiency measures can also be estimated using DEA.

The first technical efficiency measure, $TE1(x, y, b) = D_o(x, y, b \text{ constant})$, equals the ratio of actual desirable output to maximum desirable output holding inputs and undesirable output constant. It can be calculated for firm k' as the reciprocal to the following linear programming problem:

$$\begin{aligned}
\frac{1}{TE1(x,y,b)} &= \max_{\lambda,z} \ \lambda : \\
x_{k'n} &\geq \sum_{k=1}^{K} z_k x_{kn}, \ n = 1, \ldots, N, \\
\lambda y_{k'm} &\leq \sum_{k=1}^{K} z_k y_{km}, \ m = 1, \ldots, M, \\
b_{k'j} &= \sum_{k=1}^{K} z_k b_{kj}, \ j = 1, \ldots, J \\
z_k &\geq 0, k = 1, \ldots, K. \qquad (6.7)
\end{aligned}$$

Using the inputs/outputs of the 7 electric utilities found in Table 6.2 and the DEA technology given by (6.6) we can estimate $TE1(x, y, b)$ for firm G ($k = 7$) as the reciprocal to the following LP problem:

$$\begin{aligned}
\frac{1}{TE1(x,y,b)} &= \max_{\lambda,z} \ \lambda : \\
100 &\geq 100z_1 + 100z_2 + 100z_3 + 100z_4 + 100z_5 + 100z_6 + 100z_7 \\
8 &\geq 8z_1 + 8z_2 + 8z_3 + 8z_4 + 8z_5 + 8z_6 + 8z_7 \\
1.5 &\geq 1.5z_1 + 1.5z_2 + 1.5z_3 + 1.5z_4 + 1.5z_5 + 1.5z_6 + 1.5z_7 \\
\lambda 27 &\leq 15z_1 + 28z_2 + 38z_3 + 43z_4 + 46z_5 + 42z_6 + 27z_7 \\
5 &= 1z_1 + 2z_2 + 3z_3 + 4z_4 + 5z_5 + 6z_6 + 5z_7 \\
z_1 &\geq 0, z_2 \geq 0, z_3 \geq 0, z_4 \geq 0, z_5 \geq 0, z_6 \geq 0, z_7 \geq 0. \qquad (6.8)
\end{aligned}$$

The right-hand side of (6.8) consists of the observed inputs, desirable outputs, and undesirable outputs of the $K = 7$ electric utilities. The left-hand side consists of the inputs and outputs of the particular firm ($k' = 7$) under consideration. The performance measure $TE1(x, y, b)$ can be calculated for the other six electric utilities by resolving problem (6.8) six additional times, each time substituting the inputs and outputs of the firm under consideration on the left-hand side.

The second efficiency measure, $TE2(x, y, b)$, equals the ratio of minimum undesirable output to actual undesirable output. It can be estimated for firm k' as the following linear programming problem:

$$\begin{aligned} &TE2(x, y, b) = \min_{\lambda, z} \lambda : \\ &x_{k'n} \geq \sum_{k=1}^{K} z_k x_{kn}, \; n = 1, \ldots, N, \\ &y_{k'm} \leq \sum_{k=1}^{K} z_k y_{km}, \; m = 1, \ldots, M, \\ &\lambda b_{k'j} = \sum_{k=1}^{K} z_k b_{kj}, \; j = 1, \ldots, J, \\ &z_k \geq 0, k = 1, \ldots, K. \end{aligned} \tag{6.9}$$

Note that LP problem (6.9) minimizes λ with $TE2(x, y, b) = \lambda$, whereas LP problem (6.8) problem maximizes λ with $\lambda = \frac{1}{TE1(x,y,b)}$.

Using the inputs/outputs of the $K = 7$ electric utilities found in Table 6.2 and the DEA technology given by (6.6) we can estimate $TE2(x, y, b)$ for firm G ($k = 7$) as the following LP problem:

$$\begin{aligned} &TE2(x, y, b) = \min_{\lambda, z} \lambda : \\ &100 \geq 100z_1 + 100z_2 + 100z_3 + 100z_4 + 100z_5 + 100z_6 + 100z_7 \\ &8 \geq 8z_1 + 8z_2 + 8z_3 + 8z_4 + 8z_5 + 8z_6 + 8z_7 \\ &1.5 \geq 1.5z_1 + 1.5z_2 + 1.5z_3 + 1.5z_4 + 1.5z_5 + 1.5z_6 + 1.5z_7 \\ &27 \leq 15z_1 + 28z_2 + 38z_3 + 43z_4 + 46z_5 + 42z_6 + 27z_7 \\ &\lambda 5 = 1z_1 + 2z_2 + 3z_3 + 4z_4 + 5z_5 + 6z_6 + 5z_7 \\ &z_1 \geq 0, z_2 \geq 0, z_3 \geq 0, z_4 \geq 0, z_5 \geq 0, z_6 \geq 0, z_7 \geq 0. \end{aligned} \tag{6.10}$$

As in our previous examples, the right-hand side of (6.10) represents the technology and consists of all the linear combinations of inputs, desirable outputs, and undesirable outputs of the $K = 7$ firms and the left-hand side equals the inputs/outputs of the firm under consideration ($k' = 7$).

6.4 Using Excel to Estimate Technical Efficiency

When only two outputs are produced and all firms use the same inputs, the graphical approach developed in Chapter 5 can be used to identify efficient and inefficient producers. However, when firms use varying quantities of inputs to produce two or more outputs DEA (data envelopment analysis) becomes a useful tool to estimate the technical efficiencies of the firms. This section illustrates how $TE1(x, y, b)$ and $TE2(x, y, b)$ can be estimated using the solver routine available in the spreadsheet program Excel. For the most part, the approach outlined here closely follows that used by Joe Zhu [264].

As an example of the method we consider the G-20 countries (somewhat misnamed) which include the central banks of 19 countries plus the European central bank. We evaluate the nineteen countries in the G-20 for their efficiency in producing the desirable output of real GDP and two jointly produced undesirable outputs, CO_2 emissions and SO_2 emissions. These countries use labor, capital, and energy inputs. An Excel spreadsheet in Figure 6.1 reports the countries, outputs, and inputs for the year 2005. Real GDP and capital are denominated in purchasing power parity US dollars and labor equals the number of employed workers. These variables are taken from the extended Penn World Tables [165]. Emissions of CO_2 are in millions of metric tons and the energy input is in quadrillion BTUs, both taken from the Energy Information Administration: International Energy Statistics. Finally, SO_2 emissions are in gigagrams and are taken from the NASA Socio-economic Data and Applications Center.

In column I of the spreadsheet shown in Figure 6.1 we add the variables z_k which will form the linear combinations of the outputs and inputs. We arbitrarily choose starting values of $z_k = 1$, for the $k = 1, \ldots, 19$ countries in our sample. When we use the Excel solver routine these intensity variables are optimally chosen for each country. In the cells E23 to E28 enter the sum of the products of the intensity variables (z_k) and the outputs and inputs. These cells are the right-hand side of the constraints in (6.7) and form the DEA reference technology:

Cell E23=SUMPRODUCT(B2:B20,$H$2:$H$20)
Cell E24=SUMPRODUCT(C2:C20,$H$2:$H$20)
Cell E25=SUMPRODUCT(D2:D20,$H$2:$H$20)
Cell E26=SUMPRODUCT(E2:E20,$H$2:$H$20)
Cell E27=SUMPRODUCT(F2:F20,$H$2:$H$20)
Cell E28=SUMPRODUCT(G2:G20,$H$2:$H$20)

We begin by estimating technical efficiency for Argentina ($k = 1$). In cell G23 we arbitrarily choose some starting value for λ, say $\lambda = 1$. Then, in cell C23 to C28 we enter the following formulas corresponding to the left-hand side of (6.7) for the country (Argentina $k = 1$) being evaluated:

Cell C23=$G$23*INDEX(B2:B20,B22,1)

Cell C24=INDEX(C2:C20,B22,1)
Cell C25=INDEX(D2:D20,B22,1)
Cell C26=INDEX(E2:E20,B22,1)
Cell C27=INDEX(F2:F20,B22,1)
Cell C28=INDEX(G2:G20,B22,1)

Note that in Cell B22 we entered the k value for Argentina where k serves as an index and ranges from $k = 1$ to $k = 19$. In cell C23 we entered our initial starting value for λ multiplied by the Excel index function, INDEX(B2:B20,B22,1). The index function has three arguments: first is the range of cells where the y_1 values are found, B2:B20; second is the cell where the index function is located, B22; and third is the increment in the index function, 1. This index function will be useful when we write a short Visual Basic program that Excel can use to solve for technical efficiency for every country in our sample. The remaining cells, C24 to C28, correspond to the undesirable outputs and inputs for the country under consideration, in this case Argentina. When cell B22 is changed to 2, the INDEX function recognizes that $k = 2$, so Australia is now the country for which λ will be calculated. In addition, since $\lambda = \frac{1}{TE1(x,y,b)}$ we add a constraint restricting cell G23 to be greater than 1 in the **Solver Parameters** box. In cells D23 to D28 we insert the appropriate inequality symbols although those cells will not be part of the program.

Our next step is to set up the Excel Solver routine. If your computer does not have the Solver already installed you can click on the Excel Options link located on the start-up menu, then click Add-Ins, Go, and finally check the box Solver Add-In. Figure 6.2 shows the initial **Solver Parameters** box. The target cell is where λ is located, G23, and since $\frac{1}{TE1(x,y,b)}$ is a maximization (Max) problem we check the Max circle.

We want to maximize λ by changing the value in cell G23 and changing the intensity variables which are in cells H2 to H20. The **Solver Parameters** box should be set up as in Figure 6.3. Next, we need to add six constraints that define the DEA technology: the desirable output constraint, two undesirable output constraints, and three input constraints. To add these constraints click on the **Add** box in the Solver Parameters box. Figure 6.4 shows the constraint dialog box where the desirable output constraint has been added to the model. You should add the five other constraints to complete the model taking care to impose equality constraints for the two undesirable outputs and inequality constraints ($\geq$) for the three input constraints. Next, click on **Options** in the Solver Parameters box and then check the box corresponding to **Assume Non-negative** as shown in Figure 6.5. The final set up of the **Solver Parameters** box is shown in Figure 6.6. The final step is to click the **Solve** button.

The estimated $\lambda = 1.00$ for Argentina indicates that Argentina produces on the production possiblities frontier. We can estimate the λs for the other countries by changing cell B22 to 2, 3, $\ldots$, 19 one by one and rerunning the Solver routine. However, this is time consuming and so it is easier if we write a Visual Basic (VB) program to have Excel do the work for us.

	A	B	C	D	E	F	G	H
1	Country	y1	b1	b2	x1	x2	x3	z
2	Argentina	3.908E+11	151.864	132.438	5.539E+11	1.847E+07	2.955	1.00
3	Australia	7.767E+11	412.614	2522.070	1.312E+12	1.050E+07	5.650	1.00
4	Brazil	1.578E+12	370.757	1437.824	2.140E+12	9.599E+07	9.356	1.00
5	Canada	1.180E+12	623.750	2024.243	1.724E+12	1.748E+07	14.172	1.00
6	China	6.146E+12	5463.704	32673.438	1.378E+13	7.572E+08	67.915	1.00
7	France	1.949E+12	413.962	544.530	2.892E+12	2.892E+07	11.356	1.00
8	Germany	2.606E+12	847.359	572.840	4.247E+12	4.092E+07	14.422	1.00
9	India	2.789E+12	1181.398	6274.514	4.018E+12	4.219E+08	16.331	1.00
10	Indonesia	7.889E+11	330.795	1534.986	1.468E+12	1.118E+08	4.647	1.00
11	Italy	1.734E+12	471.885	406.510	3.260E+12	2.483E+07	8.079	1.00
12	Japan	4.178E+12	1241.262	833.660	9.153E+12	6.638E+07	23.135	1.00
13	Mexico	1.271E+12	397.806	2145.153	1.802E+12	4.283E+07	6.731	1.00
14	Russia	1.764E+12	1587.526	5974.767	2.016E+12	7.419E+07	27.922	1.00
15	Saudi Arabia	4.648E+11	401.927	1378.780	8.420E+11	8.482E+06	6.525	1.00
16	South Africa	3.305E+11	432.413	2476.742	4.715E+11	1.736E+07	5.120	1.00
17	South Korea	1.095E+12	493.801	395.000	2.932E+12	2.368E+07	9.332	1.00
18	Turkey	7.065E+11	230.898	1483.947	8.452E+11	2.656E+07	3.734	1.00
19	UK	2.068E+12	583.064	686.460	2.497E+12	3.061E+07	9.845	1.00
20	US	1.258E+13	5999.245	13106.420	1.960E+13	1.506E+08	100.282	1.00
21								
22	k=	1	Country		Constraints		λ	
23		y_{k1}	3.908E+11	<=	4.44E+13	sum $z_k y_{k1}$	1.00	
24		b_{k1}	151.864	=	21636.027	sum $z_k b_{k1}$		
25		b_{k2}	132.438	=	76604.323	sum $z_k b_{k2}$		
26		x_{k1}	5.539E+11	>=	7.556E+13	sum $z_k x_{k1}$		
27		x_{k2}	1.847E+07	>=	1.969E+09	sum $z_k x_{k2}$		
28		x_{k3}	2.955	>=	347.50959	sum $z_k x_{k3}$		
29		λ	1.00	>=	1			

Figure 6.1: Spreadsheet DEA Model.

To enter the VBA-editor press *Alt-F11*. Then, click on Tools/References and check the box for SOLVER in the References-VBAProject window as in Figure 6.7. A reference to the SOLVER.XLAM now appears as in Figure 6.7. Note, to get this operation to work it might be necessary to first select the **Data** tab located on the spreadsheet menu. Next, on the VBA-editor toolbar select Insert/Module and a window appears where the program code will be written.

Next, type in the code that appears in Figure 6.8. The line Sub TE() defines a subroutine called TE that will calculate the reciprocal of technical efficiency for each country. Comment lines begin with an apostrophe. The parameter k is defined as an integer in the "Dim" statement and references the country for which λ is calculated each time through the subroutine. The next line identifies the number of countries in our sample to be 19. Then, the next line tells the computer that the reference (k) for the country under consideration is in cell B22. The next two lines beginning with "Range=("I1")" tells the

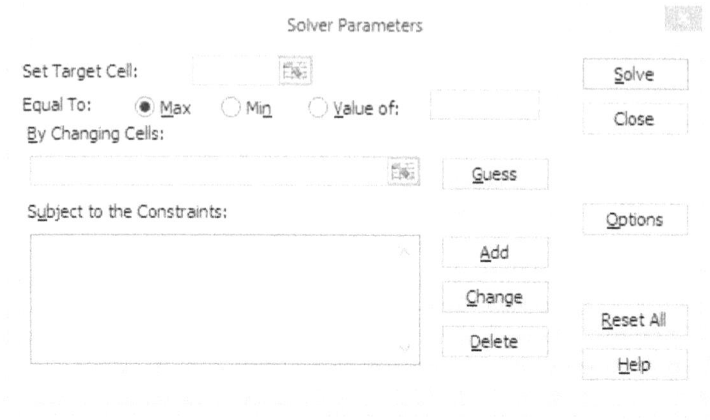

Figure 6.2: Excel Solver Parameters Box.

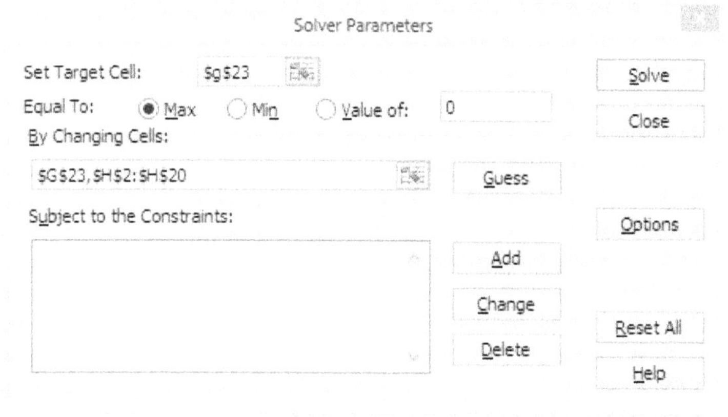

Figure 6.3: Setting up the DEA Problem in Solver.

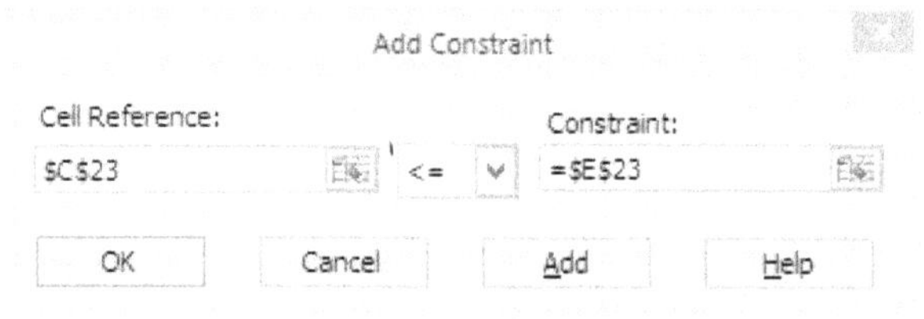

Figure 6.4: Adding Constraints in Solver.

computer to take the solution for each k found in cell G23 and place it in column I with the ".Offset=(k,0)" command telling the computer to put the solution beginning with $k = 1$ into cell I2. The last two lines of the subroutine tell the computer to loop through and run the Solver routine for the next

Figure 6.5: Constraining the Intensity Variables.

Figure 6.6: DEA Problem in Solver.

country until all countries have had λ solved for. Press $Alt - F11$ to run the subroutine.

Table 6.3 reports $\frac{1}{TE1}$ for each country. As seen by the solution values, every country except for South Korea is efficient with South Korea's value of $TE(x, y, b) = \frac{1}{1.422} = 0.703$ indicating that real GDP is only 70.3% of its potential.

The problem given by (6.10) can be easily estimated using the same subroutine, TE, but changing the objective function from a "Max" to a "Min" in the **Solver Parameters** box and modifying cells C23 to C25 as follows:
Cell C23=Index(B2:B20,B22,1)
Cell C24=$G$23*INDEX(C2:C20,B22,1)

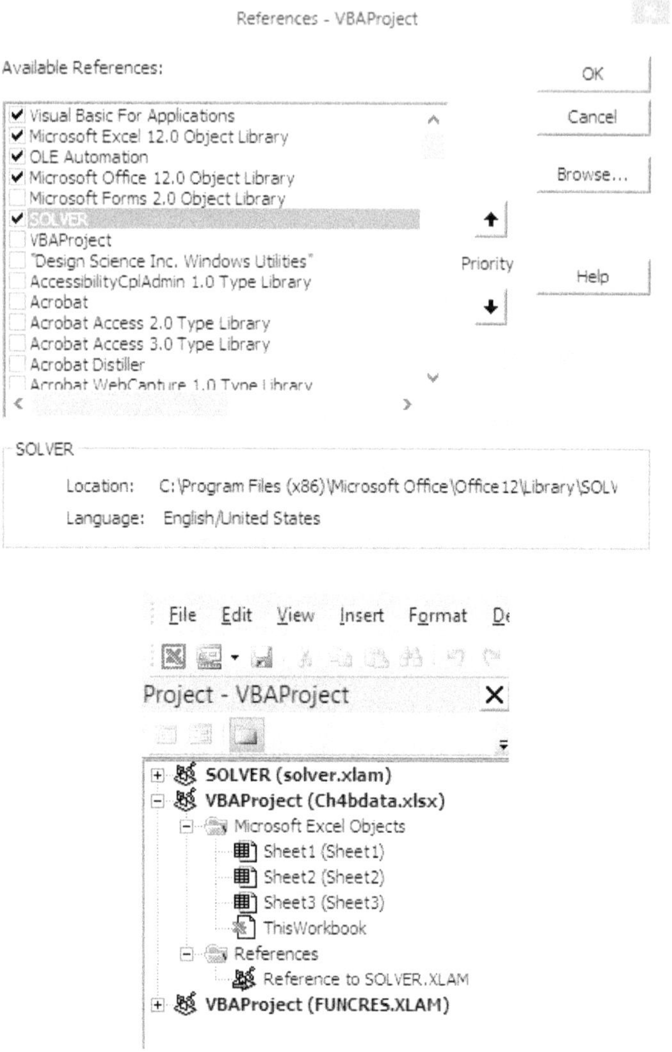

Figure 6.7: Adding the Solver as a Reference in a VBA Project.

Cell C25=$G$23*INDEX(D2:D20,B22,1)
In addition, we constrain cell $G23$ to be less than one in cell C29.

In this case, $TE2(x,y,b) = \frac{\text{minimum } b}{\text{actual } b}$ and will equal λ. After the changes to the spreadsheet and solver have been made press *Alt-F11* and run the subroutine. The estimated values of $TE2(x,y,b)$ are reported in Table 6.3. When measuring the efficiency of undesirable outputs more countries are inefficient including Canada, China, India, Indonesia, Saudi Arabia, and South Korea. Except for South Korea, these countries were efficient when esti-

```
(General)                                    TE

Sub TE()
'Declare k as an integer. This k represents the country
'under consideration and in our example ranges
'from 1 to 19.
    Dim k As Integer
    For k = 1 To 19
    Range("B22") = k
'The next statement will run the Solver routine.  The
'UserFinish is set to True indicating that the Solver
'Parameters box will not be shown.
    SolverSolve UserFinish:=True
'We want to insert the estimated lambda for each
'country in cells I2 to I20.
    With Range("I1")
    .Offset(k, 0) = Range("G23")
    End With
    Next k
End Sub
```

Figure 6.8: DEA Code in Visual Basic.

mating $TE1(x, y, b)$ but are inefficient for $TE2(x, y, b)$. The inefficiency for $TE2(x, y, b)$ but not $TE1(x, y, b)$ indicates that they are operating on the negatively sloped portion of the production possibility frontier, like Firm F depicted in Figure 5.5 of Chapter 5. These countries could actually experience some gain in real GDP if they were to reduce the two polluting outputs of CO_2 and SO_2.

Table 6.3: Technical Efficiency and Revenue Efficiency for the G-20 Countries

Country	k	$\frac{1}{TE1}$	$TE2$	$py - qb$	$R(x,p,q)$	y^*	b_1^*	b_2^*
Argentina	1	1.000	1.000	3.853E11	4.2301E11	4.290E11	164.06	538.71
Australia	2	1.000	1.000	7.612E11	8.0812E11	8.204E11	336.49	643.49
Brazil	3	1.000	1.000	1.564E12	1.7613E12	1.782E12	568.41	2,669.55
Canada	4	1.000	0.572	1.157E12	1.2618E12	1.279E12	460.38	788.06
China	5	1.000	0.344	5.941E12	1.1397E13	1.155E13	4,148.07	25,100.56
France	6	1.000	1.000	1.934E12	2.0348E12	2.059E12	673.60	1,000.76
Germany	7	1.000	1.000	2.575E12	2.8229E12	2.854E12	851.61	1,054.23
India	8	1.000	0.659	2.745E12	3.2989E12	3.335E12	979.17	2,636.89
Indonesia	9	1.000	0.612	7.666E11	9.7499E11	9.849E11	273.63	286.50
Italy	10	1.000	1.000	1.717E12	1.7172E12	1.734E12	471.88	406.51
Japan	11	1.000	1.000	4.133E12	4.6488E12	4.698E12	1,355.51	1,405.65
Mexico	12	1.000	1.000	1.256E12	1.4021E12	1.417E12	398.12	457.02
Russia	13	1.000	1.000	1.706E12	1.7056E12	1.764E12	1,587.53	5,974.77
Saudi Arabia	14	1.000	0.700	4.500E11	6.1411E11	6.223E11	225.81	389.86
South Africa	15	1.000	0.999	3.143E11	3.9586E11	4.068E11	294.41	1,216.80
South Korea	16	1.422	0.610	1.077E12	1.6918E12	1.712E12	550.52	749.03
Turkey	17	1.000	1.000	6.978E11	6.9783E11	7.065E11	230.90	1,483.95
UK	18	1.000	1.000	2.046E12	2.0465E12	2.068E12	583.06	686.46
US	19	1.000	1.000	1.236E13	1.2361E13	1.258E13	5,999.24	13,106.42

6.5 Revenue Efficiency

When the prices (p) of the desirable outputs and the charges (q) imposed on the undesirable outputs are known the output quantities that maximize revenues given inputs (x) can be determined given the DEA technology. Recall that revenues (R) are equal to $R = py - qb = \sum_m p_m y_m - \sum_j q_j b_j$. Using DEA, the maximum revenue the firm can achieve given its inputs (x), desirable output prices (p), and undesirable output prices (q) is represented as $R(x, p, q)$ and is found as the solution to the following LP problem:

$$
\begin{aligned}
R(x,p,q) &= \max_{y,b,z} \sum_{m=1}^{M} p_m y_m - \sum_{j=1}^{J} q_j b_j \text{ subject to} \\
x_{k'n} &\geq \sum_{k=1}^{K} z_k x_{kn}, \; n = 1, \ldots, N, \\
y_m &\leq \sum_{k=1}^{K} z_k y_{km}, \; m = 1, \ldots, M, \\
b_j &= \sum_{k=1}^{K} z_k b_{kj}, \; j = 1, \ldots, J, \\
z_k &\geq 0, k = 1, \ldots, K.
\end{aligned} \tag{6.11}
$$

In (6.11), maximum revenues are found by choosing the desirable outputs and undesirable outputs subject to the restrictions that the desirable and undesirable output choices can be feasibly produced given the firm's inputs. Note that in (6.11) the observed inputs for firm k' enter on the left-hand side of the input constraints but the desirable outputs (y_m) and undesirable outputs (b_j) are variables that are chosen to maximize revenues given that those outputs are feasible.

To illustrate the method return to the data for the electric utilities given in Table 6.2 and assume that the price of electricity is $p = \$1$ and that the charge imposed on SO_2 is $q = \$4$. The DEA revenue function can then be written as

$$
\begin{aligned}
R(x,p,q) = \max_{y,b,z} \ & 1y - 4b \text{ subject to} \\
100 &\geq 100z_1 + 100z_2 + 100z_3 + 100z_4 + 100z_5 + 100z_6 + 100z_7 \\
8 &\geq 8z_1 + 8z_2 + 8z_3 + 8z_4 + 8z_5 + 8z_6 + 8z_7 \\
1.5 &\geq 1.5z_1 + 1.5z_2 + 1.5z_3 + 1.5z_4 + 1.5z_5 + 1.5z_6 + 1.5z_7 \\
y &\leq 15z_1 + 28z_2 + 38z_3 + 43z_4 + 46z_5 + 42z_6 + 27z_7 \\
q &= 1z_1 + 2z_2 + 3z_3 + 4z_4 + 5z_5 + 6z_6 + 5z_7 \\
z_1 &\geq 0, z_2 \geq 0, z_3 \geq 0, z_4 \geq 0, z_5 \geq 0, z_6 \geq 0, z_7 \geq 0. \qquad (6.12)
\end{aligned}
$$

The solution to (6.12) is $y^* = 43$, $b^* = 4$, with $z_1 = z_2 = z_3 = z_5 = z_6 = z_6 = 0$ and $z_4 = 1$ yielding $R(x,p,q) = \$27$.

The DEA revenue function given in (6.11) can be estimated using EXCEL. We return to the data on real GDP, CO_2, and SO_2 for the 19 countries presented in Table 6.3. We assume that price of \$1 of real GDP is $p = \$1$. In 2013, the United States Government [190] revised their estimate of the damage caused by a ton of CO_2 from \$20/ton to an average of \$36/ton. The \$36/ton was an average of three estimates derived from discount rates of 5%, 3%, and 2%. We take up discounting in Chapter 11. Given that the data on CO_2 in Table 6.3 are in millions of metric tons we assume that $q_1 = \$36$ million.

Determing the damage caused by a ton of SO_2 is more problematic. As chronicled by Richard Schmalensee and Robert Stavins [209] the SO_2 trading market began to collapse in 2008 after the Circuit Court of Appeals for the District of Columbia vacated the Clean Air Interstate Rule. Allowance prices for a ton of SO_2 collapsed from a high of \$1,200/ton in 2005 to \$390/ton in 2008 to only \$0.56/ton in 2012. The price declines were not due to declining social damage of SO_2 but were instead due to the elimination of the SO_2 trading program as a means of meeting emissions' caps. We arbitrarily assume a charge for SO_2 of \$250/ton. Since SO_2 emissions reported in Table 6.3 are in gigagrams we convert the per ton charge to $q_2 = \$250$ thousand/gigagram, where 1 ton=0.001 gigagrams.

Figure 6.9 depicts the EXCEL spreadsheet data. Type in the cell location for the outputs and inputs for the country being evaluated in cells C23 to C28:
Cell C23=INDEX(B2:B20,B22,1)
Cell C24=INDEX(C2:C20,B22,1)
Cell C25=INDEX(D2:D20,B22,1)
Cell C26=INDEX(E2:E20,B22,1)
Cell C27=INDEX(F2:F20,B22,1)
Cell C28=INDEX(G2:G20,B22,1)
The output prices for p, q_1, and q_2 are in cells C29, C30, and C31. In cell G23 type in the formula for actual revenues for the country being evaluated: \$C29\$*C23–\$C30\$*C24–\$C31\$*C25. For Argentina, revenues are $R = 1 \times y - 36E6 \times b_1 - 100E3 \times b_2 = 3.853E11$.

Next, set up the Solver routine as a maximization problem, where cell G23

	A	B	C	D	E	F	G	H
1	Country	y1	b1	b2	x1	x2	x3	z
2	Argentina	3.908E+11	151.864	132.438	5.539E+11	1.847E+07	2.955	1.00
3	Australia	7.767E+11	412.614	2522.070	1.312E+12	1.050E+07	5.650	1.00
4	Brazil	1.578E+12	370.757	1437.824	2.140E+12	9.599E+07	9.356	1.00
5	Canada	1.180E+12	623.750	2024.243	1.724E+12	1.748E+07	14.172	1.00
6	China	6.146E+12	5463.704	32673.438	1.378E+13	7.572E+08	67.915	1.00
7	France	1.949E+12	413.962	544.530	2.892E+12	2.892E+07	11.356	1.00
8	Germany	2.606E+12	847.359	572.840	4.247E+12	4.092E+07	14.422	1.00
9	India	2.789E+12	1181.398	6274.514	4.018E+12	4.219E+08	16.331	1.00
10	Indonesia	7.889E+11	330.795	1534.986	1.468E+12	1.118E+08	4.647	1.00
11	Italy	1.734E+12	471.885	406.510	3.260E+12	2.483E+07	8.079	1.00
12	Japan	4.178E+12	1241.262	833.660	9.153E+12	6.638E+07	23.135	1.00
13	Mexico	1.271E+12	397.806	2145.153	1.802E+12	4.283E+07	6.731	1.00
14	Russia	1.764E+12	1587.526	5974.767	2.016E+12	7.419E+07	27.922	1.00
15	Saudi Arabia	4.648E+11	401.927	1378.780	8.420E+11	8.482E+06	6.525	1.00
16	South Africa	3.305E+11	432.413	2476.742	4.715E+11	1.736E+07	5.120	1.00
17	South Korea	1.095E+12	493.801	395.000	2.932E+12	2.368E+07	9.332	1.00
18	Turkey	7.065E+11	230.898	1483.947	8.452E+11	2.656E+07	3.734	1.00
19	UK	2.068E+12	583.064	686.460	2.497E+12	3.061E+07	9.845	1.00
20	US	1.258E+13	5999.245	13106.420	1.960E+13	1.506E+08	100.282	1.00
21								
22	k=	1	Country		Constraints		R(x,p,q)	
23		y_{k1}	3.908E+11	<=	4.44E+13	sum $z_k y_{k1}$	3.853E+11	
24		b_{k1}	151.864	=	21636.027	sum $z_k b_{k1}$		
25		b_{k2}	132.438	=	76604.323	sum $z_k b_{k2}$		
26		x_{k1}	5.539E+11	>=	7.556E+13	sum $z_k x_{k1}$		
27		x_{k2}	1.847E+07	>=	1.969E+09	sum $z_k x_{k2}$		
28		x_{k3}	2.955	>=	347.50959	sum $z_k x_{k3}$		
29		p=	1.00					
30		q_1=	36000000					
31		q_2=	250000					

Figure 6.9: Spreadsheet for Maximizing Revenues.

is to be maximized by changing the values of the intensity variables in cells H2 to H20 and the values of the desirable and undesirable outputs in cells C23, C24, and C25. The output and input constraints defining the technology appear in the constraint box. Figure 6.10 depicts the new Solver settings.

We write a new VBA subroutine to solve the revenue maximization problem for each country. This subroutine is listed in Figure 6.11 and is called REV(). This subroutine places maximum potential revenues for each country in column I and the optimal values of real GDP (y^*), CO_2 emissions (b_1^*), and SO_2 emissions (b_2^*) in columns J, K, and L, and actual revenues ($py - qb$) in column M.

Table 6.3 reports actual revenues ($py - qb$), maximum potential revenues ($R(x, p, q)$), optimal real GDP (y^*), optimal CO_2 emissions (b_1^*), and optimal SO_2 emissions (b_2^*). Note that each country's actual revenues are less than or equal to maximum potential revenues. Italy, Russia, Turkey, the United Kingdom, and the United States all maximize revenues given their inputs. The

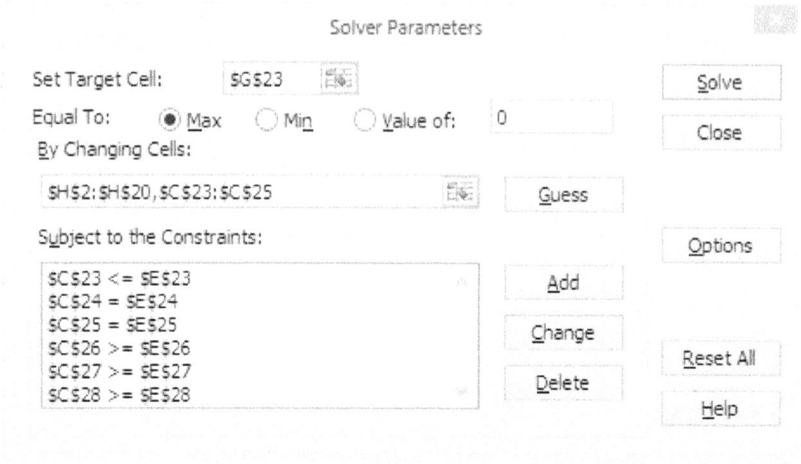

Figure 6.10: Solver for Revenue Maximization.

```
(General)                                                    REV

Sub REV()
'Declare k as an integer. This k represents the
'country under consideration and in our example
'ranges from 1 to 19.
    Dim k As Integer
    For k = 1 To 19
    Range("B22") = k
'The next statement will run the Solver routine.
'The UserFinish is set to True indicating that
'the Solver Parameters box will not be shown.
    SolverSolve UserFinish:=True
'We want to insert the estimated maximum revenue
'for each country in cells I2 to I20.
    Range("I" & k + 1) = Range("G23")
'Now, we want to insert the optimal values for
'y, b1, and b2 into the appropriate country row.
    Range("J" & k + 1) = Range("C23")
    Range("K" & k + 1) = Range("C24")
    Range("L" & k + 1) = Range("C25")
  Next k
End Sub
```

Figure 6.11: VBA Program to Maximize Revenues.

other countries could increase revenues by either becoming more technically efficient or by changing their mix of desirable and undesirable outputs. To take one example consider China, which could increase real GDP from 6.146E12 to 1.1397E13 while reducing CO_2 from 5643.7 million megatons to 4148.1 megatons and reducing SO_2 emissions from 32673 gigagrams to 25101 gigagrams. Such a change in outputs would satisfy the condition of strong sustainability defined in Chapter 5. In contrast, Argentina could increase revenues by increasing real GDP from 3.908E11 to 4.23E11 but only by increasing both CO_2 emissions from 151.8 million megatons to 164 million megatons and increasing SO_2 from 132 gigagrams to 538 gigagrams. This change in the three outputs leads to an increase in revenues but would only satisfy the condition of weak sustainability as defined in Chapter 5.

6.6 Limits of the DEA Approach

The DEA approach to estimating technical efficiency and maximum revenues can provide policy makers and firms useful information about whether policy changes are enhancing welfare given alternative notions of sustainability but several caveats apply. First, the DEA approach assumes that the failure to maximize revenues or any deviation of actual outputs and inputs from the technological frontier is due to either allocative inefficiency or technical inefficiency. In fact, measurement error in the data, random events, and luck can all play a role in producers not realizing technical efficiency or maximizing revenue. Second, the revenue maximization problem requires knowledge of the prices of the desirable outputs and the charges to be imposed on the undesirable outputs. The optimal charges on the undesirable outputs are often unknown because the undesirable outputs are not traded in markets leaving policy makers and firms in the dark. As we did above for the charge on SO_2 emissions, educated guesses can sometimes replace market prices but those guesses might also be a source of error. Third, the DEA models presented in this chapter all assume a constant returns to scale technology. It is possible and even likely that some environmental technologies are subject to increasing returns to scale or decreasing returns to scale. In these cases the gains in efficiency or revenues are biased by the constant returns to scale assumption.

6.7 Problems

1. The table below depicts the observed input/output combinations for four national forests. Each national forest uses x = acres to pro-

duce y_1 = visitors, y_2 = board feet of timber sold (in 1000s), and b = acres with severe erosion problems. Write the system of equations down that define the DEA production technology assuming constant returns to scale.

k	x	y_1	y_2	b
1	1,000	100	25	2
2	1,500	200	30	5
3	2,000	200	50	10
4	2,500	400	40	12

2. The table below depicts the observed input/output combinations for ten producers who use two inputs, x_1 = labor and x_2 = acres of forest, to produce a desirable output, y = board feet of timber (in 1000s) and undesirable output, b = tons of silt, that washes into streams in the watershed.

k	x_1	x_2	y	b
1	1	2	3	1
2	3	1	5	3
3	7	5	7	7
4	9	4	10	10
5	5	8	13	8
6	11	8	20	12
7	15	11	25	14
8	20	5	15	18
9	18	4	18	12
10	10	12	10	4

(a) Set up an EXCEL spreadsheet and program a macro that will allow you to estimate $\frac{1}{TE1}$ and $TE2$ for each firm.

(b) Identify the efficient producers which produce on the production possibility frontier.

(c) Suppose the ten firms sell the board feet of timber in a competitive market at price p = \$20. In addition, for each ton of topsoil that washes into the watershed they are fined at the rate of q = \$5 per ton. Calculate actual revenues $(py - qb)$, maximum potential revenues, $R(x, p, q)$, and each firm's revenue efficiency=$\frac{\text{actual revenues}}{\text{maximum revenues}}$.

7

Public Goods and Bads

I have never known much good done by those who affected to trade for the public good.

Adam Smith, 1776 *The Wealth of Nations* [222]

I have no other view than to promote the public good, and am unambitious of honors not founded in the approbation of my Country.

George Washington, Letter to Henry Laurens, Valley Forge, 31 January 1778 [225, p.504]

7.1 Characteristics of Public Goods and Bads

Many environmental amenities such as climate, clean air, clean water, and litter-free highways are public goods. Environmental disamenities associated with the production of undesirable outputs—pollution—are public bads. Public goods and bads have two characteristics that distinguish them from private goods. First, public goods and bads are nonrival in consumption. That is, one person's consumption of the good does not reduce the benefits available to others or for public bads, the costs borne by others. For example, Sid's enjoyment of a beautiful sunset or nice weather does not reduce the benefits available to Alice from the sunset or nice weather. For a public bad, like litter on the highway, the eyesore the litter imposes on Sid does not reduce the potential for the litter to be an eyesore to Alice. Public goods and bads are also nonexcludable in that once it is produced it is prohibitively costly or impossible to keep Sid, Alice, and others from enjoying the benefits or bearing the costs. At the margin, Sid and Alice might have different willingness to pay for the public good, but they both consume the same quantity. In contrast, a private good is rival in consumption and excludable. When Sid drinks a soda, the benefits he receives from that particular soda are not available to Alice: the soda is rival in consumption. In addition, prices serve as an exclusionary mechanism for private goods. If Alice does not have enough money to cover the price of the soda, she is legally excluded from consuming it.

Very few goods meet the strict characteristics of nonrivalry and nonexcludability. Environmental amenities like clean air or diverse ecological systems have public good qualities to some degree. However, environmental amenities are sometimes subject to some degree of rivalry or exclusiveness, even if they are better thought of as "public" rather than "private" goods. Figure 7.1 provides a two dimensional representation of alternative degrees of nonrivalry and nonexcludability for different goods.

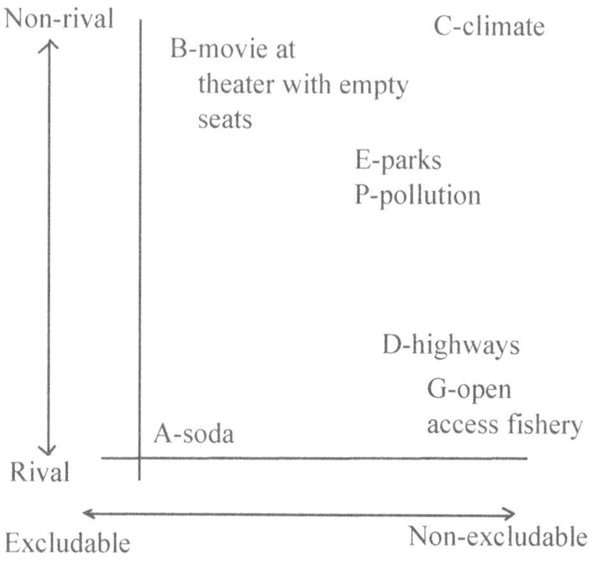

Figure 7.1: Degrees of Rivalry and Excludability for Public–Private Goods.

Point A corresponds to a pure private good like a can of soda which is both rival in consumption and excludable. Point B corresponds to a good for which it is easy to exclude individuals but for which there is a high degree of nonrivalry. For example, attending a movie at a theater is nonrival when the theater does not sell all available tickets. In such a case more people can be added to the theater without reducing the benefits of watching the movie to those already there; that is, as long as the marginal person does not wear a top hat, sit right in front of you, and talk throughout the movie. However, the movie theater can exclude people by requiring people to buy a ticket before they gain admitance. Point C represents a pure public good, like climate, which is both nonrival and nonexcludable. Point D represents goods such as highways or roads where exclusion is difficult, but there is some degree of rivalry. For instance, highways are rival when congested. During rush hour, when another vehicle enters an already congested highway, the average speed slows for the vehicles already on the highway. However, with the exception of toll roads, most highways in the United States are nonexcludable. At times, parks can also become congested such as on holiday weekends. Although some parks charge vehicle or per person admission fees, many city and state parks do not charge fees and are nonexcludable local public goods. Point G represents a common property resource such as an open access fishery or pasture shared by more than one shepherd. Common property resources are rival in that as more shepherds graze sheep or more fishermen fish, the productivity of the pasture or fishery falls. By their nature, though, common property resources are nonexcludable.

Now, think of the kinds of environmental amenities and disamenities that individuals and society consume. The amenities include services like clean air, which is nonrival in consumption, but might be excludable for some, depending on the prevailing winds, elevation, and whether the location is in a basin. Pollution, an undesirable public bad, might be local in nature, like a toxic waste site, or it might be global in nature such as the effects of carbon dioxide emissions on climate.

7.2 Efficiency Conditions for Public Goods and Bads

For private goods, individuals all face the same prices but they usually consume different quantities. To derive the aggregate or market demand function for private goods it is necessary to sum up all the individual quantities demanded at alternative prices as we did in Chapter 2. The equilibrium condition for private goods, $p = MC$, is consistent with an efficient allocation of resources. The price the individual pays represents the marginal benefits of the last unit of the good consumed. The marginal rate of substitution, MRS, represents the amount of one good an individual is willing to give up for an additional unit of another good and is another term for the marginal benefits an individual receives from consuming a good. For most goods and services, both public and private, as individuals consume more of the good, the amount of another good they are willing to forego diminishes. The diminishing marginal rate of substitution leads people to buy less at a higher relative price than they would at a lower relative price, everything else constant.

In contrast to private goods, everyone consumes the same quantity of the public good, but individuals generally have different willingness to pay. How can we determine the efficient quantities of public goods and undesirable public bads? Recall that in Chapter 2 we defined an indifference curve as the alternative quantities of two goods that yield the same level of utility. When individuals consume desirable and undesirable goods an indifference curve is defined similarly—the alternative quantities of the desirable good and undesirable good that yield the same level of utility. However, in this case, as an individual consumes more of an undesirable good they must be compensated with successively larger amounts of the private good. Therefore, an indifference curve will have a positive and increasing slope and, furthermore, the map of indifference curves will be such that as one moves to the northwest individual utility will increase. Such a map of indifference curves is given in Figure 7.2. The slope of the indifference curve is the individual's marginal rate of substitution, $MRS = \frac{\triangle y}{\triangle b}$. The MRS equals the amount of extra private good an individual must receive in compensation if they are to consume an extra unit of the undesirable public bad. Alternatively, the MRS equals the amount of private good an individual is willing to forego if the undesirable public bad is reduced by one unit.

In two classic papers Paul Samuelson [206, 207] worked out the efficiency conditions for public goods. We adapt his approach here to graphically depict the equilibrium conditions for a private good (y) and a public bad (b) that are consumed by two individuals whom we will call Sid (S) and Alice (A). The economy's production possibilities are given in Table 7.1 and depicted in Figure 7.3.

Table 7.1: Efficient Provision of a Public Bad

PPF			Alice's u_A Indifference Curve			Leftover Private Good for Sid	Slope of Leftover	Sid's
y	b	MRT	y_A	b	MRS_A	y_S	Curve	MRS_S
0	0	x	45.8	0	x	−45.8	x	x
15	1	15	46.8	1	1.0	−31.8	14.0	3.0
29	2	14	48.5	2	1.7	−19.5	12.3	3.5
42	3	13	50.9	3	2.4	−8.9	10.6	4.0
54	4	12	54.0	4	3.1	0.0	8.9	4.5
65	5	11	57.8	5	3.8	7.2	7.2	5.0
85	7	10	66.8	7	4.5	18.2	5.5	5.5
103	9	9	77.2	9	5.2	25.8	3.8	6.0
119	11	8	89.0	11	5.9	30.0	2.1	6.5
133	13	7	102.2	13	6.6	30.8	0.4	7.0
145	15	6	116.8	15	7.3	28.2	−1.3	7.5
155	17	5	132.8	17	8.0	22.2	−3.0	8.0
163	19	4	150.2	19	8.7	12.8	−4.7	8.5
169	21	3	169.0	21	8.4	0.0	−6.4	9.0
173	23	2	189.2	23	10.1	−16.2	−8.1	9.5
176	26	1	221.6	26	10.8	−45.6	−9.8	10.0

To determine the efficient quantities of the public bad and the private good we fix the utility of Alice at some arbitary level, say u_A. The alternative combinations of a private good and a public bad that yield the same level of utility for Alice are given in Table 7.1 with her indifference curve plotted in Figure 7.3 as the curve labeled u_A. Given Alice's utility some amount of private good is leftover for Sid. For instance, if Alice were to consume $y_A = 89$ and b=11 then the production possibility frontier shows that Sid would be able to consume a maximum of y_B=30 and he would also consume the same

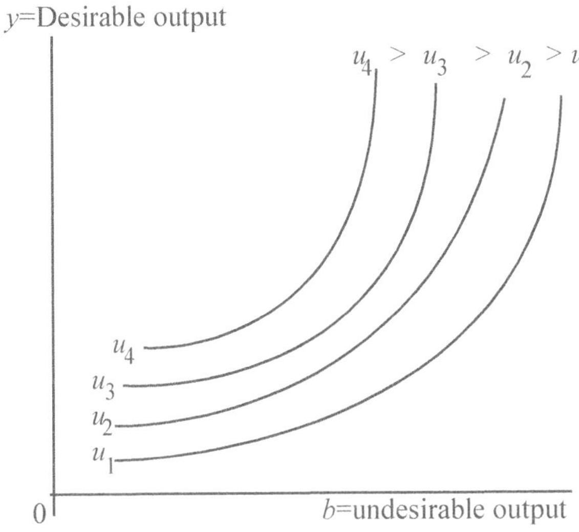

Figure 7.2: Indifference Map for Desirable and Undesirable Outputs.

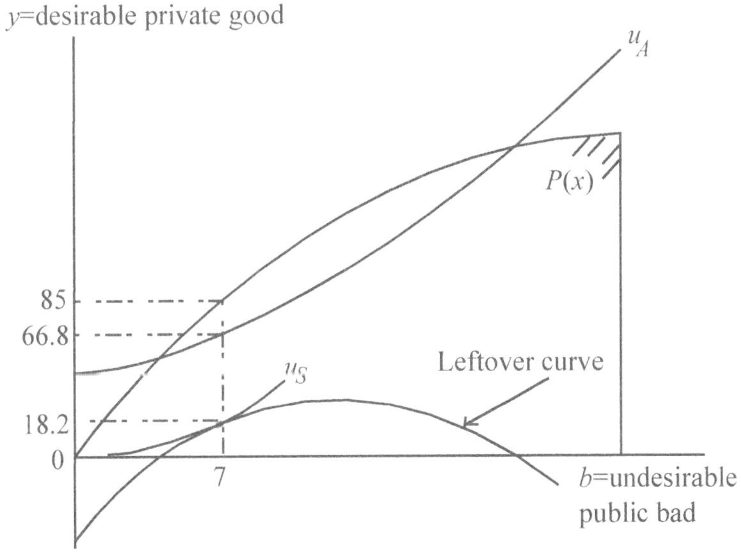

Figure 7.3: Efficient Quantities of Public Bad and Private Good.

amount of public bad, b=11. The problem facing Sid is to choose the amounts of leftover private good and public bad to maximize his own utility. That is, Sid will want to be on the highest indifference curve given the amounts corresponding to the leftover curve. Such an indifference curve corresponds to u_S which is tangent to the leftover curve. At the point of tangency the slope of Sid's indifference curve is his marginal rate of substitution.

Given Alice's and Sid's marginal rates of substitution the efficient quantities of the private good and public bad are y=85 and b=7. The private good is distributed such that Alice receives y_A=66.8 and Sid receives y_S=18.2. At the margin, if Alice is to consume another unit of the undesirable public bad she must be compensated by 5.5 extra units of private good and Sid must be compensated by receiving 4.5 extra units of private good. Together, the sum of Alice's and Sid's willingness to accept compensation totals 10 units of private good and that just equals the extra units of private good that can be produced by allowing production of the undesirable good to increase by 1 unit. From Table 7.1 we can see that at the efficient quantities of public bad and private good the slope of the leftover curve just equals the marginal rate of transformation minus Alice's marginal rate of substitution: $MRS_S = MRT - MRS_A$. Rearranging yields $MRS_A + MRS_S = MRT$.

Letting the private good be the numeraire—a good which has a price of $p = \$1$—the individual's marginal rate of substitution equals their willingness to pay. Similarly, the marginal rate of transformation is the marginal money cost of reducing the public bad by one unit or, alternatively, the marginal money damage from allowing the public bad to increase by one unit.

For public goods the aggregate willingness to pay equals the sum of the individual marginal rates of substitution. Assuming $i = 1, ...N$ individuals, the aggregate willingness to pay for the public good is $\sum_{i=1}^{N} MRS_i$. When the aggregate willingness to pay for a public good is greater than the marginal cost of provision, $\sum_{i=1}^{N} MRS_i > MC$, efficiency is enhanced by more of the public good being produced. If, instead, the aggregate willingness to pay is less than marginal cost, $\sum_{i=1}^{N} MRS_i < MC$, less of the public good should be produced. The socially efficient quantity of public goods corresponds to the quantity where the sum of the marginal rates of substitution equal marginal cost: $\sum_{i=1}^{N} MRS_i = MC$.

Consider the individual demands for a public good such as the services provided by a healthy ecosystem which includes clean air, clean water, temperature, plants, and wildlife. Table 7.2 gives the marginal rate of substitution (willingness to pay) schedules for Alice and Sid, the aggregate willingness to pay schedule, and the marginal cost of providing increments of ecosystem services. Whereas Alice and Sid had to be compensated with larger and larger increments of private good for every unit increase in the public bad, their individual marginal rates of substitution for a public good are decreasing. Thus, as they consume more of the public good their willingness to pay for an extra unit decreases.

Suppose the amount of ecosystem services currently provided is y=1. For

Table 7.2: Individual and Aggregate Demand for a Public Good

y	Alice's WTP MRS_{Alice}	Sid's WTP MRS_{Sid}	Aggregate WTP $=MRS_{Alice}+MRS_{Sid}$	MC
1	18	35	53	18
2	16	32	48	20
3	14	29	43	22
4	12	26	38	24
5	10	23	33	26
6	8	20	28	28
7	6	17	23	30
8	4	14	18	32
9	2	11	13	34
10	0	8	8	36
11	0	5	5	38
12	0	2	2	36

the second unit of ecosystem services, which is jointly consumed, Alice is willing to pay 16 and Sid is willing to pay 32 for an aggregate WTP of 48. Since the MC of providing the second unit of ecosystem services is only 20, efficiency is enhanced by the increase. The socially efficient quantity of ecosystem services is y=6. Note that at the socially efficient quantity Alice and Sid have different willingness to pay: Alice's WTP=\$8 and Sid's WTP=\$20. However, they both consume the same quantity of ecosystem services.

Figure 7.4 illustrates the vertical summation of the WTP curves to obtain aggregate willingness to pay. Since a public good is nonrival and nonexclusive, Alice and Sid consume the same quantity but have different willingness to pay. In this case we sum the demand curves (or MRS curves) vertically to obtain aggregate willingness to pay given by the bold line. When Alice's MRS=0 at y=10, aggregate willingness to pay corresponds with Sid's MRS schedule. The equilibrium condition is $MRS_A + MRS_S = MC$ and corresponds to y_E in Figure 7.4. Like private goods, increases in the MC of provision reduce the efficient quantity. On the demand side, anything that increases an individual's WTP increases the efficient quantity of the public good.

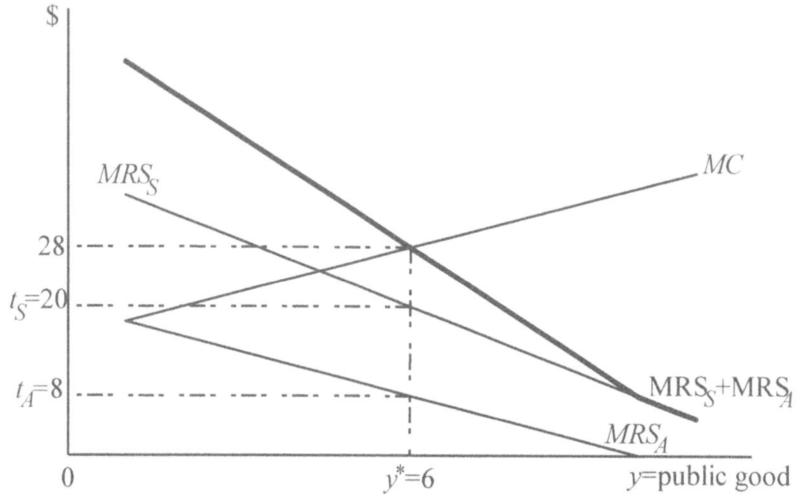

Figure 7.4: Graphical Illustration of the Public Goods Equilibrium.

7.3 Political Process and Lindahl Equilibrium

Private markets tend to underallocate resources for the provision of public goods because of the free-rider problem. Alternatively, the private market tends to produce too much of public bads. The free-rider problem arises because of the nonexclusive character of public goods and bads. Since a public good is available to all once it is provided, individuals have an incentive to free ride on the contributions of others. This kind of behavior tends to reduce the quantity of the public good that is privately produced. Many college students encounter the free-rider problem in so-called group projects, where there is sometimes a tendency for one student to do all the work with the other students coming along as free-riders. In the market place, the presence of free riders reduces the available profits to firms from providing the good, causing those firms to search for other, more profitable, ventures.

Since private markets tend to underallocate resources for the provision of public goods government often assumes the role of public good provider. However, a problem remains. How can government decision makers determine each individual's marginal rate of substitution so that efficient amounts of public goods can be provided? Since public goods are not bought and sold in markets it is difficult to determine what individuals are willing to pay. Instead, individual preferences for public goods have to be revealed in some other manner, a topic we take up further in Chapter 9.

One way that individuals express their preferences for public goods is through voting. Imagine a hypothetical issue on the ballot. Suppose the issue is whether or not a tax should be imposed to finance a new park or to pay for the mitigation of climate change. How will an individual vote on this issue? Public choice theory assumes that rational individuals will choose to vote so as to maximize their utility. In this case the citizen thinks about the costs they will incur and the benefits they will receive from the new park or from climate change mitigation policies. If the individual thinks they will receive more in benefits than what it costs them in taxes they will vote yes. On the other hand, if the perceived marginal benefits are less than the extra taxes they will vote no. We will assume that individuals who are indifferent will either not vote or flip a coin. Let the marginal rate of substitution (MRS) be the measure of marginal benefits from the public good and let the tax price, t, equal the marginal cost to the individual for a one unit increase in the public good. The individual's voting decision rule will then be:

$$\text{if } MRS > t, \text{ vote yes}$$
$$\text{if } MRS < t, \text{ vote no}$$
$$\text{if } MRS = t, \text{ indifferent.}$$

Under pairwise (two issues, thumbs-up or thumbs-down) majority rule voting an issue will pass if it receives 50% plus one vote. Individuals have single-peaked preferences over a set of policies or quantities of the public good when each individual has a most preferred quantity and as the quantity of the public good offered moves away from their most preferred quantity their utility declines. When preferences are single peaked, the median voter will determine the outcome of the election. Since candidates are trying to win the election they will seek to capture the median voter's vote. Other voting schemes, such as two-thirds majority rule or three-fifths majority rule, work similarly with the candidate trying to amass a supramajority of the vote. However, two major problems arise with majority rule voting: it tends to be redistributive and it does not necessarily ensure an efficient level of public good provision.

Redistribution occurs when those individuals with $MRS < t$ vote no on an issue and the issue passes; those individuals are made worse off and income is redistributed from them to those who voted yes because they had $MRS > t$. Such redistribution can alter the efficient level of public good provision. Inefficiencies can also arise with democratic voting. Since government officials generally do not know each individual's WTP for the public good they often devise tax schemes where the taxes individuals pay are divorced from their WTP. In these situations inefficient projects—those with $\sum MRS < MC$—sometimes attract a majority of votes and efficient projects—those with $\sum MRS > MC$—do not receive a majority of votes. Therefore, although private firms will tend to underproduce public goods, there is no guarantee that the choice made via collective decision making will be efficient.

A Lindahl equilibrium occurs when at the efficient level of public good provision each individual pays a tax price equal to their willingness to pay. As will be seen in end of chapter problem 2, income redistribution may alter the efficient level of public good provision. Therefore policy-makers need to account for both efficiency and equity in policies to provide public goods.

Furthermore, there is no guarantee that the majority rule outcome will correspond with the efficient level of public good provision, i.e., the level where $\sum_{i=1}^{N} MRS_i = MC$. Therefore, majority rule voting might result in too many or too few resources allocated for public goods.

Citizens have alternatives to the ballot box when making decisions about the level of public goods they desire. One alternative to the ballot box is that citizens might vote with their feet by moving to jurisdictions that provide a level of public goods more in line with their preferences. For example, if Sid does not like the level of clean air he can search for a city or locale that offers a level of clean air more in line with his preferences. As a second alternative sometimes individuals can purchase private substitutes for the public good. For instance, if public hunting or fishing opportunities are not sufficient, an individual or a group of individuals can acquire fishing access or purchase land and establish habitat that meets their demands. Alternatively they might pay a private landowner for the privilege of using their land. Finally, individuals can augment the level of public goods by donating their time or money to the particular cause they are concerned with. Many individuals choose to pick up trash off the sides of roadways or give money to environmental groups such as The Sierra Club, Audubon Society, The Nature Conservancy, etc., to increase the level of public goods beyond the level provided by government.

7.4 Fiscal Federalism

Various levels of government—city, county, state, federal, or international—can provide public goods. The appropriate choice of government depends on the degree to which the benefits of providing the public good spillover to noncitizens. Campaigns against street litter are probably most appropriately carried out by local governments. Since the Great Salt Lake is located entirely within the state of Utah, state management of the lake should be effective. In contrast, Lake Michigan borders Wisconsin, Michigan, Indiana, and Illinois, so attempts to govern water quality by one individual state might be less effective. A similar problem occurs for the countries that surround the Mediterranean Sea.

Since climate change is global in scope, effective policy might entail cooperation and coordination by all governments. One big problem facing such an international effort is that in 1648 the Treaty of Westphalia recognized a system of sovereign states so that international laws can only be imposed on sovereign states with their consent [185]. Therefore, a kind of unanimity voting is required to use an international governing body to solve the problems associated with a global public good such as climate. Given more than 195 sovereign countries in 2012, the free-rider problem will be a large barrier to overcome. However, agreement among a smaller number of countries, such as the United States, countries in the European Union, China, and India, might be possible. Since these countries account for more than half of the world's carbon emissions, such a coalition might prove fruitful. However, side-payments might have to be made between countries to grease the political wheels of exchange. Such side-payments from rich countries to poorer developing countries allowed CFCs to be rapidly phased out following the Montreal Protocol.

7.5 Dominant Assurance Contracts

The free-rider problem seems to indicate that unless they are coerced, individuals will tend to free ride on other people's contributions to public goods. Moreover, if all individuals are expected to free ride, then even those individuals who are against free riding might choose not to contribute to the public good to avoid wasting their time and money. For instance, if expected floodwaters are to be contained a temporary sandbag dam might have to be built six feet high and require all citizens of the low-lying area to contribute to the effort. Even if some individuals bag sand to build the temporary dike, their efforts will fail if any resident chooses to free ride. Do any mechanisms exist to help overcome the problem of free riding and provide individuals with incentives to truthfully reveal their willingness to pay for public goods?

A **dominant assurance contract** can sometimes be a way of overcoming the free-rider problem for certain public goods [231]. We assume there are N citizens each of whom place a maximum value of V on a public good. An entrepreneur wishes to produce a public good, but needs to cover the costs of provision, C. The entrepreneur knows the value that each citizen places on the public good. Suppose there is some subgroup of K citizens where $K \leq N$, who are willing to voluntarily pay for the public good. To break even, the entrepreneur needs each of the K citizens to each pay at least S for the public good where $S = \frac{C}{K}$. If the amount $S \geq V$, then no citizen will voluntarily contribute to the public good and the entrepreneur will abandon the project. Thus, we consider the case where $S \leq V$: each individual will be asked to pay an amount S which is less than their maximum willingness to pay V. The entrepreneur gives each citizen the opportunity to contribute to the public good by paying S. If $X \geq K$ citizens agree to the contract the public good is produced and each citizen who contributes receives a payoff (consumer surplus) equal to $V - S$. However, the free-rider problem assumes that nonpaying citizens also gain from the public good and, since they do not contribute S, the nonpaying citizens receive a payoff of $V > V - S$. However, those same citizens receive a payoff of zero if the public good is not produced.

To induce potential free riders to contribute to the public good the entrepreneur can offer the following contract to each of the N citizens: if $X \geq K$ citizens agree to contribute to the provision of the public good each of the X

citizens will pay S and if $X < K$ citizens agree to the contract then each of the X citizens receives a payment from the entrepreneur equal to F. When the entrepreneur provides the public good their expected profits are $E\pi = K \times S - C$. Thus, the entrepreneur wants to choose K and S so as to maximize profits subject to the constraint that each individual receives a small level of consumer surplus: $V - S \geq 0$.

Profits increase as K increases so the entrepreneur wants to set $K = N$ so that all individuals contribute. Furthermore, profits increase as S increases, so the entrepreneur chooses $S = V$ which satisfies $V - S \geq 0$. How will the citizen voters respond to the entrepreneur's choice of $K = N$ and $S = V$? If any particular citizen rejects the offer they receive a payoff of zero, since $X = N - 1 < K$ and the public good is not provided. However, if the citizen accepts the contract they receive F if someone else rejects the contract and they receive $V - S \geq 0$ if all other citizens accept the contract. Therefore, each citizen has the incentive to accept the contract. Thus, the free-rider problem is overcome and the public good is provided privately.

Consider the Black Smoke Company (BSC) which has a factory in the town of Smelterville. Each of the $N = 100$ citizens of Smelterville are willing to pay $V = \$10$ for cleaner air. The BSC could filter its emissions at a cost of \$500 so it is efficient to allocate resources for the provision of cleaner air since $100 \times \$10 > \$500$. The BSC proposes a contract such that $S = \$9.99$, $F = \$1$, and $K = 100$. That is, if all 100 citizens agree to pay \$9.99, BSC will install the filter and each citizen will earn a consumer surplus equal to $\$0.01 = \$10 - \$9.99$. If fewer than 100 citizens agreee to the contract then those citizens who do agree will earn a payoff equal to \$1 and the citizens who do not agree will earn zero. Since earning either \$1 or \$0.01 are both preferred to earning zero, each of the 100 citizens agrees to the contract, the filter is installed, and the BSC earns profits of $\$499 = \$9.99 \times 100 - \$500$.

In our example, the public good is provided by a single firm, but if competition exists among various polluters for cleaning up pollution, the process of competition will reduce the expected profits that the public good provider can earn to zero. For instance, with costs of $C = \$500$, the process of competition will force the required citizen contribution down to $S = \frac{C}{N} = \frac{\$500}{100} = \$5$. Thus, the process of competition allows each citizen to earn a surplus of \$5 as the gains from trade accrue to the consumer rather than the producer [231].

As another example, suppose that some environmental group would like to raise \$2.5 million to acquire a piece of land that helps provide habitat for some endangered species. Members might be sent a letter encouraging them to donate \$25 and that 100,000 donations are necessary for the land acquisition to occur. Those who choose to donate receive a gift of wildlife calendar or knapsack or some other item. Furthermore, their donation is returned if not enough members donate but they get to keep the gift.

7.6 Climate as a Public Good

Emissions of greenhouse gases contribute to climate change and are perceived by many observers to be a global public bad. Although much scientific research suggests that humans are contributing to climate change, the free-rider problem and the problem of imposing international treaties on sovereign nations make effective policy prescriptions difficult to achieve. Furthermore, countries like China and India quite likely perceive the marginal benefits of greater economic growth to exceed their share of the marginal costs of climate change and are thus willing to emit more greenhouse gases in exchange for higher standards of living. Thus, even if countries could be bound by international treaties, economic efficiency might still allow emissions to increase.

The free-rider problem can sometimes be overcome in small groups, but the larger the group, the greater the incentive for some members to free ride. The "Green Paradox" suggests that attempts by some countries to try and reduce their own emissions can be offset by higher emissions in other countries if energy suppliers expect lower future energy prices. Overcoming the free-rider problem is also made more difficult by the fact that greenhouse gases tend to remain in the atmosphere for long periods, so that any attempts to reduce the flow of emissions today will not be felt for more than 100 years.

Even if the free-rider problem could be overcome, there is still uncertainty about the costs and benefits of mitigating climate change. Estimates about the costs of mitigating climate change range between 1% and 12.9% of GDP [2] and reflect that uncertainty. Furthermore, the costs of doing nothing are likely to be borne by poor countries as people in more rich countries can better adapt to climate change.

Climate change involves a host of problems related to its public good (bad) characteristics including intergenerational equity, distributional equity between and within countries, and uncertainties about the costs and benefits of climate mitigation policies. Finally, if climate mitigation policies are to be implemented, questions about how much greenhouse gas emissions should be cut and when they should be cut must be resolved. The last 200 years have seen incomes increase by 15 to 23 times in developed countries. Much of that increase has been due to technological progress, so that waiting to implement policies might allow technological fixes to be implemented at lower cost. Of course, to the extent that technological progress depends on research and development, which in turn have public goods' characteristics, another public goods' problem arises.

7.7 Questions and Problems

1. Bill lives in Trashtown where all the residents except Bill throw litter onto the sides of roads. Bill is willing to pay \$100 per year to have 1 ton of litter removed from the roads of Trashtown. The other residents are willing to pay \$0 for the same amount of litter removed. The marginal cost of reducing litter by 1 ton is \$75.

   (a) Is it efficient to clean up 1 ton of litter each year?

   (b) Is litter removal in Trashtown a public good?

   (c) Given your answers to the two previous questions, how can an efficient quantity of litter removal be attained?

2. The table below depicts the marginal rate of substitution schedules for three individuals: April (A), Beckett (B), and Clyde (C). April has annual income of $I_A = \$10,000$. Beckett has annual income of $I_B = \$20,000$. Clyde has annual income of $I_C = \$30,000$.

y	MRS_A	MRS_B	MRS_C	$\sum MRS_i$	$\sum MRS_i'$	MC
1	85	120	170			300
2	80	115	165			300
3	75	110	160			300
4	70	105	155			300
5	65	100	150			300
6	60	95	145			300
7	55	90	140			300
8	50	85	135			300
9	45	80	130			300
10	40	75	125			300

   (a) Complete the aggregate willingness to pay, $\sum MRS_i$ schedule.

   (b) What is the efficient level of public good provision if the marginal cost of provision is constant at $MC = 300$?

   (c) Suppose each individual has to pay a tax price equal to $t = 100$. Describe the redistributive consequences of this tax scheme at the efficient level of public good.

   (d) Suppose each individual pays a tax price (t_i) in proportion to their income: $t_i = \frac{I_i}{\sum_{i=1}^{N} I_i} \times MC$. Calculate each person's tax price.

      i. April's tax price, $t_A =$__________.

      ii. Beckett's tax price, $t_B =$__________.

      iii. Clyde's tax price, $t_C =$__________.

(e) Given the tax prices, calculate April's, Beckett's, and Clyde's most preferred level of public good.

i. April's most preferred $y =$__________.

ii. Beckett's most preferred $y =$__________.

iii. Clyde's most preferred $y =$__________.

(f) Describe the redistributive consequences of the proportional tax system at the efficient level of public good provision.

(g) What are the Lindahl tax prices for April, Beckett, and Clyde at the efficient level of public good provision?

(h) Now suppose each individual has income of $I = \$20,000$ and an MRS schedule equal to Beckett's. Complete the new aggregate willingness to pay schedule, $\sum MRS_i'$, and determine the socially efficient quantity.

(i) As income was equalized through redistribution, what happened to the efficient level of public good provision?

3. The Wild Rapid River offers summer recreational opportunities for canoers, fishers, and people who float in innertubes (tubers), but access limits use to only 10 out of 100 miles of the river. The three groups of people have willingness to pay functions given as

$$WTP_{canoers} = 2000 - 10 \times Miles$$
$$WTP_{fishers} = 3000 - 5 \times Miles$$
$$WTP_{tubers} = 1000 - 15 \times Miles$$

The marginal cost of providing access to an additional mile is \$5000 per mile.

(a) What is the efficient number of miles to protect?

(b) What are the Lindahl tax prices for each group?

(c) How does the Wild Rapid River differ from a private good? Is the Wild Rapid River a pure public good?

4. The 500 residents of BrrrCold, Minnesota all like to do the same thing: sit in front of their heaters and keep warm. Unfortunately, their heaters generate pollution which makes each resident worse off. Each resident has a utility function that takes the form: $u_i = H_i - 3b - 0.05b^2$, where the private good, H_i, is the amount of heat consumed by the ith resident and the public bad, b, is the amount of pollution generated. The equation describing each individual's indifference curve is found by solving the utility function for H_i: $H_i = u_i + 3b + 0.05b^2$. Taking the derivative of the indifference curve with respect to pollution (b) yields the individual's marginal rate of substitution: $MRS_i = 3 + 0.1b$. Each unit of pollution allows an extra 2000 units of heat to be generated, so the marginal rate of transformation is $MRT = 2000$.

(a) What is the efficient level of pollution in the town of BrrrCold?

(b) At the efficient level of pollution, what is the MRS_i for the typical resident? Interpret what this MRS_i means.

(c) How much heat does each resident get to consume at the efficient level of pollution?

(d) At the efficient equilibrium what is the typical resident's utility?

(e) Show that the typical resident of BrrrCold loses utility if 1 more unit of pollution is produced along with a corresponding greater amount of heat.

(f) Show that relative to the efficient equilibrium, the typical resident of BrrrCold loses utility if 1 less unit of pollution is produced along with a corresponding lesser amount of heat.

5. For each of the following goods and services identify whether the good or service can best be thought of as a public good or a private good. If you think the good or service is a public good, is it a local, national, or international public good? If the good or service is a public good, is it subject to congestion? If you think it is a public good, are there instances when it might be excludable for some or all people?

(a) Canoe trip on the National Wild and Scenic Current River.

(b) Cape Girardeau builds a new water treatment plant that reduces the amount of chemicals in drinking water.

(c) Congress designates 50 million additional acres as wilderness.

(d) State of Missouri passes a law that reduces the amount of sulfur and nitrogen emissions from electric power plants.

(e) City of St. Louis reduces the amount of sewage discharged into the Mississippi River.

(f) The United States and Canada agree to cut greenhouse gas (carbon dioxide and methane) emissions by 50% from their current levels by 2020.

(g) Missouri designates 5000 acres along the Mississippi River in Southeast Missouri as a wildlife refuge for migratory birds (ducks and geese).

(h) Joe Peabody eats lunch at McDonald's.

8

Externalities and Common Property Resources

Even the most ardent environmentalist doesn't really want to stop pollution. If he thinks about, and doesn't just talk about it, he wants to have the right amount of pollution. We can't really afford to eliminate it—not without abandoning all the benefits of technology that we not only enjoy but on which we depend.

Milton Friedman, *There's No Such Thing as a Free Lunch* [82, p.16]

8.1 Is the Price Right?

Sid, like many other college students, takes a last look around his dorm room and then heads out the door for spring break from Big State University. As he starts the car he notices that the gas gauge is on Empty so he heads across the street to Cousin Carl's Convenience Store to fill up. He notices the price of \$3.49[9] per gallon and then quickly slides his debit card in and out of the gas pump and exchanges \$35 for 10 gallons of gas. Two weeks earlier, Cousin Carl, in anticipation of the start of Spring Break, increased his order for fuel with the distributor. In turn, the distributor had ordered more gas from the refinery and the refinery manager had delayed the scheduled maintenance on Well No. 7 so that more oil could be placed into the pipeline. Instead of shortages of gasoline and long lines at Cousin Carl's and other convenience stores around the country, thousands of college students filled up and left town for a week without having to stand in line for gas.

Not only had participants in the market process anticipated the increased gasoline demands of college students leaving for Spring Break, they had also stocked sun tan oil, chips, beer, and frozen pizzas in their stores for the spring break travelers. Remembering the influx of students from past years, fast food joints along the highways leading to the beaches and ski resorts had made sure french fries, hamburgers, and tortillas would not run out as the students stopped for food. In fact, no government official had to assure enough gasoline or food, although the state patrol was reminded of the potential for increased traffic. In fact, Cousin Carl and the other gas station owners did not even need to know whether it was Sid who bought the gas or Ruth and Ed who were on their way to Ed's mother's ninety-fifth birthday party. Instead, producers and consumers responded to changes in relative prices by increasing production and consumption of some items and reducing production and consumption of other items.

Although prices generally account for most of the costs and benefits associated with consumption and production, those prices do not always reflect the true costs and/or benefits of the exchange. Sometimes costs and benefits spill over to third parties who are not part of the voluntary exchange. These spillover costs and benefits arise because of a lack of private property rights to some resources. Recall from Chapter 5 that undesirable outputs are jointly produced by-products of desirable output production. In driving to the beach, Sid's car fouls the air. If no one owns clean air, Sid and other college students like him do not pay the full cost of their actions, and in turn, the market system and price mechanism do not necessarily guarantee that the Pareto efficient amount of production and consumption occur. Accounting for these spillovers or externalities is the subject of this chapter.

8.2 Trade with Spillover Costs and Benefits

Voluntary trade between two people tends to make both parties better off, or, at the least, leaves no one worse off. If someone were made worse off, they would not voluntarily agree to the trade. In voluntary market exchanges the two parties that trade compensate each other. When Sid goes to the gas station and puts ten gallons of gasoline in his tank his action makes the gas station owner worse off because now the owner has ten fewer gallons of gas in her storage tank. But, Sid compensates the owner by paying her $3.50/gallon. Both Sid and the gas station owner are made better off by the exchange. Sid reveals that he would rather have ten gallons and less money in his pocket and the gas station owner reveals that she would rather have the money and ten fewer gallons of gasoline.

An **externality** or spillover or third-party effect occurs when one economic agent's actions affects the well-being of others and no compensation is made. Continuing with the gasoline example, no externality occurs until Sid starts his car and pulls away from the gas station. However, as soon as Sid continues his journey, the gasoline is burned and generates pollution. If Sid does not compensate others for the pollution he generates, an externality occurs.

Externalities can be either beneficial or harmful. Positive externalities extend benefits to third parties while negative externalities impose costs on third parties. When positive externalities exist people produce and consume too little of the good or service in private market transactions. In contrast, when there are negative (harmful) externalities imposed on a third party due to production and consumption private individuals acting in their own self-interest will tend to overproduce and overconsume goods and services.

As an example of a positive externality, suppose that you get a flu shot or some other immunization against contagious disease. If you get a flu shot not only do you have a reduced probability of getting the flu, but other people that you are in contact with them through socializing, in class, at work, etc., should also have a reduced probability of getting the flu. Your consumption of the flu shot makes you and other people better off, but very seldom will those other people compensate you for your action; nor will you tend to account for others' well-being when deciding whether to get a flu shot.

Pollution is an example of a negative externality. In the manufacturing production process, clean air and water are used up and replaced by dirty air, dirty water, and other kinds of waste. Joint production of desirable and undesirable outputs can, when charges are not imposed on the undesirable outputs, impose third party costs. When negative externalities are present too much of the good is produced and consumed in the private market relative to the socially optimal amount. Not all pollution is bad, though, in an economic sense. Clearly, one way to reduce air pollution would be for individuals to stop driving and take up walking or riding a bike. But these kinds of actions would entail a huge change in lifestyles and well-being for many people. It might even be the case that people forced to give up travel would have less willingness to pay for environmental services like parks, recreational lakes, and hiking trails, since the absence of the automobile would limit their access to those areas.

Why do externalities arise? Usually they result from a lack of well-defined property rights over some resource. Loud music in dormitories occurs because no individual "owns" the property right to "quiet." Air pollution and water pollution occur when no one owns clean air or clean water. When property rights are well defined, the owner of the resource can make others pay for the use of that resource, so third-party effects are avoided. Property rights also provide an incentive to owners to use the resource in its highest and best use as they seek to maximize the value of the resource.

Have you ever wondered why people litter the highway, but seldom throw trash in their own yard? Well, if you throw trash in your yard, the value of your land and house decline. But when people litter the highway, it does not reduce the value of their own resources. This is not to say that individuals won't trash their own yards. Some people do. However, people who engage in that kind of behavior must also bear the costs of their actions.

8.3 Efficiency in the Presence of Externalities

When there are no externalities or public goods, markets will allocate resources up to the point where the marginal benefit of the last unit consumed

(MB) equals the marginal cost (MC) of producing that unit. When negative externalities are present the process of production and consumption generates external costs and these costs, borne by third parties, are termed the marginal external costs (MEC). These costs represent the damage done to a third party who is not part of the trade and represent the amount someone would charge for the use of the resource if they owned the property rights to it. Similarly, when positive externalities are present due to either production or consumption, third parties receive marginal external benefits (MEB).

When negative externalities are present, social efficiency requires that resources be allocated up to the point where the marginal benefits of consumption equal the marginal private costs plus the marginal external costs of production: $MB = MC + MEC$. For instance, when electric utilities burn coal to drive steam turbines and generate electricity they have to pay for the coal, but generally do not pay for the clean air they use up when coal is burned. In this case the marginal external cost occurs because of the absence of private property rights to the air. If someone (individual, firm, or group of citizens) owned the property rights to clean air, they could charge for its use and depletion and the marginal cost of production would be greater. The higher costs cause the price to rise and rational producers (and consumers) seek alternatives that are cheaper and use less clean air.

When positive externalities are generated by either production or consumption too little of the good is produced and consumed at the private equilibrium relative to the social optimum. Think of the flu shot again. If the marginal cost of providing a flu shot is $25, a rational individual like Bob who expects to receive more than $25 in benefits (reduced suffering from the flu) will choose to get the flu shot. In contrast, someone else like Carl, who expects to receive only $24 in benefits, would rationally choose not to get the flu shot. However, people who interact with Carl also benefit when Carl gets a flu shot—those other people have a reduced probability of getting the flu. Suppose the external benefits that spill over to the people who interact with Carl amount to $2. In this case, society would be better off if Carl got the flu shot since the marginal social benefits equal $26; the $24 in private benefits to Carl plus the $2 in external benefits to other people more than offset the $25 marginal cost. For social efficiency in the presence of positive externalities, resources should be allocated up to the point where the sum of the private marginal benefits and marginal external benefits just offset the marginal cost of production: $MB + MEB = MC$.

As an example of a negative externality consider the city of Snowtown. The city council of Snowtown has contracted with a private snow removal firm to plow and salt the roads during the winter months. The cost of the salt, the wear and tear on the trucks used to spread the salt, and the salaries of the truck drivers are all part of the private marginal cost (MC) of using salt as a deicer. In addition to paying for the salt, salaries, and truck maintenance, there are environmental damages from salt as snow melts and rains wash salt into rivers and streams where it causes harm to fish and invertebrates and increases the costs of using the river as drinking water for downstream towns. The use of salt makes roads safer to drive on and that increased safety is reflected in salt's marginal benefits (MB). However, the marginal benefits of increased salt use decline as more is used since salt tends to be spread on the most heavily traveled streets first before it is spread on secondary residential roads and streets. Table 8.1 depicts the marginal benefits, marginal costs, and the marginal external costs associated with environmental damage for Snowtown.

Table 8.1: Benefits and Costs of Salt Use in Snowtown

y=Tons of Salt	MB	MC	MEC	MSC
1	150	60	10	70
2	140	65	15	80
3	130	70	20	90
4	120	75	25	100
5	110	80	30	110
6	100	85	35	120
7	90	90	40	130
8	80	95	45	140

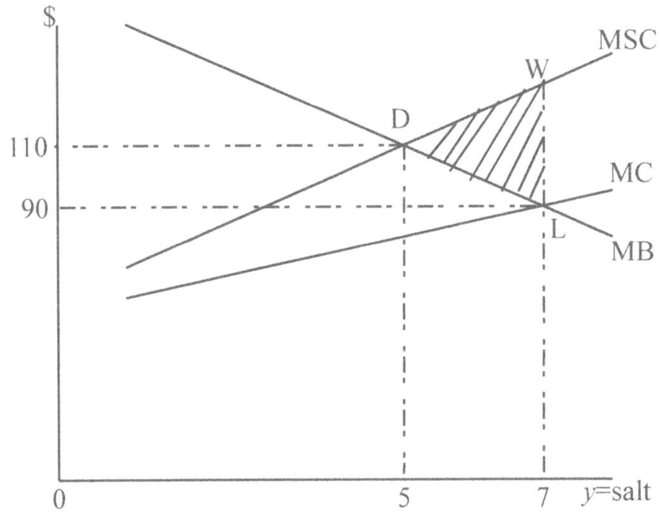

Figure 8.1: Deadweight Loss of a Negative Externality.

If the external costs of salt use are ignored the private equilibrium quantity is $y = 7$ tons where $MB = MC$. The true marginal social costs (MSC) of production equal the sum of the private marginal costs and the marginal external costs. The socially efficient quantity of salt is $y = 5$ where $MB = MSC$. The deadweight loss associated with the private equilibrium is the lost value due to overproduction and overconsumption of salt. Although the sixth unit of salt has a private marginal benefit of 100, its marginal social cost is 120. If the sixth ton of salt is dumped on the streets it results in 20 in lost value. Similarly, although the seventh unit of salt has a marginal benefit of 90, its use would impose a marginal social cost of 130 resulting in an additional 40 in lost value. The total deadweight loss equals the 20+40=60. Figure 8.1 graphs the data from Table 8.1. The triangle DWL approximates the deadweight loss from ignoring the external cost of salt use.

The following conditions generate a socially optimal allocation of resources in the presence or absence of externalities:

No negative externalities, no positive externalities: $MB = MC$.
Negative externalities exist, no positive externalities: $MB = MC + MEC$.
No negative externalities, positive externalities exist: $MB + MEB = MC$.
Both negative and positive externalities exist: $MB + MEC = MC + MEC$.

Social efficiency requires that the allocation of resources be such that the marginal social benefits ($MB+MEB$) from the last unit consumed just offset the marginal social costs ($MC + MEC$) of producing the last unit of output. When external costs are present the private equilibrium quantity tends to be greater than the socially optimal quantity. When external benefits are present the private equilibrium quantity tends to be less than the socially optimal quantity. Sometimes consumption and production might generate both external costs and external benefits. The use of antibiotics against bacterial infections can generate a positive externality by reducing the threat of the infection spreading but also a negative externality since each succeeding generation of the bacteria is more resistant making the use of the antibiotic less effectual.

8.4 Common Property Resources

As another example of a positive externality, consider a farmer who plants wildlife feed plots along the edge of a field to enhance hunting or viewing opportunities for wildlife. Since wildlife tends to ignore fences, not only will the farmer benefit, but other wildlife watchers and hunters also benefit. We often think of wildlife as a common property resource (CPR). Common property resources are nonexcludable, but rival in consumption. A resource is nonexcludable when it is prohibitively costly or impossible to exclude users. A resource is rival, when one person's use reduces the amount available to others. Viewing wildlife while on a photo-safari is a nonrival use of wildlife, but hunting constitutes a rival use of wildlife. Common property resources include buffalo and beaver, fisheries, oil and gas wells, and aquifers. Common property resources are sometimes overexploited as users ignore the costs of their actions on other users.

Consider the following case where the five citizens of Fish City are all fishers who can either choose to fish in a commonly owned lake or work at a privately owned fish farm. Let N_F equal the number of citizens who choose to work at the fish farm and let N_L equal the number of citzens who choose to work at the lake, where $N_L + N_F = 5$. If the workers choose the fish farm their average product will be 2 fish per worker, $AP_F = 2$, and the total number of fish caught at the farm is $y_F = AP_F \times N_F$. Alternatively, if the workers fish at the lake their average product is $AP_L = 6 - N_L$; as more people fish at the lake, the average number of fish caught per person falls. The total number fish caught at the lake is $y_L = AP_L \times N_L$.

Each person chooses where to work based on their average product. If $AP_F > AP_L$ then the person chooses to work at the farm and if $AP_F < AP_L$ the person chooses to work at the lake. Table 8.2 illustrates the alternative production possibilities for the people of Fish City.

Table 8.2: Common Property Resource Problem

N_F	N_L	AP_F	AP_L	y_F	y_L	$y = y_F + y_L$
5	0	2	x	10	0	10
4	1	2	5	8	5	13
3	2	2	4	6	8	14
2	3	2	3	4	9	13
1	4	2	2	2	8	10
0	5	2	1	0	5	5

Suppose all five citizens work at the farm. Any one of those five workers has an incentive to switch to the lake because they could catch 5 fish at the lake whereas they can only catch 2 fish at the farm. A second worker also has the incentive to switch from the farm to the lake since they would catch 4 fish at the lake while only sacrificing 2 fish from their farm production. A third worker also has an incentive to switch since they would catch 3 fish at the lake but only sacrifice 2 fish from the farm. A fourth worker would be indifferent between switching from the lake to the farm since their average product is the same at both. Thus, the workers, acting independently and ignoring the effects of their actions on others, will align themselves such that $N_F = 2$ and $N_L = 3$ with 13 total fish caught or $N_F = 1$ and $N_L = 4$ with only $y = 10$ fish caught. Clearly the allocation of citizens between farm and lake is inefficient since $y = 14$ fish could be caught if $N_F = 3$ and $N_L = 2$. The inefficiency arises because each person who fishes at the lake imposes a negative externality on others who also fish the lake. Although the second lake fisher catches 4 fish they reduce the amount of fish caught by the first lake fisher by 1 (from 5 to 4). The third lake fisher catches 3 fish but reduces the amount of fish caught by the first two lake fishers by 2. If a fourth person chose to fish the lake they would catch 2 fish but reduce the quantity caught by the first three fishers by 3 fish. A fifth fisher would catch 1 fish but reduce the quantity caught by the first four fishers by 4 fish, resulting in a net loss of 3 fish.

Table 8.3: External Costs in the CPR Problem

N_L	F_L	$MB_L = \frac{\triangle y_L}{\triangle N_L}$	MC_L	MEC_L	MSC_L
0	0	x	x	x	x
1	5	5	2	0	2
2	8	3	2	1	3
3	9	1	2	2	4
4	8	−1	2	3	5
5	5	−3	2	4	6

Table 8.3 reports the private marginal benefits of fishing at the lake, the private marginal cost of fishing at the lake which equals the foregone fish from the farm, and the marginal external cost of fishing at the lake. The MEC of fishing at the lake equals the reduction in fish caught by the previous fishers when another fisher arrives and fishes. Social efficiency requires that as long as $MB \geq MSC$ then more people should fish the lake. Thus, it would be socially efficient for two citizens to fish the lake. If more than two people fish the lake a deadweight loss occurs as total fish production falls with the addition of another fisher at the lake.

Chapter 1 introduced the prisoners' dilemma faced by two shepherds who graze sheep in a commonly owned pasture. In Table 1.2 we showed that the game theoretic stable equilibrium for the two shepherds was to both graze 10 sheep, even though that number of sheep would exceed the carrying capacity of the pasture. Similarly, Table 8.3 shows that when individuals act independently too many people fish the lake resulting in a lower overall harvest than would occur if the individuals could coordinate their actions. One way to ensure that each individual acting in their own self interest also advances the group's welfare would be to establish private property rights to the CPR (common property resource). In the absence of private property rights is coordination toward the common goal possible among the users, or must the problem be solved by an outside planner or regulator who likely has less information about the problems with the CPR than the individual appropriators? Several problems must be overcome in solving the CPR problem. First is the problem of commitment. The users must somehow agree to limit their own use of the CPR. Second, the users need to be able to monitor each other to ensure compliance.

In 2009, Elinor Ostrom won the Nobel Prize in economics for her work studying the common property resource problem. Her research examined groundwater basins near Los Angeles, California, communal mountain pastures and forests in Switzerland and Japan, irrigation ditches in Spain, and an inland fishery near Alanya, Turkey. In studying the CPR problem, Ostrom found that successful long-enduring and self-monitoring systems tended to have some common characteristics[191]:

1. Boundaries and who may use the CPR are clearly defined.
2. Restrictions regarding when, where, by what method, and how much of the resource can be used are related to local conditions.
3. Most individuals affected by the operational rules can participate in modifying the operational rules. It should not be expected that the rules are formed perfectly the first time, but, instead, the rules evolve and an important organizing rule is for the users to specifically allow the rules to evolve.
4. Monitors, who actively audit CPR conditions and appropriator behavior, are accountable to the appropriators or are the appropriators.
5. Users who violate the operational rules are subject to graduated sanctions by the officials or other appropriators.
6. Appropriators and their officials have timely access to low-cost local arenas to resolve conflicts between monitors and appropriators or between various appropriators.
7. The rights of appropriators to devise their own institutions and rules are not challenged by external governmental authorities.
8. Appropriation, provision, monitoring, enforcement, conflict resolution, and governance activities are organized in multiple layers of nested enterprises.

If one or more of the characteristics are not met, the possibility that the CPR will be overexploited increases. Ostrom reports on an ocean fishery in Mawelle, Sri Lanka, which had established a set of rules governing the rotation and catch limits of fishers, but were unable to prevent entry, leading to overexploitation. In two other Turkish fisheries in the Bays of Bodrum and Izmir, fishers who belonged to heterogenous groups were unable to organize themselves. In addition, a high demand for fish from nearby cities changed local conditions and made it profitable for fishers to go for the quick profit at the expense of a sustainable fishery. Finally, as group size increases, efforts at self-monitoring become more difficult as the free-rider problem becomes greater in large groups.

8.5 Internalizing Externalities through Taxes and Subsidies

Various policies can be tried to move the economy in the right direction when externalities exist. These alternatives include taxes and subsidies, defining property rights, imposing quantity controls, negotiation, and appeal to the social contract. Taxes can be imposed on goods that generate a negative externality and subsidies can be given to producers or consumers who generate positive externalities. Private property rights such as marketable pollution permits are used in the electric power industry to control pollution and individual transferrable quotas are used in fisheries to prevent overfishing. Sometimes quantity controls are imposed. If the production of a good or service is such that the marginal social cost is always greater than the marginal benefit an outright ban on production is efficient.

The first policy option we consider is the use of taxes or subsidies to help control negative or positive externalities. In the case of a negative externality, imposing a tax equal to the marginal external cost of production gives the producer the incentive to cut production thereby reducing the external costs. In contrast, when a positive externality occurs a subsidy given to the consumer equal to the marginal external benefit encourages greater consumption. These kinds of taxes or subsidies help to internalize the externality and are referred to as Pigouvian taxes and subsidies after the economist A.C. Pigou. The idea behind the **Pigouvian tax** is simple. To internalize a negative externality, a tax equal to the MEC of production/consumption is imposed. People try to avoid paying taxes whenever possible. One way to avoid paying the tax is to reduce the activity that generates the negative externality. Sometimes business firms will be able to shift part or all of the tax to consumers in the form of higher prices, but this tax shifting is okay. The law of demand tells us that consumers will respond to the higher price by reducing quantity demanded. When consumers buy less of the polluting product, the negative externality is reduced. In fact, if we were to impose price ceilings to keep firms from shifting the tax to consumers the policy would likely be sub-optimal because consumers would continue to demand the same quantity in the absence of higher prices.

Figure 8.2 illustrates the case where electric utilities use clean air as an input in the production of electricity. Unless clean air is privately owned, electric utilities will tend to use too much clean air in producing the equilibrium quantity of electricity, y_E. Using clean air and dumping NO_x, SO_2, and CO_2 into the air generates negative externalities in the form of acid rain and greenhouse gases. Acid rain stunts tree growth, acidifies streams leading to fewer insects (fish food) and fish, and causes limestone buildings and statues to deteriorate. Carbon emissions contribute to climate change. If production and consumption occur up to the private equilibrium, y_E, the external costs of acid rain and greenhouse gases generate a deadweight loss equal to the area of the triangle DWL. Imposing a per unit tax equal to the marginal external cost of acid rain and greenhouse gases causes the electric utility to internalize the externality and reduce its use of clean air. This per unit tax equals the vertical distance between the MSC and MC curves: distance WL or DE. The tax revenue equals the per unit tax (t) multiplied by the quantity produced and consumed and equals the shaded rectangle P^*DEF. In theory, this tax revenue might be given back to consumers in the form of a lump-sum tax cut, but the price of electricity will still rise to P^* which gives consumers the incentive to reduce consumption of electricity.

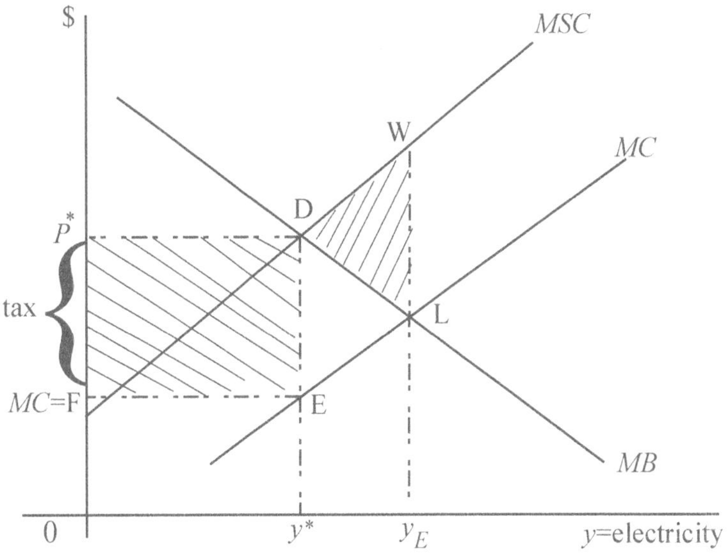

Figure 8.2: Pigouvian Tax to Internalize a Negative Externality.

When external benefits are present the Pigouvian solution would entail granting a per unit subsidy, s, to consumers, where $s = MEB$. This subsidy encourages greater consumption and production moving the economy toward the socially efficient quantity, y^*. Figure 8.3 illustrates the Pigouvian subsidy for a farmer who also likes to fish, hunt, and observe wildlife. The farmer, instead of using all of her land in farming, leaves an unplowed buffer zone around the edge of fields and in river bottoms. The buffer zone helps prevent erosion fostering better fish habitat and also provides wildlife habitat for birds and deer. The private landowner only invests in the buffer zone to the extent that they derive private marginal benefits from that zone. However, marginal external benefits accrue to third parties, such as downstream landowners who also get higher water quality or neighbors who also like to hunt or view wildlife, since wildlife seldom respect property lines. The subsidy (WE or DL) is paid for each additional increment to the buffer zone and the shaded area P^*WEF equals the total subsidy given to the landowner. Triangle DWL represents the deadweight loss due to underconsumption and underproduction of the buffer zone .

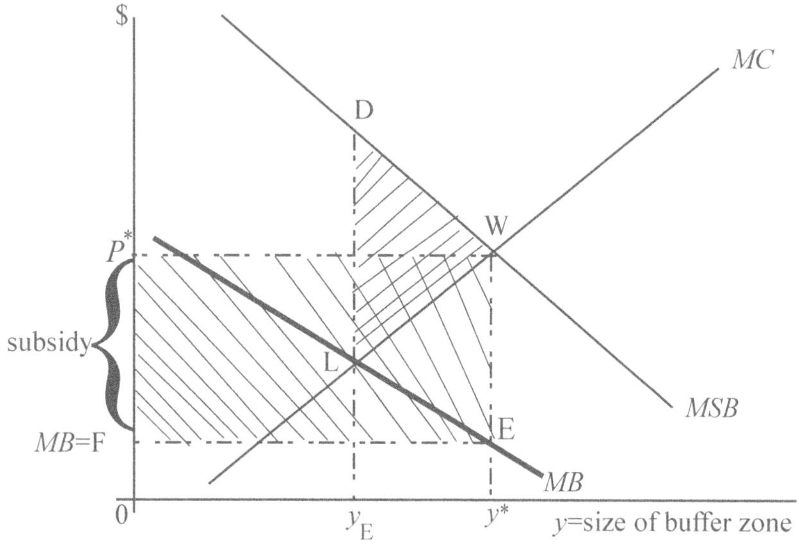

Figure 8.3: Pigouvian Subsidy to Internalize a Positive Externality.

8.6 Using Property Rights to Internalize External Costs and Benefits

As an alternative to Pigouvian taxes and subsidies, a second policy option would be to extend property rights to the resource that is commonly owned. The institution of private property rights gives the owner the incentive to employ the resource in its highest and best use. It also means that any economic agent who wants to use the resource must pay the resource owner a price at least as great as the value of the resource that is used up. Well-defined property rights are comprehensively assigned, exclusive, transferable, and secure. However, the assignment of property rights will affect the distribution of income. As we saw in Chapter 7, the socially efficient allocation of resources will often depend on the distribution of income, so that the assignment of private property rights can alter the socially optimal level of pollution.

Consider two firms, each of which uses a different mix of inputs to produce a desirable output, such as electricity, and a jointly produced undesirable output, such as SO_2 emissions. Table 8.4 below lists the production possibilities for the two firms. In the absence of property rights to clean air, each firm generates the maximum amount of electricity which they can sell for $1 per unit. The marginal cost of abatement equals the value of foregone electricity production necessary to reduce pollution by one unit.

Suppose the government decides that the optimal amount of pollution is not 18 units, but only 10 units. To achieve the optimal quantity they impose a quantity control that requires each firm emit no more than 5 units of pollution. Efficiency requires that the marginal abatement cost (MAC) be the same for both firms. However, if both firms abate 4 units of pollution, the MAC to firms A and B are not equal. Instead, firm A can abate a 5th unit of pollution for only $5, whereas it costs firm B $8 to abate the 4th unit. Therefore, the same amount of pollution (8 units) can be abated more cheaply by allowing firm A to abate more and firm B to abate less. In fact, the cost efficient way of abating 8 units of pollution would be for firm A to abate 6 units and firm B to abate only 2 units. Here we note that the marginal cost of abating the last unit of pollution is $6 regardless of which firm abated the 8th unit. In addition, the marginal cost of abating a 9th unit of pollution is $7 regardless of whether A abates the 7th unit or B abates the 3rd unit. Here we note that in the context of our production model abatement means producing less of the desirable output. If production of the desirable output and jointly produced undesirable output occurs, and, then, the undesirable output is somehow cleaned up, more inputs are needed.

A system of marketable emission permits is one method of promoting cost efficiency. In such a system the government issues (or auctions) 10 permits and firms bid for those permits with permits going to the highest bidder. If firms want to dump a unit of SO_2 into the air they must buy a permit. The maximum price a firm would pay for a permit equals their cost savings from not having to abate the marginal unit of pollution. For instance, if firm A does not buy any permits they have to abate 9 units of pollution by not producing any desirable output. The cost savings (value of electricity) to firm A from not having to abate the 9th unit equal $9. Thus, firm A would be willing to pay up to $9 for the first permit. Similarly, firm B would be willing to pay up to $13 for their first permit. In fact, the permit demand schedules for the two firms equals their MAC schedule turned upside down as shown in Table 8.5.

Table 8.6 gives the market demand for permits which equals the sum of the individual firm demands. For instance, if the price is $13 per permit Firm A will not purchase any permits and Firm B will only want to purchase 1 permit for a total market demand of 1 permit. If the price of a permit is $9, market demand is 6 permits as Firm A will buy 1 permit and Firm B will buy 5 permits at that price.

Since the supply of permits is fixed by the government at 10, the equilibrium price of a permit is $7. At $7, Firm A buys 3 permits and abates 6 units of pollution. Similarly, at $7, Firm B buys 7 permits and abates 2 units of

Table 8.4: Pollution Abatement Costs for Two Firms

Firm A				Firm B			
SO_2 Abated	$b = SO_2$	$y =$ Elec.	MAC $= p \times dy/db$	SO_2 Abated	$b = SO_2$	$y =$ Elec	MAC $= p \times dy/db$
0	9	45	x	0	9	81	x
1	8	44	$\$1 = \$1 \times (45 - 44)$	1	8	76	$\$5 = \$1 \times (81 - 76)$
2	7	42	$\$2 = \$1 \times (44 - 42)$	2	7	70	$\$6 = \$1 \times (76 - 70)$
3	6	39	$\$3 = \$1 \times (42 - 39)$	3	6	63	$\$7 = \$1 \times (70 - 63)$
4	5	35	$\$4 = \$1 \times (39 - 35)$	4	5	55	$\$8 = \$1 \times (63 - 55)$
5	4	30	$\$5 = \$1 \times (35 - 30)$	5	4	46	$\$9 = \$1 \times (55 - 46)$
6	3	24	$\$6 = \$1 \times (30 - 24)$	6	3	36	$\$10 = \$1 \times (46 - 36)$
7	2	17	$\$7 = \$1 \times (24 - 17)$	7	2	25	$\$11 = \$1 \times (36 - 25)$
8	1	9	$\$8 = \$1 \times (17 - 9)$	8	1	13	$\$12 = \$1 \times (25 - 13)$
9	0	0	$\$9 = \$1 \times (9 - 0)$	9	0	0	$\$13 = \$1 \times (13 - 0)$

Table 8.5: Bid Prices for Permits

Firm A		Firm B	
Permit	Bid	Permit	Bid
1	$9	1	$13
2	$8	2	$12
3	$7	3	$11
4	$6	4	$10
5	$5	5	$9
6	$4	6	$8
7	$3	7	$7
8	$2	8	$6
9	$1	9	$5

Table 8.6: Market Demand for Permits

Price	Firm A's Demand for Permits	Firm B's Demand for Permits	Market Demand
$13	0	1	1
$12	0	2	2
$11	0	3	3
$10	0	4	4
$9	1	5	6
$8	2	6	8
$7	3	7	10
$6	4	8	12
$5	5	9	14
$4	6	9	15
$3	7	9	16
$2	8	9	17
$1	9	9	18

pollution. This price also equals the marginal cost of abating the 7th unit of pollution for Firm A or the 3rd unit of pollution for Firm B.

If Pigouvian taxes were used instead of permits, what tax rate would ensure that only 10 units of pollution are emitted instead of the current 18? Similar to the permit system a tax equal to $7 per unit of pollution would give Firm A the incentive to abate the first 6 units of pollution since the marginal cost of abatement is less than the tax and they would be indifferent between abating the 7th unit or paying the tax. Similarly, with a tax equal to $7 per unit Firm B would have the incentive to abate the first two units of pollution and they would be indifferent to abating the 3rd unit of pollution or paying the tax. Thus, setting the tax equal to the marginal cost of abating the 9th unit would give Firm A the incentive to abate 6 units and Firm B to abate 2 units resulting in only 10 units of pollution emitted between the two firms.

While property rights' schemes can sometimes be less costly than command and control regulations, the number of permits is still determined via the political arena. In setting the number of permits regulators must account for regional differences in the damages that pollution causes, the potential for firms to substitute an unregulated pollutant for the regulated one, and for changing demands for an amenity that occurs in nonpolluted areas. Consider the case of permits for wastewater discharges. Since amenity values of water for fishing, boating, swimming, and wildlife are not priced in markets, a system of effluent permits for wastewater discharges might have the effect of concentrating water pollution in certain watersheds and not in others. The reasoning is that as water pollution increases in one watershed, the amenity value of clean water increases in the nonpolluted area increases, so that discharging effluents in the nonpolluted watershed would have a higher marginal cost in terms of foregone recreational benefit [56].

8.7 Uncertainty and the Choice between Taxes and Permits

When regulators face uncertainty about firms' marginal abatement costs the efficacy of taxes and permits might diverge [248]. The marginal benefits of abatement equal the damages averted from not emitting the last unit of pollution. Suppose the regulator knows the true marginal benefits of abatement, given as MB_1 in Table 8.7, but is uncertain about the marginal costs. If there is a 50–50 chance of marginal costs being high (MC_{High}) versus low (MC_{Low}) the regulator might average the two marginal costs schedules to obtain the expected marginal cost of abatement shown as EMC in Table 8.7. Given marginal benefits (MB_1) and expected marginal costs (EMC) the regulator might think it best to abate 8 units of pollution. This level of abatement can be achieved by issuing 7 permits or by imposing a tax of 15 which is equal to the expected marginal cost of abating the 8th unit of pollution.

Suppose the true marginal cost of abatement is MC_{High}. If the regulator issues 7 permits then 8 units of pollution are abated, but the socially optimal level of abatement is only 6 units of pollution corresponding to where $MB_1 = MC_{High}$. Thus, too few permits are issued and too much pollution is abated resulting in a deadweight loss illustrated in Figure 8.4 as the area of the triangle EFH. If instead, the regulator had imposed a tax equal to 15, the producer with marginal costs of abatement equal to MC_{High} would choose to abate only 2 units of pollution, which is too little abatement. In the case of the tax, the deadweight loss would be represented by the triangle ACH.

Suppose instead, the true marginal cost of abatement is MC_{Low}. Given MB_1 it would now be optimal to abate 10 units of pollution. However, if the regulator equates MB_1 and EMC then they incorrectly think that 8 units of pollution should be abated. If the regulator issues 7 permits too little pollution is abated resulting in a deadweight loss equal to the area of the triangle EGL. If the regulator were to impose a tax of 15, then in the low marginal cost world, the producer would choose to abate 14 units of pollution rather than abate the socially optimal amount of 10 units resulting in a deadweight loss equal to triangle JKL.

Table 8.7: Uncertainty and the Choice between Permits and Taxes

Pollution Abated	Permits	Flat MB_2	Steep MB_1	EMC	MC_{High}	MC_{Low}
0	15	x	x	x	x	x
1	14	18.5	29	8	14	2
2	13	18.0	27	9	15	3
3	12	17.5	25	10	16	4
4	11	17.0	23	11	17	5
5	10	16.5	21	12	18	6
6	9	16.0	19	13	19	7
7	8	15.5	17	14	20	8
8	7	15.0	15	15	21	9
9	6	14.5	13	16	22	10
10	5	14.0	11	17	23	11
11	4	13.5	9	18	24	12
12	3	13.0	7	19	25	13
13	2	12.5	5	20	26	14
14	1	12.0	3	21	27	15
15	0	11.5	1	22	28	16

Now, consider the case where the marginal benefit schedule is flatter, given by MB_2 in Table 8.7. Again, the regulator equates MB_2 with EMC, determines that 8 units of pollution should be abated, and then issues 7 permits or imposes a tax equal to 15. If the true marginal cost is high then the optimal level of abatement is 4 units. Under a permit program where the regulator issues 7 permits and requires abatement of 8 units, the deadweight loss is given by the triangle EFH in Figure 8.5. If marginal costs are low the optimal level of abatement is 12 units. If the regulator issues only 7 permits and requires 8 units of abatement the deadweight loss would be given by triangle EGL.

Suppose instead that a tax equal to 15 is imposed. In the high marginal cost world, the producer would choose to abate only 2 units which is less than optimal abatement of 4 units resulting in a deadweight loss equal to triangle ACH. In the low cost world, a pollution tax equal to 15 would cause the

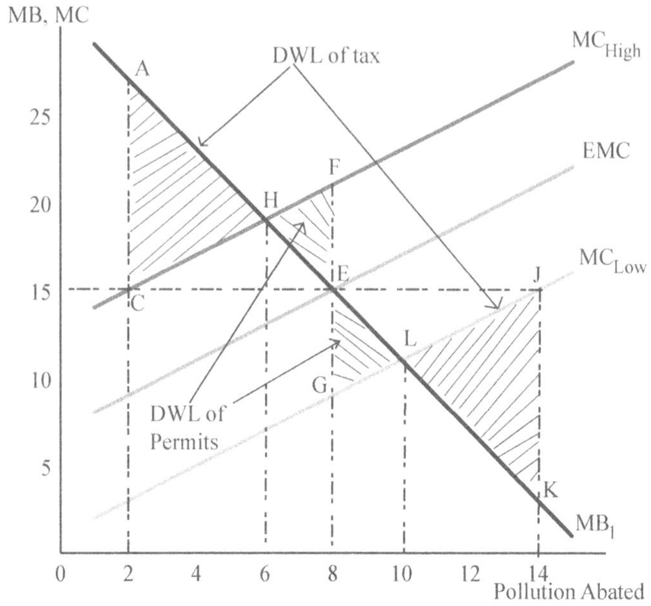

Figure 8.4: Uncertainty and the Choice between Permits and Taxes: Absolute slope $|MB|$ > slope MC.

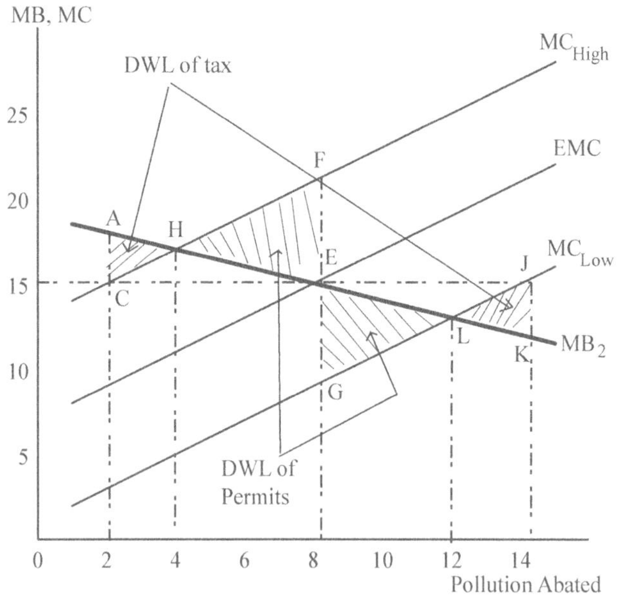

Figure 8.5: Uncertainty and the Choice between Permits and Taxes: Absolute slope $|MB|$ < slope MC.

producer to abate 14 units of pollution, which is more than optimal abatement of 12 units resulting in a deadweight loss equal to triangle LJK.

To summarize, when the absolute value of the slope of the marginal benefit schedule is greater than the slope of the marginal cost schedule, a quantity control such as "cap and trade" yields a smaller deadweight loss than a tax. In Table 8.7 the slope of the marginal cost schedule equals 1, regardless of whether marginal costs are high MC_{High}, low MC_{Low}, or average EMC. The absolute slope of MB_1 equals 2 and the absolute slope of MB_2 equals 0.5. In Figure 8.4 we assumed that the marginal benefit schedule was MB_1. Here, we observe that the deadweight loss of a quantity control is smaller than the deadweight loss of a tax. In contrast, when MB_2 is used as the marginal benefit schedule as in Figure 8.5, the deadweight loss of a quantity control is larger than the deadweight loss of a tax.

To understand, remember that the benefit of abating an extra unit of pollution equals the damage avoided from reduced exposure. If the marginal benefit curve is steep, then there is a threshold effect around the socially efficient quantity of abatement. If too little abatement occurs the marginal damages are high relative to the marginal costs and a quantity control would be more efficient. In contrast, if the marginal benefit schedule is relatively flat the marginal damages around the threshold are similar, so it is more efficient to impose a tax, which has a smaller deadweight loss when costs are uncertain.

8.8 Negotiation and Bargaining

Another policy option is to allow third parties to resolve the externality through negotiation. For example, suppose I like to smoke cigarettes and you feel harmed by my smoking. If I own the air in the room I can smoke whenever and as much as I want. However, you will be able to induce me to smoke less if you pay me an amount that is at least as great as the value of lost utility I would incur from not smoking. If I value smoking the next cigarette at $0.25, you could pay me $0.26 and I would not smoke the cigarette. On the other hand, suppose you own the property rights to the air in the room and I value smoking a cigarette at $0.25. If smoking the cigarette imposes less than $0.25 damage, you should be willing to take $0.25 and allow me to smoke. This process of negotiation has the potential to make both of us better off since it is a voluntary transaction.

The process described above is known as the Coase Theorem, for the Nobel prize-winning economist Ronald Coase [43]. Coase used as an example a factory emitting smoke which imposed costs on its neighbors. However, Coase pointed out that a regulation that prohibits the factory from emitting smoke imposes a cost on the factory. Coase [43, p. 2] argued that "The real question that has to be decided is: should A be allowed to harm B or should B be allowed to harm A? The problem is to avoid the more serious harm."

As an example of how the Coase Theorem might work consider the runoff from a chicken farm that pollutes a local stream. The chicken farm benefits from the runoff in the form of enhanced profits (or lower costs). Local fishers bear the cost of the farm's runoff in the form of degraded fishing. Figure 8.6 depicts the marginal benefits to the farm $MB(b)$ from alternative levels of pollution (b). If the amount of water pollution is b_1, the height of the $MB(b)$ curve at b_1 represents the marginal benefit and the area under the $MB(b)$ curve from 0 to b_1 equals the total benefit to the farmer from allowing runoff. As the amount of runoff increases, the total benefits from runoff increase, but the marginal benefits become smaller since less extra desirable output is produced for each one unit increase in pollution. Let the maximum amount of runoff that can occur be represented by $\bar{b}$. The total benefits from that level of runoff are the sum of the areas A and B and the marginal benefit of the last unit of pollution equals zero.

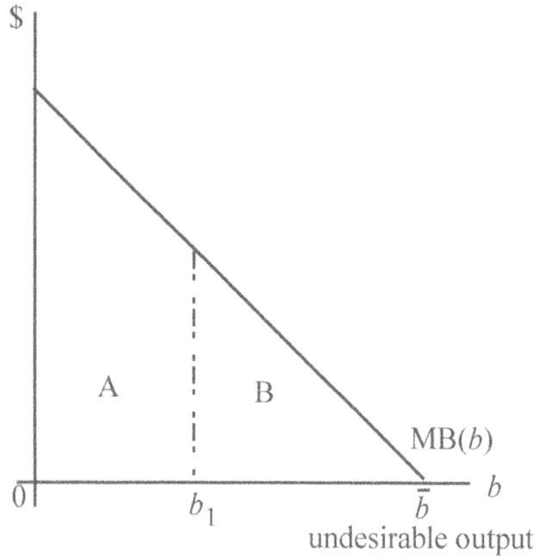

Figure 8.6: Marginal Benefits from Runoff.

Although the chicken farm gains benefits from its runoff, local fishers bear the costs of pollution in the form of degraded fishing opportunities. The marginal cost (damage) of the pollution is depicted in Figure 8.7. The marginal cost of pollution equals the vertical distance from the horizontal axis to the marginal cost curve at a given quantity of pollution. For a given level of runoff, say b_1, the total cost or damage to the fishers equals area under the $MC(b)$ curve from 0 to b_1: area C. As the amount of runoff increases the total cost to the fishers increases as does the marginal cost. At the maximum level of pollution, $\bar{b}$, the total cost of the runoff equals the sum of the areas C and D.

The height of the $MB(b)$ curve gives the marginal benefits to the farm from pollution and it also represents the marginal cost curve to the farmer if the farmer is required to abate pollution. If the farmer is polluting the

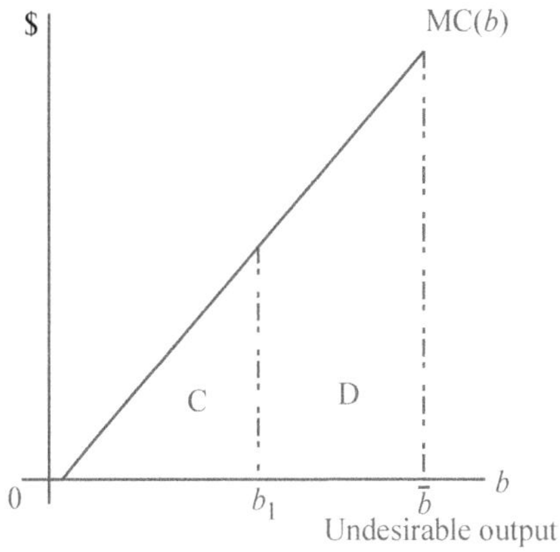

Figure 8.7: Marginal Costs from Runoff.

maximum amount, $\bar{b}$, the total cost to the farmer from reducing pollution, say to b_1, is represented by area B in Figure 8.6. Similarly, any abatement of pollution by the farmer represents a benefit to the fishers. In Figure 8.7 if abatement occurs from $\bar{b}$ to b_1 the fishers gain total benefits equal to area D and the height of the $MC(b)$ curve at b_1 represents the marginal benefit to the fishers from abating the last unit of runoff.

Figure 8.8 depicts both the $MB(b)$ and $MC(b)$ curves to the farmer and the fisher. Suppose the chicken farmer owns the property rights to the local stream and can choose to allow any amount of runoff. Given this scenario, the chicken farmer will increase chicken production and its associated runoff up to the point where the last unit of runoff gives zero in marginal benefits, $\bar{b}$. However, at $\bar{b}$, the marginal cost of runoff borne by the fishers is positive. Thus, the fishers could compensate the farmer to reduce the amount of runoff into the stream. For instance, at $\bar{b}$, the cost to the fishers from the last unit of runoff might be \$25 and so the fishers would be willing to pay up to \$25 to induce the farmer to reduce runoff by one unit. On the other side of the potential exchange, the farmer would gladly accept \$25 to reduce runoff since the marginal benefit of the last unit of runoff was zero. In fact, as long as the marginal cost of runoff to the fisher is greater than the marginal benefit to the farmer, the fisher can more than compensate the farmer for reducing runoff. The socially optimal level of runoff occurs when $MB(b) = MC(b)$ at b^* with price c^*. When the farmer owns the property rights, and is compensated by the fishers for reducing the runoff, the farmer bears a cost equal to triangle Q, but the benefits accruing to the fishers from having cleaner water to fish equals the areas Q+R+S, for a net social gain of R+S.

Now, consider the case where the fishers own the property rights to the river. Given those rights, the fishers want $b = 0$ unless they are compensated for the degradation of the fishery caused by the runoff. At $b = 0$, the marginal

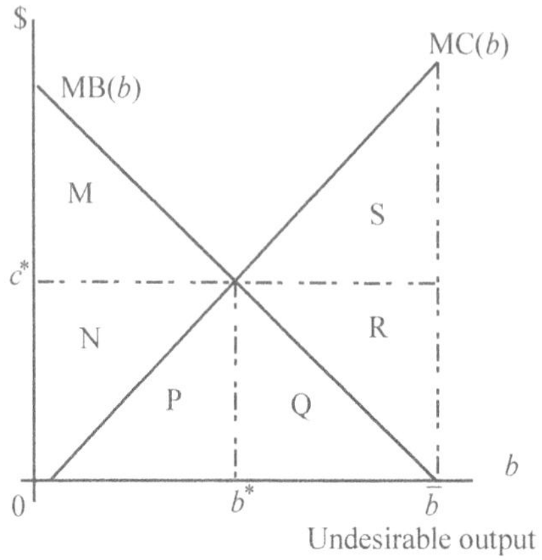

Figure 8.8: Optimal Level of Pollution.

cost of runoff to the fishers is low relative to the marginal benefit to the farmer. In this case, the farmer could more than compensate the fishers for the damages caused by runoff. At the socially optimal level of runoff, b^*, the fishers would bear a total cost equal to area P, but the farmer would gain up to the area M+N+P, for a net social gain of M+N.

If trade between the fishers and farmers takes place at the equilibrium price of pollution, c^*, the net gains from farmers compensating fishers for the right to pollute are divided such that farmers receive a gain equal to area M while fishers receive a gain equal to area N. If instead, farmers own the property rights and charge c^* for every unit of runoff they reduce, the gain accruing to the fishers from compensating the farmer to reduce pollution would equal area S with the farmer receiving a gain equal to R. In some cases, one party or the other might exercise greater bargaining power or skill. In such a case, the party with the greater bargaining skills can extract the entire net gains from trade, either M+N when the fisher owns the property rights, or R+S when the farmer owns the property rights.

In our example with the farmer and the fisher there are only two parties and for a given assignment of property rights either the fisher compensates the farmer or the farmer compensates the fisher. However, as the number of parties or people that gain benefits and bear costs increases, the potential number of side payments or "bribes" increases.

Table 8.8 illustrates the geometric increasing relation between people and the number of side payments between those people necessary for the Coasian process to generate an equilibrium outcome. Because of the problems associated with large group negotiations, groups of like-minded individuals sometimes find it advantageous to form coalitions. Groups such as The Nature Conservancy, the Sierra Club, the Audubon Society, and the National Wildlife Federation are composed of members who give money and time to promote

Table 8.8: Negotiation and Side Payments

People	Side Payments
2	1
3	3
4	6
5	10
6	15
25	325
100	4,950

various environmental objectives. The Nature Conservancy purchases land that it considers worthy (by its own standards) for preservation. The Sierra Club on the other hand is much more aggressive on the political side of the environmental advocacy movement. Other groups, such as those who were part of the Sagebrush Rebellion, formed coalitions against environmentalists and pressed for local control over the use of western public lands. Local interests lobbied for continuing mining, timber, and ranching that were allowed by the Multiple Use-Sustained Yield Act (MUSY) signed into law by President Dwight Eisenhower in 1960. Through a process of bargaining, litigation, lobbying, and changing preferences, the various coalitions of interest reached a kind of Coasian equilibrium [167] whereby the federal government continued to control western public lands, but ranching and timber activities were allowed to continue.

In the early 1990s wolves were reintroduced into Yellowstone National Park. The expected benefits of the wolf program were to provide viewing opportunities for people and provide a predator for elk that had overpopulated and overgrazed the Greater Yellowstone ecosystem. Cattle ranchers, fearing that wolves would also prey on sheep and calves, opposed the program. The potential conflict between environmentalists who wanted wolves reintroduced and ranchers who opposed the wolves provided an opportunity for negotiation. The Bailey Wildlife Foundation Proactive Carnivore Conservation Fund is overseen by the environmental group Defenders of Wildlife. The Fund helps compensate ranchers when wolves kill livestock. Two programs implemented with Fund support include providing electric fences for pens for sheep and range riders who help monitor cattle herds and fire cracker shells when wolves approach herds. Both programs reduce the costs borne by ranchers due to wolves. In addition, environmentalists and ranchers alike want to maintain intact ranches that are not subdivided into 20-acre ranchettes. Such subdivision is more likely when ranchers' costs rise due to wolves preying on sheep and cattle.

8.9 Social Contract

The last way of correcting for externalities can best be thought of as a preventative measure: the social contract. The social contract is a set of unwritten rules and, in a sense, a code of ethics that people try and abide by. Certainly when a person is traveling down the road and no one is around, they could at no cost to their self throw out the trash from their lunch bought at McDonald's. No one would know and it would save them from having it stink up their car. However, some people choose not to litter, even though they could get away with it. Similarly, even without compensation, some people choose to plant wildlife plots and build brush-piles for wildlife. Still other people choose to take the welfare of strangers into account when making decisions that would likely cause third-party effects. For example, some fishers practice catch-and-release fishing, which improves fishing, even when they might not ever fish an area again and could legally take a limit.

The social contract applies to states or countries as well. The United States is certainly big and powerful and can choose to abide by or reject various environmental treaties. As one of the largest consumers of energy in the world, what responsibilities do US citizens have to citizens of other countries for reducing pollution and the effects of global warming? The social contract sometimes draws a very thin line between what is legal and what is socially acceptable. However, some people argue that we should never make ethics into law (and probably could not anyway), because ethics are what you practice when no one is watching.

8.10 Pigouvian Tax to Reduce Global Warming

The IPCC (Intergovernmental Panel on Climate Change) presents evidence from scientists in many fields that burning coal and oil have increased CO_2 emissions by 70% from 1970 to 2004. Other emissions, including methane (CH_4) and nitrous oxides (NO_x), have also increased. Such GHG (greenhouse gas) emissions are thought to be a source of global warming. Adding a sense of urgency, the IPCC reports that ten of the warmest years since 1850—when widespread recording of temperature began—have occurred since 1998.

Government policies such as corn-based ethanol and higher fuel efficiency standards for vehicles are likely to be of little real value in reducing emissions and serve more as political window dressing in an election year. Instead, resource users should pay a charge or tax equal to the climate damage caused by consuming GHG intensive products, such as electricity and transportation services. European countries have adapted a cap and trade system to GHG emissions. With a cap in place, firms that want to emit more GHGs must buy permits in the market. The permit costs are eventually reflected in higher prices which send a signal to producers to search for less GHG intensive production methods and signal consumers to buy less GHG intensive products. However, policy-makers can only guess at what the proper cap should be. Instead, a system of taxes per unit of CO_2 equivalent might be imposed, where the tax equals the marginal social damage caused by GHG emissions. Current estimates put the social damage of an additional ton of GHG emissions at about \$36 with a range between \$12 and \$58 depending on the rate of discount. Such a tax could be increased or decreased as more information becomes available on the contribution of GHG emissions to climate change.

According to IPCC scientist and economist Ross McKitrick, "climate models predict that, if greenhouse gases are driving climate change, there will be a unique fingerprint in the form of a strong warming trend in the tropical troposphere, the region of the atmosphere up to 15 km altitude, over the tropics from 20 degrees North to 20 degrees South [169, p. 708]." McKitrick [169, 170] proposes a T3 tax on CO_2 equivalent emissions tied to the degree of warming in the tropical troposphere, where T3 stands for tropical troposphere temperature. McKitrick argues that skeptics of climate change who believe that there has been little warming will be pleased that the tax will be small if global warming does not occur and will decline to zero if cooling occurs. Global warming activists will be pleased that if warming rapidly proceeds, the tax will rise sharply, sending a strong price signal to investors and consumers to adapt cleaner technologies.

8.11 Regional Differences in the Damages of Pollution

So far, our analysis of permit trading has treated emissions coming from various sources as equivalent. In fact, the damages caused by emissions will generally depend on the location of the firm, the size of population affected by the emissions, and the demographic makeup of the population. Emissions that occur in densely populated cities with young children will typically have higher damages than an equivalent level of emissions in a sparsely populated rural area with an elderly population. Recent work by Muller and Mendelsohn [179] estimated the marginal damages of six pollutants from 10,000 different sources. The pollutants were fine particulate matter, coarse particulate matter, nitrogen oxides, ammonia, volatile organic compounds, and sulfur dioxide. Using a \$2 million value of a statistical life, the marginal damages of an additional ton are reported in Table 8.9.

Although there are differences in the marginal damages of pollution between rural and urban areas, the marginal damages also vary across geographic regions of the country. For instance, sulfur dioxide emissions have a relatively lower marginal damage in western states which have alkaline soils than in east-

Table 8.9: Marginal Damages from One Ton of Emissions.

	Percentile					
Pollutant	1st	25th	50th	75th	99th	99.9^{t}
Fine Particulate Matter	\$250	\$700	\$1,170	\$1,970	\$12,400	\$41,77
Coarse Particulate Matter	\$60	\$120	\$170	\$280	\$1,960	\$6,55
Nitrogren Oxides	\$20	\$180	\$250	\$370	\$1,100	\$1,78
Ammonia	\$100	\$300	\$900	\$2,000	\$20,620	\$59,45
Volatile Organic Compounds	\$40	\$120	\$180	\$280	\$1,370	\$4,54
Sulfur Dioxide	\$220	\$550	\$970	\$1,300	\$4,130	\$10,86

(Source: Muller and Mendelsohn [179]).

ern states with more acidic soils. For a pollutant such as sulfur dioxide, the lowest marginal damages (1st percentile) of sulfur dioxide equal \$220 per ton and at the 99.9th percentile the marginal damages are \$10,860. An example of a city in the 1st quantile is Klamath, Oregon and at the 99.9th percentile is Hudson, New York. Since the ratio of marginal damages between the 99.9th and 1st percentiles is 49.4 to 1 ($\frac{\$10,860}{\$220}$), abating 1 ton of sulfur dioxide in Hudson, New York is equivalent to abating 49.4 tons of sulfur dioxide in Klamath, Oregon. Therefore, the trading ratio between a firm in Klamath and a firm in Hudson is 49.4 to 1. Thus, a firm in Hudson seeking to increase emissions by 1 ton of SO_2 would have to acquire 49.4 permits from a firm in Klamath. Alternatively, a firm seeking to expand emissions by 49.4 tons in Klamath would only have to acquire 1 permit from a firm in Hudson. Accounting for the regional effects of pollution would give firms a greater incentive to abate pollution in areas with high marginal damage per ton and increase emissions in areas where there is a low marginal damage per ton.

8.12 Problems

1. The table below illustrates the private marginal benefits and the private marginal costs of producing a million kilowatt hours of electricity. The producer is a firm that burns coal to generate electricity. In the process of burning the coal, sulfur is emitted into the atmosphere and later combines with oxygen producing SO_2 which causes acid rain and breathing problems for those with asthma and for the general population. The electric utility does not have to pay for the damage that burning the coal causes because there are no private property rights to air.

MC	MEC	MSC	$y =$ Elec.	MB
x	x	x	0	x
10	4		10	50
15	6		20	45
20	8		30	40
25	10		40	35
30	12		50	30
35	14		60	25
40	16		70	20

(a) If resources are allocated in a private market what will be the private equilibrium price and equilibrium quantity of electricity?

(b) Complete the schedule of marginal social costs.

(c) What is the socially efficient quantity of electricity to be produced and consumed?

(d) What is the socially efficient price per unit of electricity?

(e) If we use a Pigouvian tax to internalize the negative externality, what should the tax rate equal to give producers and consumers the incentive to produce/consume at the social optimum?

(f) Comment on incidence of the Pigouvian tax. That is, how much of the tax does the producer pay and how much of the tax does the consumer pay? Why is it efficient to make consumers pay at least some of the tax?

2. The table below reports the private marginal benefit (MB) schedule and private marginal cost (MC) schedule for timber products. Individuals gain marginal benefits from timber to the extent that they use wood for housing, furniture, firewood, etc. The private marginal costs of timber include the costs of hiring lumberjacks, truckers, and operating sawmills. The timber production process generates positive externalities in the form of recreational services as people gain forest access from former logging roads and negative externalities as the roads reduce the migratory routes of some wildlife. In addition, timber companies reduce biodiversity as they plant a monoculture of fast-growing trees of the same age, rather than a mix of trees suitable for an area. The marginal external benefits associated with timber production are assumed to be constant at every level of output and are MEB=25. The marginal external costs associated with timber production are also constant at every level of output and equal MEC=50.

(a) Ignoring the external effects identify the private market equilibrium price and quantity.

MC	MEC	MSC	y = Millions of Board Feet	MSB	MEB	MB
x	x	x	0	x	x	x
25			5			100
40			10			90
55			15			80
70			25			70
85			30			60
100			35			50
115			40			40
130			45			30
145			50			20
160			55			10

(b) Complete the MEB, MSB, MEC, and MSC schedules.

(c) If we account for the marginal external benefits, but not the marginal external costs, what will be the optimal price and quantity?

(d) If we account for the marginal external costs, but not the marginal external benefits, what will be the optimal price and quantity?

(e) If we account for both the marginal external benefits and the marginal external costs, what will be the socially optimal price and quantity of timber production and consumption?

(f) Accounting for both the MEB and MEC devise a tax or subsidy scheme that will generate the socially optimal price and quantity of timber production and consumption.

3. The table below gives the total benefits (in dollars) schedule to Nick O. Teen Fiend from smoking cigarettes and the total cost schedule (in dollars) to Hack A.(Asthma) Tack from Nick's smoking.

   (a) Complete the marginal benefit schedule (MB) to Nick and the marginal cost schedule (MC) to Hack.

   (b) If Nick owns the property rights to the air and there is no possibility for negotiation between Nick and Hack concerning the number of cigarettes Nick will smoke, how many cigarettes will Nick smoke? What cost will Hack incur?

   (c) Now suppose that Nick still has the property rights to the air but we allow Hack to negotiate with Nick about the number of cigarettes Nick will smoke. How many cigarettes will Nick smoke? What will be the price received by Nick and paid by Hack to keep Nick from smoking?

Nick TB	MB	Number of Cigarettes	Hack MC	TC	Net Surplus $= TB - TC$
0	x	0	x	0	0
16		1		5	
31		2		11	
45		3		18	
58		4		26	
70		5		35	
81		6		45	
91		7		56	
100		8		68	
108		9		81	
115		10		95	

(d) Instead of Nick, suppose that Hack has the property rights to the air in the dorm room and no cigarettes are smoked by Nick. Is this allocation efficient? Explain.

(e) If we allow Nick to make side payments to Hack what will be the equilibrium price that Nick will pay and Hack will receive and the equilibrium quantity of cigarettes smoked by Nick?

(f) Compare your answers in (c) and (e).

4. Two different groups—members of The Fishers, Liars, and Drinkers Club and The Bad Water Company—would like to use water in the local Whiskey River. The members of the FLD Club would like to fish in Whiskey River and The Bad Water Company would like to discharge their waste water into the river. The FLD Club currently receives $20,000 in recreational benefits from fishing. The Bad Water Co. would save $25,000 each year if it could discharge their waste into the river.

(a) Is it efficient to ban waste water discharges into Whiskey River?

(b) If the FLD Club owns the property rights to the river is a Pareto improving exchange possible between Bad Water and FLD?

(c) Suppose instead that Bad Water owns the property rights to the river. A scientist/engineer learns of a method of reducing the pollution in waste water such that after treatment and discharge, the river would still be fishable. The cost of installing the waste water treatment is $14,000. Is a Pareto improving exchange possible between Bad Water and FLD?

5. Ford and GM make cars and pollute the air. The table below depicts the hypothetical marginal costs for Ford and the total costs to GM to reduce polluting emissions from new automobiles.

Ford				GM			
$b =$ Pollution	Pollution Abated	MC	TC	$b =$ of Pollution	Pollution Abated	MC	TC
5	0	x	0	5	0	x	0
4	1	10		4	1		5
3	2	20		3	2		15
2	3	30		2	3		30
1	4	40		1	4		50
0	5	50		0	5		75

(a) Complete the TC of abatement schedule for Ford and the MC of abatement schedule GM.

(b) Before any pollution policy is implemented assume that each firm is producing a given number of autos such that 5 tons of pollution are emitted. If a regulation is imposed requiring each firm to abate 2 units of pollution (reduce emissions by 2) what will be the total cost of the regulation?

i. Cost to Ford=
ii. Cost to GM=
iii. Total cost=

(c) How many tons of emissions will there be after the regulation?

(d) Instead of a regulation requiring each firm to abate two units of pollution, suppose the government decides to sell six emission permits to the highest bidder. Complete the schedule of highest bids for Ford and GM. Remember, if a firm buys an emissions' permit they will not have to abate a unit of pollution. If they do not buy any pollution permits the emissions they may legally emit equals zero. To get the bid price figure out what each firm will save by not having to abate the marginal unit of pollution.

Ford		GM	
Permit	Bid	Permit	Bid
1		1	
2		2	
3		3	
4		4	
5		5	

(e) Out of the six permits, how many will Ford purchase? How many will GM purchase?

(f) What is the minimum total cost of cleaning up four units of pollution with the program of marketable emission permits?

(g) Why is the total cost of the marketable emission program less than the total cost of the regulatory scheme for cleaning up pollution?

(h) Complete the market demand schedule for permits. Determine the equilibrium price of permits given a supply equal to six permits.

Price	Market Demand Ford+GM	Demand by NC	New Market Demand Ford+GM+NC
$50		0	
$40		0	
$30		1	
$25		2	
$20		3	
$15		4	
$10		5	
$5		6	

(i) Suppose another buyer in addition to Ford and GM enters the market. Call this buyer The Nature Conservancy (an environmental group). The Nature Conservancy's demand schedule is given in the third column. Complete the new market demand schedule that includes the three buyers: Ford, GM, and The Nature Conservancy.

(j) The new equilibrium price of permits is ________ and at that price Ford will buy ________ permits, GM will buy ________ permits, and The Nature Conservancy will buy ________ permits.

6. The table below gives the marginal cost schedules for Greasy Bros. and the Stale Air Co. to clean up pollution.

   (a) What is the most cost-effective means of abating four units of pollution? That is, how much should Greasy Bros. abate and how much should Stale Air Co. abate?

   (b) What is the most cost-effective means of abating seven units of pollution?

   (c) What is the most cost-effective means of abating nine units of pollution?

   (d) What principle is at work in determing how much pollution each firm abates?

7. This problem investigates whether taxes or quantity restrictions are more efficient when the regulatory authorities are uncertain about

Greasy Bros.		The Stale Air Co.	
Pollution Abated	MC	Pollution Abated	MC
0	x	0	x
1	5	1	10
2	8	2	20
3	10	3	30
4	14	4	40
5	20	5	50
6	30	6	60
7	35	7	70

the marginal cost (MC) of cleanup. Let the marginal benefit function for pollution abatement be given as $MB = 28 - 2 \times A$, where A is the level of abatement. The regulator is uncertain whether firms have high abatement costs or low abatement costs but they know there is a 50% probability that firms have high abatement costs and a 50% probability that firms will have low abatement costs. If firms have high abatement costs the marginal cost of abatement is $MC_H = 7 + A$ and if firms have low abatement costs the marginal cost of abatement is $MC_L = 1 + A$.

(a) Given the probabilities for high MC or low MC, derive the expected marginal cost (EMC) of abatement function.

(b) Given the marginal benefit function and the EMC function calculate the expected efficient level of abatement A^* and the expected efficient tax (T^*) on pollution that would induce firms to abate A^*.

(c) Compare the absolute values of the slopes of the MB and EMC of abatement. Which schedule, MB or EMC, has the greatest slope in absolute value?

(d) Suppose the regulator uses a quantity control requiring firms to abate A^* units of pollution. Also, assume that the true marginal cost of abatement is MC_H.

   i. Calculate the level of abatement that would be socially efficient given MC_H.

   ii. Calculate the deadweight loss of the quantity control set at A^* given the socially efficient level of abatement.

(e) Instead of a quantity control, suppose the regulator imposes a tax equal to the amount you calculated in (b). If the firm's true marginal cost of abatement is MC_H how much pollution will the firm choose to abate? Given the socially efficient level of abatement calculate the deadweight loss of the tax.

(f) Redo parts (c) and (d) assuming that the true marginal cost of abatement is MC_L.

(g) If the marginal benefit schedule is $MB = 16 - 0.5 \times A$ instead of $MB = 28 - 2 \times A$, what will happen to the deadweight loss of the tax? the deadweight loss of permit scheme?

8. Use EXCEL to solve the following problem. In the small fishing village of Bridgeport all workers are fishers. These fishermen can fish in a lake that is an open access CPR or they can work at a private fish farm. Let the number of fishers who fish the open-access lake be N_L and the number of fishers who work at the fish farm be N_F. The aggregate supply of labor is perfectly inelastic and equal to 10 so $N_L + N_F = 10$. The number of fish caught at the lake depends on how many people fish the lake. The average product of each fisher at the lake is given as $AP_L = 13 - N_L$ and the total harvest at the lake is given as $y_L = AP_L \times N_L = (13 - N_L) \times N_L$. Fishers who work at the fish-farm harvest 3 fish per day, $AP_F = 3$, and the total harvest at the fish farm is $y_F = AP_F \times N_F = 3 \times N_F$. The goal is to allocate labor to the lake and to the fish farm so as to maximize the total harvest of fish. That is, choose N_F and N_L to maximize $y = y_L + y_F$.

To solve this problem set up an Excel spreadsheet as in the following table:

The Bridgeport Fishery

N_L	N_F	y_L	y_F	AP_L	AP_F	MEC_L
0	10			x	x	x
1	9					x
2	8					
3	7					
4	6					
5	5					
6	4					
7	3					
8	2					
9	1					
10	0					

(a) Complete the spreadsheet and determine the optimal N_L and N_F.

(b) In the absence of any government intervention a worker will choose to fish the lake if their average product, AP_L, is greater than their average product at the fish farm, AP_F. Given this incentive, how many workers will choose to fish at the lake?

(c) At the lake, each fisher after the first imposes an external cost on previous workers because the average catch declines. Construct the column for the marginal external cost of fishers at the lake, MEC_L.

(d) How high should the tax (in terms of fish) be to induce an efficient number of lake fishers, N_L?

9. Use Table 8.9 to determine the trading ratios of Nitrogen Oxides (NO_x) between each pair of cities in the table below.

City	Percentile	Trading Ratio with Klamath
Klamath, OR	1st	1 to 1
Tyler, TX	25th	
Bradley, AR	50th	
Florence, TX	75th	
Cuyahoga, OH	99th	
Hudson, NY	99.9^{th}	

Self-test

1. In each of the following situations determine whether the market exchange is generating a positive externality, a negative externality, or neither.

(a) An individual builds a $1M house in a neighborhood of $100,000 houses.

(b) A woman loves her man-friend so much that she decides to marry him despite the fact that he smokes. Twenty years later, she dies from the effects of second-hand smoke.

(c) A farmer applies nitrogen fertilizer to his field. Heavy rains cause nitrogen runoff into a local stream. The high nitrogen levels cause an algal bloom leading to a fish kill that ruins the fishing trip of two university professors.

(d) A farmer plants wildflowers and a feed plot by the side of a county road. Drivers passing by the field enjoy seeing the flowers and wildlife.

(e) Cross-country skiers in Northern Colorado enjoy using the Colorado State Forest near Cameron Pass for their winter enjoyment.

(f) A smoker pays his office mates $25 per week (they would accept

$22) so he can smoke in the office rather than have to go outside to puff.

(g) St. Joe Lead Corp. in Herculaneum, Missouri deposits toxic substances into the air as a consequence of their smelting process.

(h) You build a fishing pond on your own property, post no-trespassing signs, and enjoy catching big bass and bluegill for years to come.

(i) You participate in the Missouri Stream Team and clean up trash in the Castor River every spring.

2. The current allocation of resources for a canoe livery on the Current River is such that the private marginal cost of providing a day trip for two people is $20 and the private marginal benefit is $25. On any given summer day there are approximately 200 people floating in 100 canoes on a given river stretch. Each additional canoe reduces the well-being of others by approximately $0.04. Are the following True or False statements?

   (a) It would improve social well-being if fewer canoes were on the river.
   (b) The current number of canoes is socially optimal.
   (c) The marginal social benefits of adding an additional canoe (and two people) are greater than the marginal social costs.
   (d) Given $MB = \$25$, no more than 125 canoes should be on the river.

3. Determine whether the following are True or False statements. As the number of third parties affected by a market transaction increases,

   (a) we can expect the transaction costs of negotiation via the Coase theorem to decrease.
   (b) it will be easier to use the market mechanism via the Coase theorem to reduce the third party effects.
   (c) the transaction costs of negotiation via the Coase theorem will increase and the third-party effects (i.e., the externality) will be less likely to be internalized through private negotiation.
   (d) calls for government intervention to correct the third-party effects will become less pronounced.

4. In each of the following cases determine whether more, less, or the same amount of the good should be produced and consumed.

   (a) Market for electricity: MB=$0.30, MC=$.025, MEC=$0.07.
   (b) Market for mass transport: MB=$0.75, MC=$0.80, MEB=$0.12.
   (c) Market for food: MB=$1, MC=$1.

(d) Market for trash removal: MB=$7.50, MC=$7, MEC=$0.50.
(e) Market for smoking: MB=$0.10, MC=$0.08, MEC=$0.01.
(f) Market for wildlife viewing: MB=$15, MC=$20, MEB=$7.

5. In Northern Colorado some people enjoy viewing elk while others enjoy hunting elk. During the fall there is a large concentration of elk near the town of Estes Park, which is on the eastern border of Rocky Mountain National Park. Some people argue that elk hunting imposes a large external cost on elk viewers, while elk hunters simultaneously argue that elk hunting extends external benefits to society.

   (a) What rationale would you expect to hear from those who argue that elk hunting imposes external costs?
   (b) What rationale would you expect to hear from those who argue that elk hunting extends external benefits?
   (c) Suppose elk hunting generates external benefits of MEB=$75 per elk and external costs of MEC=$85 per elk. The private marginal benefits from elk hunting are MB=$250 and the private marginal cost from elk hunting is MC=$230. Is it socially efficient to harvest another elk? Explain.

6. We are interested in cleaning up pollution at the lowest possible cost. Identify each statement as either True or False.

   (a) The firm with the lowest marginal cost of abatement cleans up the next unit.
   (b) Efficiency is achieved when the amount of pollution cleaned up is such that the marginal cost of abatement is equal for all firms.
   (c) If Firm A has total cost of cleaning up 5 units of pollution of $50 and Firm B has total cost of cleaning up 5 units of pollution of $100, Firm A should clean up all 5 units of pollution.
   (d) The firm that pollutes the most should clean up the most.

7. Every time a person throws a beer can or Styrofoam cup out a car window along a highway society bears $0.25 in damage due to the unaesthetic aspects of litter. There is only a 1 in 1000 chance of any individual being caught littering. How high does the fine have to be to promote an efficient quantity of littering? Does the punishment fit the crime?

9

Valuing Environmental Resources

Earth gets its price for what Earth gives us; ...

James Russell Lowell, "The Vision of Sir Launfal"

9.1 Introduction

Since prices for environmental amenities and disamenities are seldom directly observed, policymakers must rely on various indirect methods to infer the benefits from environmental amenities and the costs of disamenities. While voting can be used to make social decisions, economic efficiency requires that we compare aggregate willingness to pay for public goods with the marginal costs of production. The assumptions required for willingness to pay to be revealed by the median voter tend to be restrictive. Alternatively, if production or consumption generates externalities policymakers need to know the marginal external costs or marginal external benefits.

Several methods have been devised to reveal individual and aggregate willingness to pay for changes in environmental amenities. One method requires observing the travel patterns of individuals as they enjoy recreational activities at lakes, forests, or wilderness areas. As distance to the recreational area and therefore the travel cost changes, how do the number of visits change? How do people change their planned number of visits to a recreational area as its quality changes? Answers to these questions allow an indirect measure of willingness to pay to be revealed. A second method constructs what are called hedonic prices for nonmarket characteristics. The hedonic price method has commonly examined the effects of various environmental amenities on housing prices. Houses in neighborhoods with cleaner air, scenic views, and access to national forests should sell at a premium relative to similar houses in neighborhoods without those amenities. Examining the effects of environmental variables on workers' wages—the hedonic wage method—is done similarly. Economic theory predicts that workers require a wage premium when they are exposed to greater environmental risks. Holding other determinants of wages constant, the wage premium provides a measure of the individual's assessment of the cost of exposure to the environmental hazard. With respect to environmental amenities there is strong rational reason why Bill Gates located

Microsoft in Redmond, Washington rather than Grand Forks, North Dakota: the wages required by software engineers living in Redmond are significantly less given the surrounding recreational opportunities offered by Puget Sound, the Cascades, and Coastal mountain ranges.

A third way of determining willingness to pay is the contingent valuation method (CVM). The CVM directly asks respondents what they are willing to pay or how they would vote on a referendum for a change in environmental quality. By varying the amount that different groups of respondents are asked to pay for a change in environmental quality the demand function for environmental quality can be recovered. A fourth method of eliciting willingness to pay is the weak complementarity approach. This method measures the change in consumer's surplus for a private good due to a change in environmental quality (the public good). Finally, shadow-pricing methods attempt to infer values for environmental amenities/disamenities by examining the structure of production and recovering marginal costs or the marginal rate of transformation, which in equilibrium should equal the marginal value accruing to consumers.

9.2 Travel Cost Approach

One of the first preference revelation schemes to be developed was the travel cost approach. The travel cost approach imputes the price-quantity reactions of consumers by observing their actual behavior with respect to travel costs to a particular site, the number and duration of trips taken to a site, and the quality of the public good at the particular site.

The following example will help clarify. Suppose we make observations on the number of trips individuals make to two popular recreational lakes which have different water quality. We will assume that the individuals come from three different cities, each with the same population, although the assumption of constant population can be easily relaxed. The cost of travel to the lake depends upon the distance the city is from the lake and the opportunity cost of travel. For instance, we might estimate the opportunity cost of travel as the average hourly wage earned in the city multiplied by the number of hours of travel to the recreational lake and add the costs of fuel and vehicle use. Consider two lakes with varying water quality. The hypothetical costs and number of visits for individuals residing in the three cities are given in Table 9.1.

The plot of the travel cost per visit (C) against the # of visits (V) is shown in Figure 9.1 for those going to the lake with water quality of $WQ = 1$ and for those going to the lake with $WQ = 2$. Connecting the points one can obtain the linear willingness to pay function for those going to the lake with $WQ = 1$ as $C = 310 - V$, denoted by demand curve $D(WQ = 1)$. A similar

Table 9.1: Travel Costs from Three Cities

City	Travel Cost	Water Quality	# of Visits
A	\$10	1	300
B	\$60	1	250
C	\$100	1	210
A	\$10	2	310
B	\$60	2	260
A	\$100	2	220

linear willingness to pay function for those going to the lake with $WQ = 2$ is $C = 320 - V$, denoted by demand curve $D(WQ = 2)$. To determine an

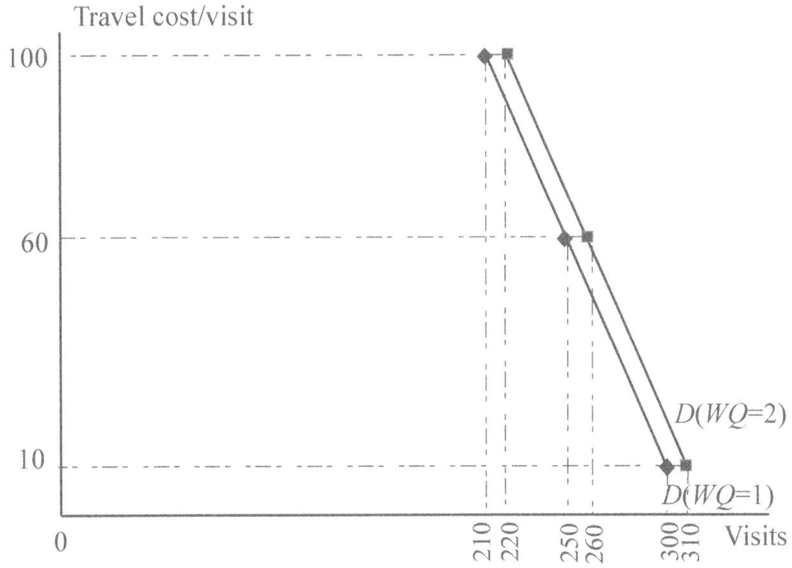

Figure 9.1: Illustration of the Travel Cost Model.

individual's willingness to pay for an improvement in water quality we need to calculate the change in consumer's surplus due to the change in water quality. Then, summing over all individuals we can derive the aggregate willingness to pay for a change in water quality. First consider the individuals in city A, who incur travel costs of \$10 to either lake. Consumers' surplus for each of those individuals when water quality is $WQ = 1$ is $CS = 0.5 \times (310 - 10) \times 300 = \$45,000$. When $WQ = 2$, consumers' surplus for those same individuals is $CS = 0.5 \times (320 - 10) \times 310 = \$48,050$. We can think of the change in consumers' surplus as a money measure of willingness to pay for the water quality improvement. For the people of city A, the change in consumers' surplus due to the one unit change in water quality is $\triangle CS = \$3,050$.

For the individuals who live in city B, travel costs are \$60 to either lake. The change in consumers' surplus due to the change in water quality is

$$\triangle CS = \underbrace{0.5 \times (320 - 60) \times 260}_{\text{New CS}} - \underbrace{0.5 \times (310 - 60) \times 250}_{\text{Old CS}} = \$2,550.$$

For the individuals who live in city C, travel costs are \$100 to either lake. The change in consumers' surplus due to the change in water quality is

$$\triangle CS = \underbrace{0.5 \times (320 - 100) \times 220}_{\text{New CS}} - \underbrace{0.5 \times (310 - 1000) \times 210}_{\text{Old CS}} = \$2,150.$$

Summing over all individuals yields the aggregate willingness to pay for the improvement in water quality: Aggregate $WTP = \$3,050 + \$2,550 + \$2,150 = \$7,750$.

9.3 Hedonic Price Approach

The hedonic price approach is similar to the travel cost approach. As one application of the hedonic pricing researchers examine the relation between housing prices and any associated environmental amenities. For example, suppose we observe two houses that are identical in all respects except that house A is located next to a garbage dump and house B is not. House A sells for \$20,000 less than house B. If the houses are identical in all respects then the difference in the two house prices should reflect the costs of the dump. One can think of other examples. How much higher are house prices next to popular recreation areas or in areas with clean air and clean water supplies compared to identical houses that do not have ready access to recreation areas or do not have access to clean air and clean water supplies? What is the premium attached to a house price in a location with a Mediterranean climate compared to an identical house with climate extremes in January and July? What is the premium for houses with scenic views or ocean access?

The hedonic price approach tries to determine the housing price premium (discount) due to an environmental amenity (disamenity). A key to accurately inferring the premium or discount is the maintenance of the *ceteris paribus* assumption. Again, suppose that there are two identical houses (same square footage, same style, etc.) but house A is located next to a garbage dump and house B is not. House A sells for a price which is \$20,000 lower than house B, but in addition to house A's proximity to the garbage dump it is also in a lower quality school district than the one where house B is located. If we ignore the differences in quality of the two school districts we would incorrectly infer that the entire \$20,000 difference in the two housing prices is due to the dump. Instead, part of the premium that must be paid for house B is due to the lack of the dump and part of the premium is due to the superior school district. Therefore, to infer willingness to pay for a particular environmental amenity, we have to control for the physical characteristics of the houses and all other environmental and neighborhood amenities like climate, school quality, crime rates, and fire protection to name a few. Also, we note that it is usually the land component of the housing price that captures the amenity value.

As a simple illustration of the hedonic housing approach let us suppose that a researcher has accurately determined that housing prices (H) depend on the square footage of the house ($SQFT$), whether the house is brick or frame ($BRICK$), the number of dollars the local school district spends per pupil ($EDUC$), and the number of days in the year that air quality in the neighborhood is above a given threshold (AIR) level. If the house is brick then the indicator variable $BRICK$ equals one; otherwise $BRICK$ equals zero. Higher values for AIR indicate better air quality. The estimated linear housing price function takes the form:

$$H = 10,000 + 100 \times SQFT + 2 \times EDUC + 15000 \times BRICK + 75 \times AIR \quad (9.1)$$

The change in the house price due to a one unit change in air quality equals the slope coefficient for the variable AIR: $\frac{\triangle H}{\triangle AIR} = 75$. Therefore, the willingness to pay for an improvement in air quality is revealed by the hedonic house price equation. In this case, homeowners value a unit increase in air quality at \$75.

The linear housing price function is just one of many functional forms. One problem with the linear form is that individual preferences tend to be subject to diminishing marginal rates of substitution. However, the change in housing prices revealed by the linear hedonic price equation assigns a value of \$75 to increases in air quality, no matter if air quality is low or high. If air quality is subject to diminishing marginal rates of substitution, then the value of a one unit increase in air quality should be greater when air quality is low than when air quality is high.

To account for the possibility of diminishing marginal rates of substitution, hedonic price functions which are nonlinear must be examined. If the housing price function is nonlinear then to infer the marginal willingness to pay for a unit increase in air quality we need to evaluate the derivative of H with respect to AIR ($dH/dAIR$) for given values of $SQFT$, $EDUC$, $BRICK$, and AIR. We leave this problem as an end of chapter exercise.

9.4 Compensating Wage Differentials and Hedonic Wages

The hedonic price approach can also be used to measure willingness to pay by examining how wages respond to changes in environmental amenities. In this instance, the researcher would estimate a wage function with the environmental amenity as an argument, controlling for other wage determinants such as an individual's education, experience, and the risk associated with the job. Earlier we suggested that housing prices will rise due to an increase in environmental amenities. With respect to wages, economic theory predicts that individuals would be willing to accept lower wages for greater environmental amenities and would require higher wages to offset exposure to environmental disamenities [204]. Therefore, the hedonic wage equation measures the wage premium (discount) due to environmental disamenities (amenities).

The wage function can also be used to value the cost of exposure to harmful chemicals when the underlying probability distribution of acquiring some disease associated with the chemical is known. By using expected utility theory and analyzing the underlying lottery one can determine an individual's assessment of the cost of exposure. Think of a simple lottery where one draws a card from a deck of fifty-two. If the card is red the individual wins \$2 and if the card is black the individual loses \$1. Since there is an equal (50-50) chance of drawing a red or a black card the expected value of the lottery is $EV = 0.5 \times \$2 + 0.5 \times -\$1 = \$0.50$.

An individual who is risk neutral would willingly pay \$0.50 for this lottery. On the other hand, a risk-taker would pay more than \$0.50 for the lottery. Finally, a risk-averse individual would only pay some amount less than \$0.50 to participate in the lottery. Individuals exhibit risk aversion when they prefer a certain payoff over an uncertain payoff with the same expected value. The expected value of a lottery with probability *pr* of winning \$X and probability $(1 - pr)$ of losing \$Y is $EV = pr \times \$X + (1 - pr) \times \$Y$.

The relation between willingness to pay (WTP) for a lottery and the expected value of the lottery is such that a person is:

$$\text{risk-averse if } WTP < EV$$
$$\text{a risk-taker if } WTP > EV$$
$$\text{risk neutral if } WTP = EV.$$

The results of laboratory experiments suggest that most individual behavior is characterized by risk-aversion or risk-neutrality, at least for lotteries in some relevant range. However, we note that recent research in behavioral economics has found evidence that individuals sometimes exhibit "loss aversion." Loss aversion occurs when individuals exhibit risk-aversion when given the opportunity to achieve a gain in utility, but exhibit risk-taking behavior when faced with a loss. Such behavior arises when there is an endowment effect as individuals face inertia from leaving the status quo. For instance, people might be willing to pay \$25 to reduce some mortality risk by 1 in 100,000, but require payment of \$40 if the mortality risk were to increase by 1 in 100,000.

To take an example of how workers reveal their preferences for risk and the value they place on a risk reduction suppose that workers in a particular industry are exposed to the chemical toluene. We assume workers are risk averse or risk neutral and that long-term exposure to this chemical results in a 1 in 1000 chance of the worker acquiring cancer and dying. We observe that wages in this particular industry are \$2.50 per hour higher than in another occupation which requires the same human capital, experience, etc. but where workers experience no exposure to the chemical. Assuming that the individual works 40 hours per week for 50 weeks in the year, what value does the individual place on their life to make them indifferent to the lottery where they receive more annually versus dying which has a negative payoff of $\$Y$? The relevant lottery is:

$$EV = \frac{999}{1000}(\$2.50/\text{hr} \times 40\text{hrs/week} \times 50 \text{ weeks}) + \frac{1}{1000} \times -\$Y. \tag{9.2}$$

If we assume risk neutrality and set $EV = 0$ we can solve (9.2) for $\$Y$ which yields a hypothetical economic value of life equal to $\$Y = \$4,995,000$ or approximately \$5 million. If people are risk-averse instead of risk-neutral, the \$5 million is a lower bound on the revealed economic value of life.

With the lottery approach the environmental economist would work with industrial engineers and medical professionals to determine chemical exposures, the underlying risk distribution (i.e., the probabilities associated with each outcome), and the wage premium in order to impute the economic cost of the chemical exposure. Once the economic costs of exposure have been determined the regulations can be assessed to see if they pass a cost-benefit test. For instance, suppose a new regulation which reduces chemical exposure in the work place costs \$15 million to implement but is expected to save four lives. The benefits associated with the four lives saved are $4 \times \$5$ million $= \$20$ million, which more than offsets the costs of the regulation. However, if the regulation were expected to save less than three lives it would not be cost effective.

There are numerous problems in determining the value of a statistical life (VSL). First, should a risk-reduction that saves the life of a three-year-old child be treated the same as an equivalent risk-reduction that saves the life of a 70-year-old adult? Many people might argue that saving the life of a three-year-old is more valuable than the life of a 70-year-old, but three-year-old children don't vote. Second, higher income people generally have a higher willingness to pay for most goods and services. Thus, two identical individuals except for income will have different willingness to pay for a reduction in risk leading to different estimates of the value of a statistical life. Differences in willingness to pay for a risk-reduction based on income seems objectionable to some people. However, low income people might be subject to more and different kinds of risks than high income people and forcing them to pay for a risk-reduction that costs more than they value the risk-reduction prevents them from paying for a risk reduction that they value more. For instance, certain kinds of luxury cars, say a Mercedes, have various safety features that make them less risky than other kinds of cars. Forcing low income people to buy a Mercedes reduces their ability to pay for other kinds of safety reductions and makes them worse off [230].

Third, regulatory policies that reduce the risk of some uncertain event, like cancer, tend to cost money today but do not save lives until some time in the future. Should saving a life twenty or thirty years from today be treated the same as saving a life today? Fourth, in some cases regulatory policies redistribute income from those who pay more in costs than they receive in benefits to those who receive more in benefits from the regulation than they pay in costs. For instance, suppose that poor workers receive positive net benefits from some regulation that increases workplace safety at the expense of rich consumers. Even if the regulation has higher costs than benefits, some people might argue that it is justified based on equity considerations. However, regulatory policy is usually a blunt and ineffective instrument for redistributing income [230].

Fifth, people must be well informed about various risks for methods that value a statistical life to be meaningful. For instance, which has more risk? Smoking 1.4 cigarettes or living for 150 years within 30 kilometers of a nuclear power plant? In fact, both risks increase the chance of death by one in one million per year [255]. Which kills more children? Swimming pools or guns? Although the media focuses on gun deaths, swimming pools result in more deaths to children each year. Which of the following results in a greater loss of life expectancy? Lacking good social connections via friends and family or industrial air pollution? The lack of good social connections for a single person at age 55 results in a loss of life expectancy of 3.2 years for males and 1.9 years for females whereas industrial air pollution results in a loss of life expectancy of only 4 days [46]. To the extent that individuals cannot accurately assign risks, methods to recover the value of a statistical life become suspect.

In evaluating environmental and other risks it is important to recognize that reducing one type of risk might cause another type of risk to increase. For instance, reducing energy consumption might reduce the risk of global warming but increase the risk that much of the world's population will live in poverty resulting in a loss of life expectancy. Even in wealthy countries energy conservation that reduces available lighting will tend to increase automobile accidents, murders, and accidents by falls [46]. This is not to say that energy conservation is inefficient, just that all costs should be accounted for in assessing the policy.

Various government agencies use different estimates of the value of a statistical life in conducting cost-benefit analysis of regulations. In 2006, the EPA used a \$7.4 million dollar VSL. At the FDA, the VSL ranged from \$6.35 million to \$8.25 million in 2013. The Department of Transportation used a \$9.1

million VSL. In addition, the DOT assumes an income elasticity of 1.0 so that every 10% increase in income increases the VSL by 10%. A recent econometric study examining the wages and annual fatality rates for male head of household reports estimates of a VSL between $4 million and $10 million [129]. Given the range of estimates for the VSL, cost-benefit analysis should test to see whether projects that pass or fail a cost-benefit test are sensitive to the choice of the VSL.

9.5 Weak Complementarity Approach

Rather than examine the effects of environmental amenities on housing prices or wages, the weak complementarity approach measures the change in consumer's surplus for a private good due to a change in the quantity of a public good, like an environmental amenity [27, 157]. A key assumption underlying weak complementarity is that if the individual consumes none of the private good, they have no willingness to pay for the public good. As an example, consider hiking boots as a private good and hiking trails as the public good. One would expect the demand for hiking boots to be greater in areas where there are numerous hiking trails, *ceteris paribus.* If hiking boots and hiking trails are weak complements, then one can infer the willingness to pay for a new hiking trail by estimating how consumer's surplus for hiking boots changes as the number of hiking trails changes.

Figure 9.2 illustrates the weak complementarity method for the private good of hiking boots (x) and hiking trails (y). The compensated demand curve for hiking boots when the quantity of hiking trails is y_1 is represented by $D(y_1)$. Given a price of hiking boots equal to P the individual's quantity demanded of hiking boots is x_1. If miles of hiking trails increase to y_2 the individual's compensated demand curve shifts to $D(y_2)$. How much is the consumer willing to pay for the change in the quantity of the public good? At price P and public good quantity of y_1 the individual receives a consumer's surplus equal to the area PCB. A hypothetical price change of hiking boots to P' makes the consumer worse off by the amount PCB so that amount must be paid to the consumer to restore the initial level of utility. At P' the individual consumes $x = 0$ hiking boots. Thus, a change in the quantity of the public good from y_1 to y_2 does not change utility because of the assumption of weak complementarity. Now, reduce price from P' to P and the new quantity demanded of hiking boots is x_2. Since the individual is made better off due to the decrease in price they would be willing to pay up to PFP' holding utility constant. Thus, the consumer's willingness to pay for the change in the quantity of the public good is $BP'FC = PFP' - PCB$.

With respect to recreation, if the assumption of weak complementarity holds, we can determine an individual's willingness to pay for fishing lakes,

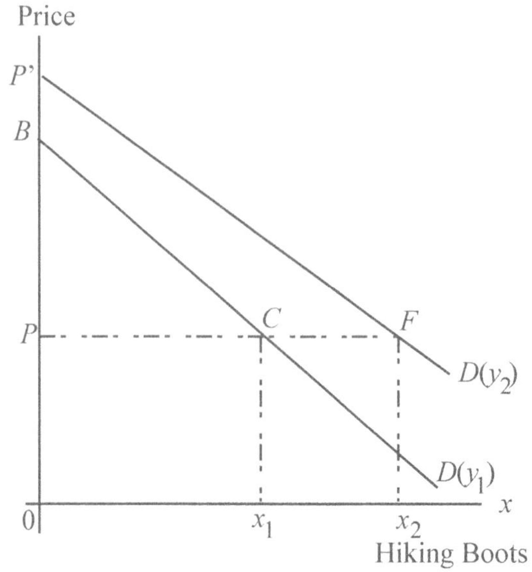

Figure 9.2: Weak Complementarity between Hiking Boots (x) and Hiking Trails (y).

hunting preserves, or ski slopes from their demands for fishing rods, guns, or ski equipment. However, we note that we used the compensated private demand curves to recover willingness to pay for the public good. In general, we only observe the uncompensated demand curve and extending the weak complementarity approach to the uncompensated private demand curve introduces bias in our estimates of willingness to pay for the public good.

9.6 Contingent Valuation Method

A recent approach to measuring willingness to pay for environmental amenities is the contingent valuation method (CVM). The CVM is a survey approach to measuring willingness to pay. People are asked whether they would vote in favor or against a policy that expands some environmental amenity if they had to pay some price. For instance, "Are you willing to pay $10 for a new public park?" To be accurate, the *ceteris paribus* assumption must be invoked so that the survey must also determine the respondent's income, tastes, age, gender, years of schooling, etc. The CVM approach is particularly useful for measuring willingness to pay for environmental amenities that people do not currently have access to. If people do not have access to a particular environmental amenity then it is impossible to use the travel cost, hedonic price approach, or weak complementarity methods to infer willingness to pay.

To take an example, suppose that the Missouri Department of Conservation would like to learn what residents of Missouri are willing to pay to establish a new state park along the National Wild and Scenic Current River. The researcher would first provide the citizen with some background information about the park. For instance, "A new state park in Texas County along the Current River has been proposed. This state park will comprise 1500 acres of previously private land that borders the river for 1 mile along its eastern bank. Camping facilities, playgrounds, showers, and laundry facilities will be available. Approximately 7 miles of hiking trails will be available and one of the trails will pass by a scenic overlook of the Current River valley."

Next, the researcher will attempt to elicit what the citizen would be willing to pay for the proposed park. Several methods are available. The **open-ended** approach would ask the respondent "What is the maximum amount you would be willing to pay in taxes to help pay for the proposed park?" Under a **bidding** approach the researcher would ask "Would you be willing to pay \$5 for the park?" If the respondent answers yes, then the researcher would increase the amount and ask the question again: "Would you be willing to pay \$10 for the park?" The last amount that the respondent accepts would be the person's willingness to pay. Under a **check off** approach, the researcher would show the respondent a series of ranges and ask the respondent to check which range their willingness to pay falls in. For example, the ranges might be: \$0-\$5; \$5-\$10; \$10-\$15; \$15-\$25; etc.

Finally, under the **referendum approach**,an approach that has been endorsed for use in legal cases, the researcher would present the respondent with some dollar amount, say, X, and ask the respondent whether they would support allocating resources for the park if they were required to pay X. For instance, suppose that at $X = \$25$, 70% of the respondents said yes in support of the new park and 30% of the respondents said no. Given these responses, a lower bound on willingness to pay is $0.70 \times \$25 + 0.30 \times \$0 = \$17.50$. The amount is a lower bound because the people who responded no to the amount $X = \$25$, might have said yes to \$24, and the people who said yes to $X = \$25$, might have been willing to pay \$30, or \$50, or some larger amount. If we somehow know or assume that nobody would be willing to pay more than \$100 for the park we could also derive an upper bound on willingness to pay as $0.70 \times \$100 + 0.30 \times \$24 = \$77.20$. Thus, our range is $\$17.50 \leq WTP \leq \$77.20$.

When using the referendum approach and varying X to each respondent, it is possible to narrow the range of willingness to pay. Suppose that four groups of people are offered an up or down vote with $X = \$75$, $X = \$50$, $X = \$25$, and $X = \$10$. For the group of people offered $X = \$75$, 25% of them vote yes and 75% vote no. For the group offered $X = \$50$, 40% vote yes. For the group offered \$25, 70% vote yes, and for the group offered $X = \$10$, 90% vote yes. Table 9.2 shows the proportion of citizens who vote yes for each X value and the marginal proportions who vote yes as the price (X) is lowered.

Table 9.2: Referendum Contingent Valuation Method

Offer Amount=X	Proportion Voting Yes	Marginal Proportion
\$75	0.25	0.25
\$50	0.40	0.15
\$25	0.70	0.30
\$10	0.90	0.20
\$0	1.0	0.10

To obtain the lower bound estimate we assume that the maximum willingness to pay is given by X for the marginal proportion who vote in favor of the park. Therefore, the lower bound is $WTP = 0.25 \times \$75 + 0.15 \times \$50 + 0.30 \times \$25 + 0.20 \times \$10 + 0.10 \times \$0 = \$35.75$. The upper bound is found by assuming that the maximum WTP for the population is some fixed amount, say \$100. In addition, it is assumed that the marginal proportion of the population who voted no at some X value would have voted yes if the offered amount X had been been \$1 less. For instance, the 25% of the people who voted yes for \$75 would have also voted yes at $X = \$100$. The marginal 15% of people who voted yes for $X = \$50$ would have been willing to pay up to \$74, and so forth. In this case the upper bound estimate is $WTP = 0.25 \times \$100 + 0.15 \times \$74 + 0.30 \times \$49 + 0.20 \times \$24 + 0.1 \times \$9 = \$56.5$. Therefore, the range is $\$35.75 \leq WTP \leq \$56.5$. As more information is elicited by changing the X values offered to various groups of respondents a narrower range of WTP is revealed by the referendum CVM approach.

A major criticism of the CVM is that what people say they will do and what they actually end up doing are often not the same. People might lie, they might try to give an answer to try and please the researcher, or they might just guess. Some studies report evidence that willingness to pay revealed from the CVM is twice as high as actual willingness to pay. A second criticism of the CVM is that the surveys are expensive to implement. Telephone and face-to-face interviews with follow-ups are generally employed with costs that often reach \$75-\$100 per interview, a figure which can be prohibitively costly.

9.7 Shadow Pricing

9.7.1 Valuing Nonmarket Outputs

As shown in Chapter 2 when individuals maximize utility subject to a budget constraint the equilibrium conditions imply that the price ratio equals the consumer's marginal rate of substitution between the two goods. Alternatively, marginal rate of substitution can be thought of as the ratio of marginal utilities of the two goods. This condition can be represented as

$$\frac{p_1}{p_2} = \frac{\triangle U/\triangle y_1}{\triangle U/\triangle y_2} = \frac{\triangle y_2}{\triangle y_1}. \tag{9.3}$$

The term $\triangle y_2/\triangle y_1$ is the consumer's marginal rate of substitution and equals the amount of good 2 they are willing to forego for an additional unit of good 1, holding utility constant. In general, we do not observe the consumer's marginal rate of substitution directly; instead, it is indirectly revealed by a person's actions. For instance, if traveling to Alaska for a trip to Denali National Park costs $2000 and traveling to Yellowstone National Park costs $1000, then people who choose Denali NP are revealing that they value that trip at least twice as much as a trip to Yellowstone NP. Alternatively, we can say the individual is willing to give up at least two trips to Yellowstone to make one trip to Denali.

Now, let's turn our attention to the producer's objective. Given inputs, producers try to maximize revenues subject to their output choices being part of the production possibility set. The marginal cost of producing one more unit of the ith good is the extra cost of producing that unit of output, $MC_i = \triangle TC/\triangle y_i$, where TC is the total cost of production. Resources are efficiently allocated when price equals the marginal cost of production and this condition must hold for all goods and services. Therefore, in equilibrium allocative efficiency requires that the price ratio equal the ratio of marginal costs: $\frac{p_1}{p_2} = \frac{MC_1}{MC_2}$. The ratio of marginal costs between two goods is termed the marginal rate of transformation (MRT) and equals the amount of good 2 that must be foregone if one more unit of good 1 is to be produced. Substituting for marginal cost yields

$$\frac{p_1}{p_2} = \frac{MC_1}{MC_2} = \frac{\triangle TC/\triangle y_1}{\triangle TC/\triangle y_2} = \frac{\triangle y_2}{\triangle y_1}. \tag{9.4}$$

Therefore, (9.3) and (9.4) imply that a Pareto efficient allocation occurs when the consumer's marginal rate of substitution between two goods equals the marginal rate of transformation in production. With a desirable output that has price p and a jointly produced undesirable output with price q the equilibrium condition is $\frac{q}{p} = \frac{\triangle y}{\triangle b}$.

Graphically, the MRT equals the slope of the production possibility frontier. Also, recall that when two desirable outputs are produced the slope of the production possibility frontier is negative; if more of one good is produced, less of the other good must be produced because of scarcity. However, when desirable goods and undesirable by-products are jointly produced, the slope of the production possibility frontier tends to be positive. In this case, if more of the desirable output is produced we must also accept more of the undesirable output. Alternatively, if we want less of the undesirable output we must forego some of the desirable output.

If the economy is in equilibrium, then the consumer's marginal rate of substitution equals the slope of the production possibility frontier. Figure 9.3 depicts this condition when two desirable goods are produced and when a desirable good and an undesirable good are produced. With public goods allocative efficiency requires the sum of the marginal rates of substitution to equal the marginal rate of transformation. If researchers and policy makers do not have access to consumer demand functions, housing prices, wages, etc., it might be possible to assign a value to the environmental amenity or disamenity by determining the marginal rate of transformation. Green accounting systems might rely on shadow pricing to impute values for environmental amenities or pollution rather than valuing those services based on the cost of the inputs used to produce them.

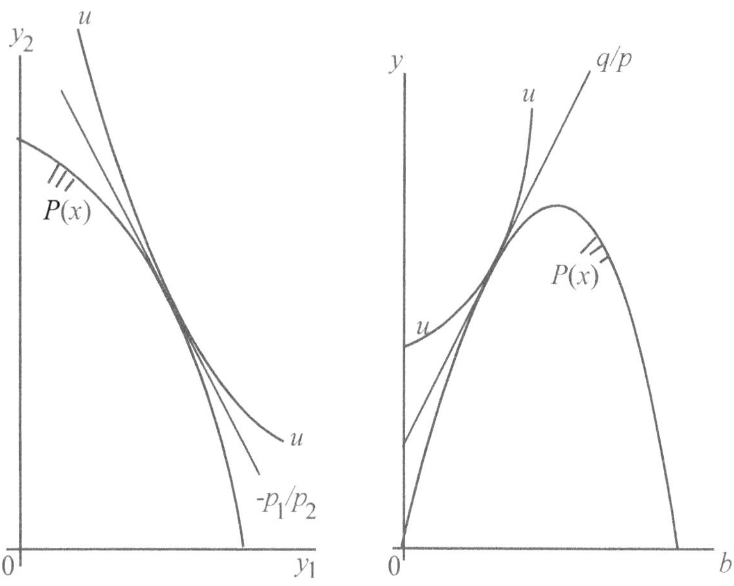

Figure 9.3: Shadow Pricing Using the Production Possibilities Frontier.

In the system of national income and product accounts there are two ways to determine national income. The **income approach** sums up all the sources of income including wages, rent, interest, and profits. On the other hand, the **expenditure approach** sums up all the spending by households (consumption spending), business firms (investment spending), government (government spending), and the foreign sector (net exports). Under a shadow pricing system "green" accounts can be derived by valuing the environmental amenities at their respective shadow prices. Using a similar method, we can also derive "brown" accounts for undesirable outputs such as pollution.

As an example of shadow pricing, let good 1 be the services produced (and received) from conservation land and let good 2 be money income and assume that the price of one dollar of money income is $p_2 = \$1$. The equilibrium condition for allocative efficiency requires that $\frac{p_1}{p_2} = MRT$. By examining the production relation and determining the MRT of private money income into public conservation land we can determine the shadow price of the conservation land as $p_1 = p_2 \times MRT$. Since y_1 represents the quantity of conservation land (in acres) then $p_1 \times y_1$ equals the shadow value we place on the aggregate quantity of conservation land. This amount should be added into the measure of national income to obtain a more accurate measure of societal welfare. Over time we can examine changes in this "green" account to judge performance and policy.

The harmful effects of pollution and other undesirable outputs can also be valued at their shadow prices. These shadow prices represent the marginal damage caused by the undesirable output. The equilibrium condition is $\frac{q}{p} = MRT$, where q is the charge to be imposed on the undesirable output depicted in Figure 9.3 as the slope of the tangent line separating the production possibility set $P(x)$ with the consumer's indifference curve. Rearranging the equilibrium condition yields the undesirable shadow price: $q = p \times MRT$. Multiplying the shadow price and the undesirable output quantity, $q \times b$, yields the amount that should be subtracted from national income. In this manner, society can assess whether it is better off with more material goods and services (say sport utility vehicles) accounting for the costs of pollution associated with producing and consuming those goods.

9.7.2 Valuing Nonmarket Inputs

A production function gives the maximum amount of a single output (y) that can be produced from one or more inputs with the N input quantities represented by $x = (x_1, x_2, \ldots, x_N)$. The marginal product of an input is the additional output that can be produced from one more unit of input. Algebraically, the marginal product of input j is $MP_j = \frac{\triangle y}{\triangle x_j}$. For instance, if a firm uses two more units of input j and output increases by 5, then $MP_j = 5/2 = 2.5$. The law of diminishing returns indicates that after some amount of input usage, output tends to increase at a decreasing rate. That is, as more of an input is employed its marginal product of an input tends to fall, holding other inputs constant.

Figure 9.4 depicts four hypothetical production functions where a watershed, labor, various purifying chemicals, and a water processing plant (capital) are used to produce gallons of potable water. In the graph we hold labor, chemicals, and capital inputs constant and allow only the size of the watershed (in 1000s of acres) to vary. The slope of the production function is the marginal product of the watershed and diminishing returns are evident in that the slope decreases as the size of the watershed increases. We let x_i represent the amount of labor, chemicals, and capital inputs. For instance, let $x_i = 1$ represent 10 units of labor, 10 pounds of chemicals, and a processing plant capable of producing 100 thousand gallons of potable water per day. Similarly, when $x_i = 2$ the firm uses twice as much labor, chemicals, and processing capacity. The production function shifts upward as more labor, chemicals, and capital inputs are used, holding the size of the watershed constant.

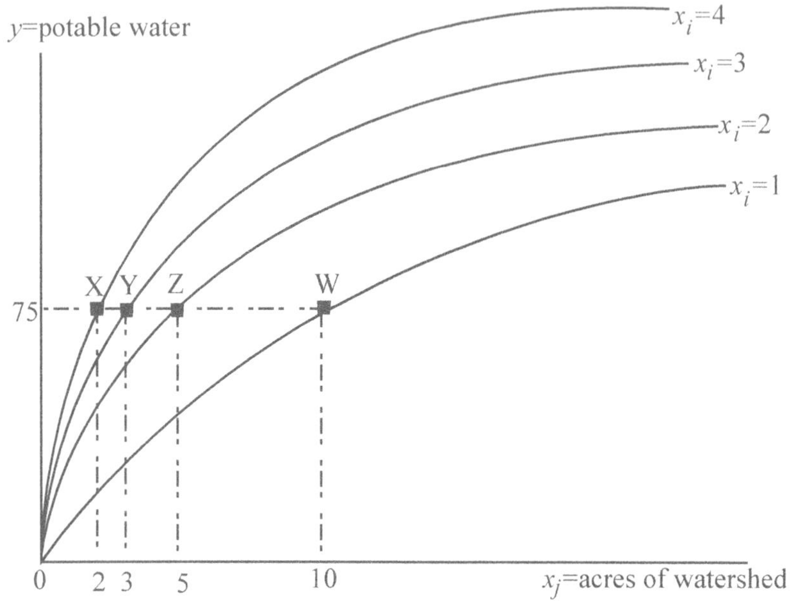

Figure 9.4: Four Production Functions.

The production technology can also be represented by the different combinations of inputs that can be used to produce a fixed level of output. In Figure 9.4 we see that four different combinations of watershed and other inputs (x_i) can be used to produce 75 thousand gallons of potable water per day. These combinations of inputs are called the input isoquant and are represented in Table 9.3.

Table 9.3: Isoquant for Producing $y = 75$ Thousand Gallons of Water

	Labor, Chemicals, and Plant $= x_i$	Watershed Acres $= x_j$	$MRTS = \left\|\frac{\Delta x_i}{\Delta x_j}\right\|$
X	4	2	x
Y	3	3	1
Z	2	5	0.5
W	1	10	0.2

As a firm uses more of one input it can reduce the use of another input. For example, suppose a water processing plant is pulling water from a nearby stream. More trees and vegetative matter in the watershed upstream from the intake pipe would allow a decrease in the amount of chemical additives and labor needed to make the water potable. Thus, the isoquant has a negative slope. Furthermore, the isoquant is convex to the origin (bowed inward) because of the law of diminishing returns. Convexity means implies that as a firm uses less of one input, say fewer acres in the watershed, it must use larger and larger amounts of another input like chemical additives to maintain the given output. The absolute value of the slope of the isoquant is called the marginal rate of technical substitution between the two inputs and equals the ratio of marginal products of the two inputs:

$$MRTS = \left|\frac{MP_j}{MP_i}\right| = \left|\frac{\Delta y/\Delta x_j}{\Delta y/\Delta x_i}\right| = \left|\frac{\Delta x_i}{\Delta x_j}\right|. \tag{9.5}$$

The marginal rate of technical substitution equals the amount that input i (chemicals, labor, and capital) can be reduced if one more unit of input j (watershed) is employed. An isoquant map gives the different combination of two inputs that can produce a given level of output, where isoquants further to the northeast yield higher levels of output. Figure 9.5 shows the isoquant map for three output levels, $y = 55$, $y = 75$, and $y = 95$. At point Z the water utility uses 2 units of x_i (labor, chemicals, capital) and 5 units of x_j (watershed) to produce $y = 75$ thousand gallons of potable water. Furthermore, as seen in Table 9.3 the marginal rate of technical substitution between input combinations Y and Z is $MRTS = 0.50$ since the firm was able to reduce x_i by 1 by increasing x_j by 2.

Now, consider the problem a producer faces in trying to minimize the cost of producing a given level of output, or, alternatively, in trying to maximize output subject to a given budget with which to hire inputs. Let the input prices be represented by w_i and w_j. Suppose the producer gets to choose the level of output. To minimize costs the producer should choose inputs that have the highest marginal product per dollar. If $\frac{MP_j}{w_j} > \frac{MP_i}{w_i}$ the firm receives more output per dollar from the next unit of the jth input than it does from the next unit of the ith input and should choose more of input j and less of input i to reduce costs. However, as more of input j is employed, its marginal product declines because of diminishing returns. Likewise, as less of input i is employed, its marginal product increases. The equilibrium occurs when the marginal product per dollar of each input is the same: $\frac{MP_j}{w_j} = \frac{MP_i}{w_i}$. Rearranging the equilibrium condition yields: $\frac{w_j}{w_i} = \frac{MP_j}{MP_i}$. The input price ratio $\frac{w_j}{w_i}$ gives the amount of input i that must be given up if another unit of input j is employed in order to hold the costs of production constant and the ratio of marginal products gives the $MRTS$.

Alternatively, suppose the firm wants to minimize the costs of producing a given level of output. In this case, it compares the $MRTS$ with the price ratio of the two inputs, $\frac{w_j}{w_i}$. If $MRTS > \frac{w_j}{w_i}$ the firm can lower costs by using more of input j and less of input i. However, Table 9.3 shows that as the firm

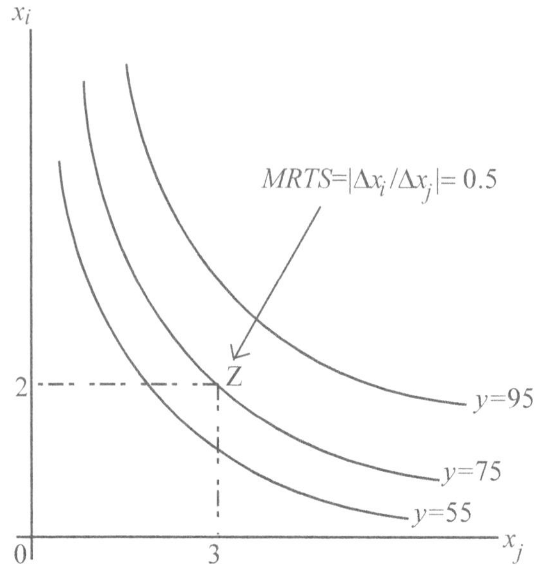

Figure 9.5: Isoquant Map.

uses more of input j the $MRTS$ falls and will eventually equal the price ratio: $MRTS = \frac{w_j}{w_i}$.

To use the equilibrium condition for inputs to determine shadow prices, imagine that the jth input (the watershed) is an un-priced input. The watershed allows the water utility to reduce its use of chemicals, labor, and capital, which in turn reduces production costs. The producer currently uses $x_i = 2$ and $x_j = 5$ to produce $y = 75$ thousand gallons of water corresponding to point Z in Figure 9.5. What is the value of the watershed in this use? The shadow price of the watershed can be found by rearranging the equilibrium condition: $w_j = w_i \times \frac{MP_j}{MP_i} = w_i \times MRTS$. For instance, if the price of the input i is $w_i =$ \$10 million, the shadow price of the 5000 acre watershed is \$5 million = \$10 million $\times 0.5$.

9.8 Problems

1. Use the travel cost approach to measure the willingness to pay for clean air improvements which increase the number of high visibility days ($Days$) at the Grand Canyon. Current conditions are such that $Days = 300$ per year. The inverse price demand function for a trip to the Grand Canyon is:

$P = 500 - 0.5 \times Q + 20 \times Days$, or rearranging and solving for quantity demanded we have $Q = 1000 - 2 \times P + 40 \times Days$.

The average price a household faces is: $P = 1000$ for a trip to the Grand Canyon. A proposed policy would increase the number of high visibility days to $Days = 330$. Holding price constant, calculate the willingness to pay for the increase in $Days$ from 300 to 330.

2. Housing prices in Cape Girardeau, MO (H_{Cape}) and Jackson, MO ($H_{Jackson}$) depend on the number of square feet in the house ($SQFT$), the age of the house (AGE), the number of dollars the local school spends per pupil ($EDUC$), and the number of acres of National Forest land that is within 50 radial miles ($ACRES$). The housing price equations for Cape Girardeau and Jackson take the form:

   $$H_{Cape} = 80 \times SQFT - 500 \times AGE + 0.5 \times EDUC + 0.25 \times ACRES$$

   $$H_{Jackson} = 80 \times SQFT - 400 \times AGE + 0.6 \times EDUC + 0.3 \times ACRES.$$

   The current value of $ACRES = 20,000$ in Cape Girardeau, primarily in the Shawnee National Forest in Southern Illinois and the current value of $ACRES = 15,000$ in Jackson, primarily in the Mark Twain National Forest in Missouri. The current averages in Cape Girardeau are $SQFT = 2,000$, $AGE = 10$, and $EDUC = 4,000$. The current averages in Jackson are $SQFT = 2,200$, $AGE = 12$, and $EDUC = 4,200$. A proposal in Congress would expand the Mark Twain Forest in Missouri by 25,000 acres of which 500 acres would be within 50 radial miles of Cape Girardeau and 1000 acres of which would be within 50 radial miles of Jackson. Use the hedonic price technique to calculate the willingness to pay for the 25,000 acre expansion of National Forest land for residents of Cape Girardeau and Jackson.

3. Housing prices in Cape Girardeau depend on the number of square feet ($SQFT$), the age of the house (AGE), education spending per pupil in the local school district ($EDUC$), and the number of acres of National Forest land within 50 radial miles ($ACRES$). The housing price equation takes the form:

   $$H_{Cape} = 80 \times SQFT - 500 \times AGE + 0.5 \times EDUC + 0.25 \times ACRES - 0.000001 \times ACRES^2.$$

   The current averages in Cape Girardeau are $SQFT = 2,000$, $AGE=10$, and $EDUC = 4,000$.

   (a) Given that $ACRES = 20,000$ calculate willingness to pay for a 500 increase in $ACRES$: $\triangle ACRES = 500$.

   (b) Suppose instead that $ACRES = 21,000$. Calculate willingness to pay for a 500 increase in $ACRES$: $\triangle ACRES = 500$.

(c) Compare your answers in (a) and (b). As the number of $ACRES$ increases what happens to willingness to pay for an extra 500 $ACRES$? What property of the utility function is this?

4. Wage rates in St. Louis, MO depend on the number of years of schooling a person has attained, YRS, their years of experience, EXP, and their exposure to hazardous waste, $WASTE$. The wage equation takes the form:

$$WAGE = 1000 + 1000 \times YRS + 500 \times EXP + 3 \times WASTE + 0.2 \times WASTE^2$$

In St. Louis, the average years of school attained is $YRS = 14$ and the average years of experience is $EXP = 10$.

(a) Given $WASTE = 100$ estimate the wage change for a policy that decreases waste to $WASTE = 90$: $\Delta WASTE = -10$.

(b) Given $WASTE = 50$, estimate the wage change for a policy that decreases waste to $WASTE = 40$: $\Delta WASTE = -10$.

(c) Compare your answers in (a) and (b). Each time the policy reduced waste by 10. Why was the change in wages different? What does this tell us about an individual's willingness to pay to avoid waste?

5. Canoe rentals (CR) and water quality (WQ) on the Current River are weak complements. The compensated demand function for canoe rentals (the private good) is:

$$Q_{CR} = 500 - P + 2 \times WQ, \text{ or, } P = 500 - Q_{CR} + 2 \times WQ,$$

where $P = \$25$ is the price of renting a canoe. Water quality on the Current River is $WQ = 1$, which corresponds to a high level of coliform bacteria due to antiquated septic systems in several upstream campgrounds and housing developments. An upgrading of septic systems in the Current River drainage would increase water quality to $WQ = 4$. Calculate the willingness to pay for the upgraded septic systems using the weak complementarity approach.

6. The county commission is considering the possibility of acquiring 100 acres to build a park and recreation area. They have hired you to carry out a contingent valuation survey to determine whether there is enough willingness to pay among county residents to justify the park. You decide to use the referendum approach in your survey and you offer three different groups of respondents amounts of $X = \$30$, $X = \$20$, and $X = \$10$. For the group offered $X = \$30$, 30% of the group votes in favor. For the group offered $X = \$20$, 60% of the group votes in favor. For the group offered $X = \$10$, 75% vote in favor.

(a) Assuming a maximum willingness to pay of \$60, calculate upper and lower bound estimates for willingness to pay for the average county resident.

(b) If the cost of acquiring and developing the land is \$350,000 and the county consists of 20,000 residents, what recommendation do you give the county commissioners?

7. Suppose we value a life at \$5 million. What is the minimum annual wage premium a worker must be paid to induce them to accept a job where their exposure to toxic waste is associated with a 5/10,000 probability of contracting cancer and dying? Assume risk neutrality.

8. Workers in Lander, Wyoming are willing to accept \$3/hour less than they would if they lived in Peoria, Illinois doing the same identical job. Lander is adjacent to national forest land in the Wind River Range. The town has a labor force of 3,000. The average worker works 50 weeks per year, 40 hours per week. There are no national forest lands adjacent to Peoria. What is the aggregate value the residents of Lander place on the national forest land?

9. The following two diagrams depict the production possibility frontiers for two desirable outputs (Figure A) and a desirable and an undesirable output (Figure B). The slope of a curve at a particular point on the curve equals the slope of a tangent line (the derivative) at that point, which in turn equals the marginal rate of transformation (MRT).

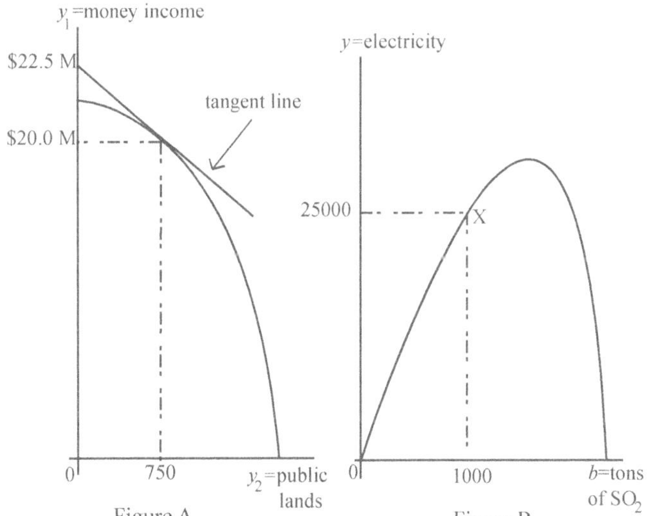

Figure A

Figure B

(a) In Figure A the current production of public conservation land in Yoknapatawpha County is 750 acres and private aggregate money income equals \$20 million. The price of money income is \$1. Calculate the marginal rate of transformation and interpret.

(b) Given your answer in (a), enter the appropriate amount in Yoknapatawpha County's "green" income account.

(c) In Figure B the current production of electricity is 25000 megawatt hours (mWh) annually. Along with the electricity, 1000 tons of SO_2 are emitted into the air. The slope at point X equals 7. The price of a mWh of electricity is \$35. Calculate the shadow price of SO_2 and interpret.

(d) Given your answer in (c), estimate the amount that should be entered into the utility's "brown" income account.

10. The table below depicts the production function for a salmon farmer who uses fish feed, x_1, and ocean pen space, x_2, to grow salmon from smolt to a marketable final output, y, which is measured as kilos of salmon.

	$x_2 = 0$		$x_2 = 1$		$x_2 = 2$		$x_2 = 3$	
x_1	y	MP_1	y	MP_1	y	MP_1	y	MP_1
0	0	x	0	x	0	x	0	x
1	0	0	20	20	30		35	
2	0	0	38		50		60	
3	0	0	54		68		81	
4	0	0	68		84		99	
5	0	0	80		98		115	
6	0	0	90		110		129	
7	0	0	98		120		140	
8	0	0	104		128		150	
9	0	0	108		134		157	

(a) Complete the columns for the marginal product of fish feed (MP_1) for the various pen sizes (x_2).

(b) In the table below record the marginal product of pen space at every level of fish feed. The marginal product of ocean pen space (MP_2) is read across the row in the table above, while the marginal product of feed (MP_1) is read down the column.

(c) Do diminishing returns hold for ocean pen space?

(d) A salmon farmer uses $x_1 = 6$ units of fish feed, $x_2 = 2$ units of ocean pen space, and the price of fish feed is $w_1 = \$10$. What is the shadow value of pen space?

(e) A salmon farmer uses $x_1 = 6$ units of fish feed, $x_2 = 3$ units of ocean pen space, and the price of fish feed is $w_1 = \$10$. What is the shadow value of pen space?

	Pen Space Increases from:		
	$x_2 = 0$ to $x_2 = 1$	$x_2 = 1$ to $x_2 = 2$	$x_2 = 2$ to $x_2 = 3$
x_1	MP_2	MP_2	MP_2
0	x	x	x
1	$20 = (20 - 0)$	$10 = (30 - 20)$	$5 = (35 - 30)$
2	38		
3			
4			
5			
6			
7			
8			
9			

(f) As more x_3 is used, what happens to its shadow price?

11. A water utility has a 1000-acre lake from which they draw water for delivery to customers. They also own land surrounding the lake on which there are trees and other vegetative matter that filter rainwater into the lake. The utility uses a charcoal/gravel bed to filter the water. The more acres of trees, the less the water needs to be filtered. The graph below depicts the isoquant between acres of woodland and pounds of chemicals for the water utility when they deliver 1 million gallons of water per day. The utility inputs consist of 250 acres of woodland surrounding the lake and 50 tons of filtering material. At that combination of inputs the marginal product of the last acre of woodland is $MP_1 = 20,000$ gallons and the marginal product of the last ton of filtering material is $MP_2 = 500$ gallons. The price per ton of filtering material is $w_2 = \$20$. Calculate the shadow price of the wooded acreage.

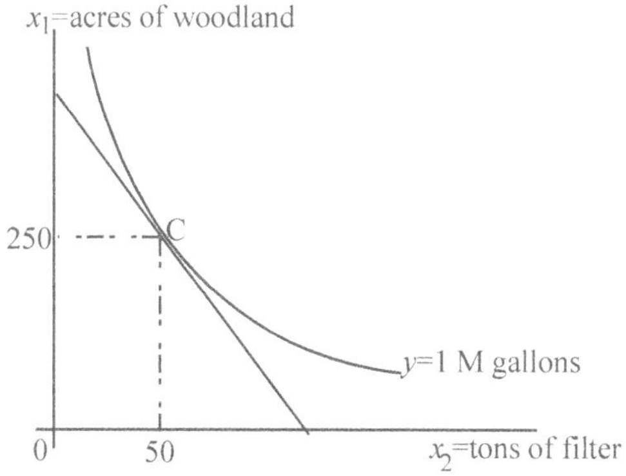

10

Endangered Species

. . . conservation will ultimately boil down to rewarding the private landowner who conserves the public interest.

Aldo Leopold, "Conservation Economics" [143, p. 202]

10.1 Introduction

Biologists have catalogued more than 1.75 million species and various forecasts suggest the number of species not yet discovered or catalogued might be as high as 12 to 13 million. Species extinctions are now thought to be occurring at a rate between one thousand and ten thousand times the normal rate [156]. Table 10.1 lists the number of identified endangered and threatened species. A species is **endangered** if it is in danger of becoming extinct throughout all or a significant part of its natural range. A species is identified as **threatened** if it is likely to become endangered in the foreseeable future.

Table 10.1: Endangered and Threatened Species

	Endangered			Threatened		
	Total	US	Foreign	Total	US	Foreign
Mammals	325	70	255	34	14	20
Birds	271	77	194	29	16	13
Reptiles	79	13	66	40	24	16
Amphibians	22	14	8	11	10	1
Fishes	83	72	11	68	67	1
Snails	26	25	1	11	11	0
Clams	67	65	2	8	8	0
Crustaceans	19	19	0	3	3	0
Insects	54	50	4	10	10	0
Arachnids	12	12	0	0	0	0
Plants	645	644	1	150	148	2

(Source: Statistical Abstract of the United States 2012, Table 387.)

In the past, meteors and climate change were the primary causes of extinction. However, recent extinctions are largely attributed to human caused habitat destruction and fragmentation, overexploitation via hunting/fishing/gathering, the introduction of alien species, pollution, and disease. These extinctions arise from a tragedy of the commons that results from a lack of well-defined property rights and from market prices that do not reflect the true marginal social costs and benefits of an action. The introduction of exotic species arises as exploration and trade occurs and specimens are introduced deliberately or accidently as stowaways or hitchhikers. For instance, bubonic plague was carried into the United States by rats on trading ships arriving from Asia. The plague became manifest in prairie dog colonies and contributed to the decline of both prairie dogs and the endangered black-footed ferret. Another example is the Burmese python which has populated the Everglades after being released into the wild by pet-owners. With no natural predator, the python population has grown rapidly and caused precipitous declines in rabbits, bobcats, and a host of birds in the Everglades.

Extinctions reduce biodiversity and cause a host of problems. Biodiversity promotes ecosystem stability and prevents fragility. For instance, although private timber companies have been diligent in replanting trees after harvest, the replanting is often of one species with all timber in a stand of one age. These monoculture forests are more susceptible to fires, pests, and disease than more diverse forests with different species and ages of trees. Biodiversity also increases the number of available food sources which provides a kind of insurance when one food source is lost due to drought or flood. Furthermore, many plants provide natural chemical compounds that are valuable in the production of marketable goods. Biodiversity also provides a library of genetic information that can be used in developing new seed varieties and drugs. Finally, biodiversity is valuable because people think it is important for ethical reasons [120].

The Endangered Species Act (ESA) was passed in 1973 to account for the social benefits of endangered species that are not valued in the market. One large problem with the ESA is that all species were to be treated as equal. That is, economic cost–benefit analysis could not be used in evaluating which species to list as endangered or threatened. Although placing prices on species might be considered crude, the budgets of the EPA and the Fish and Wildlife Service are limited and somehow choices must be made about how to spend those limited resources. To the extent that economic growth and development have threatened various species prior economic choices have influenced which species are endangered or threatened.

In 2011, state and federal agencies spent approximately $1.6 billion on endangered species, up from only $43.7 million in 1989. Table 10.2 reports spending on the top 20 species in 2011. Although there were 1196 species listed in 2011, approximately 49% of all expenditures net of land acquisition costs were spent on only 20 species. However, this percent was down from

72.7% spent on the top 20 species in 1996. In contrast, only 1% of total nonland expenditures were spent on the 565 lowest ranked species.

Table 10.2: Top 20 Listed Species in Terms of Expenditures, 2011, E=endangered, T=threatened

Rank	Species name	Status	Cumulative percent
1	Salmon, chinook—Sacramento R.	E	6.3%
2	Steelhead—middle Colombia R.	T	10.7%
3	Steelhead—Snake R. Basin	T	14.9%
4	Salmon, chinook—Snake R. spring/summer run	T	18.8%
5	Sturgeon, pallid	E	22.1%
6	Trout, bull	T	25.1%
7	Salmon, chinook—Snake R. fall run	T	27.9%
8	Salmon, chinook—Upper Columbia R. spring run	E	30.4%
9	Steelhead—Upper Columbia R. basin	T	32.8%
10	Woodpecker, red-cockaded	E	34.9%
11	Steelhead—Upper Willamette R.	T	36.7%
12	Salmon, chinook—Lower Columbia R.	T	38.4%
13	Salmon, sockeye—Snake R.	E	40.1%
14	Salmon, chinook—Upper Willamette R.	T	41.6%
15	Plover, piping—except Great Lakes watershed	T	43.0%
16	Tortoise, desert—except Sonoran Desert	T	44.4%
17	Bat, Indiana	E	45.6%
18	Salmon, chinook—Puget Sound	T	46.8%
19	Owl, northern spotted	T	48.1%
20	Salmon, coho—Southern OR-Northern CA	T	49.2%

(Source: US Fish and Wildlife Service [79].)

Table 10.3 reports federal and state expenditures for various animal classes and plant divisions in 2011. Fishes, birds, mammals, reptiles, and flowering plants comprise the top five in terms of expenditures. That fish are at the top of the list is no surprise given that fifteen out of the top twenty species listed in Table 10.2 are fish.

Table 10.3: Expenditures for Endangered Species, 2011

	Nonland Expenditures	Land Acquisition	Total Expenditures
Fishes	$785,773,850	$94,623,943	$880,397,793
Birds	$160,767,219	$126,078,104	$286,845,323
Mammals	$146,839,737	$44,224,693	$191,064,430
Reptiles	$67,709,265	$13,289,178	$80,998,443
Flowering plants	$47,477,323	$7,162,894	$54,640,217
Amphibians	$14,807,895	$4,120,579	$18,928,474
Crustaceans	$8,540,350	$9,874,668	$18,415,018
Clams	$15,362,327	$2,570,787	$17,933,114
Insects	$14,854,997	$647,825	$15,502,822
Snails	$4,799,923	$18,714	$4,818,637
Conifers and Cyads	$162,600	$3,557,000	$3,719,600
Corals	$3,308,022	$4,000	$3,312,022
Arachnids	$1,386,167	$0	$1,386,167
Ferns and Allies	$912,252	$8,400	$920,652
Lichens	$104,607	$2,067	$106,674

(Source: US Fish and Wildlife Service [79].)

10.2 Unintended Consequences of the ESA

Amendments to the ESA in 1978 allowed federal agencies to begin limited cost–benefit analysis of habitat preservation, but the favor was not extended to private landowners. This led some landowners, who might have otherwise been conservation minded, to adopt a "shoot, shovel, and shut-up" strategy to avoid the constraints of the ESA. Instead of protecting endangered species, the lack of compensation to private property owners actually led to species' decline. In contrast, compensation "would immediately make the costs of preserving species more apparent and would, therefore, spur the search for a broad range of innovative approaches[218, p. 323]."

The 1992 Rio Declaration states that "Where there are threats of serious or irreversible damage, lack of full scientific certainty should not be used as a reason for postponing cost-effective measures." Some researchers [174] argue that the ESA takes a strong narrowly defined precautionary approach and does not recognize the time dimension of risk to habitat and species. In this sense, policies that involve short-term risks to a species, but which can also lead to long-term gains are contrary to the narrowly defined precautionary principle. As an example, the Northwest Forest Plan imposed a policy of fighting all fires which led to changing forest habitat in the Pacific Northwest; habitat changed from one where there were frequent low intensity ground fires which maintained many tree stands, to an environment where high intensity crown fires have become more frequent.

Northern spotted owls are listed as endangered and are found in Pacific Northwest forests in Late Successional Reserves. These types of forests are on their way to becoming old growth forests. In developing the Northwest Forest Plan foresters argued in favor of prescribed burns to eliminate fuel. Although the prescribed burns would have increased short-term risks for owls, they had the potential to reduce the risk of high intensity crown fires and protect habitat for owls in the long-run. However, politicians and bureaucrats in the Clinton era exhibited narrow, short-term, risk averse behavior and opposed the prescribed burns. Paradoxically, the strict adherence to the precautionary principle contributed to habitat destruction and species' decline.

After passage of the ESA in 1973, private landowners were reluctant to talk with Fish and Wildlife Service employees about potential endangered species inhabiting their land, since landowners did not want to risk losing their property rights. In 1982, the ESA was modified to allow private landowners to obtain permits to engage in activities that might harm endangered species provided the landowners developed a plan to minimize and mitigate the impact of those activities. Although this modification weakened the letter of the law, it helped start a flow of information between private landowners and Fish and Wildlife Service employees about where endangered species were found and how those species might best be protected [254].

Still, other unintended consequences occurred. The red-cockaded woodpecker is found in old growth pine forests of the Sandhills and Coastal Plain regions of North Carolina. It nests in tree cavities of living pine trees that are sixty or more years of age. When this woodpecker was listed as endangered, property owners who worried about losing their property rights if the woodpecker was found on their land began cutting trees to ensure that their land would not become prime woodpecker habitat. One study found that as the distance from a known colony of woodpeckers decreased, the probability of harvesting a forest increased and the age of the trees decreased [150]. To remedy this preemptive habitat destruction, the federal government implemented the Safe Harbor program in 1995. Under the Safe Harbor program, landowners who engage in voluntary activities that benefit endangered species do not incur additional regulatory restrictions as a result of their good deeds [175].

10.3 Economics and Endangered Species

How can economics contribute to the preservation of endangered species? First, we should recognize that since resources are scarce choices have to be made regarding the use of those resources. Whether we like it or not, tradeoffs have to be evaluated regarding how much money should be spent saving species and how that money should be allocated between various species. Gardner Brown and Jason Shogren argue that "economists can frame the endangered species debate in benefit/cost terms" to see whether species' protection regardless of cost "holds a moral trump card over such policy issues as health care or the welfare of children[32, p. 15]."

Second, there will be diminishing returns to resources used to save a species. It is impossible to ensure 100% survivability of a species. Furthermore, increasing the probability of survival from 35% to 50% will, in general, be cheaper than increasing the probability of survival from 80% to 95%. Recognitition and acknowledgement of diminishing returns along with the equal marginal product per dollar principle would allow resources to be shifted toward species where a greater change in the probability of survival can be realized per dollar spent.

Third, a shift in focus toward preserving and enhancing habitat might yield greater returns than a focus on individual species. To the extent that species are interdependent, better habitat for one species, such as prairie dog colonies, can contribute to the survivability of another species like the black-footed ferret.

Fourth, economists think that producers and consumers respond to incentives and changing relative prices provide information about the costs and benefits of an action. Habitat destruction often occurs because of economic development as the various economic actors respond to changing relative prices. Reducing various price distortions, such as subsidies to agricultural interests and local developers, might indirectly enhance habitat and increase the probability of a species' survivability. More than "90% of endangered species had at least some habitat on private land and more than one third were completely on nonfederal land [32, p. 16]." Compensating landowners to reduce private takings can contribute to survivability. Since the reintroduction of wolves in the greater Yellowstone ecosystem, private ranchers have been compensated for the loss of livestock due to wolves. Similarly, subsistence farmers in Africa consider elephants pests since they trample fields. Compensating those subsistence farmers for their losses can help provide incentives to keep farmers from looking the other way when poachers take elephants.

As discussed in the previous section, unintended consequences arise in formulating policies to protect endangered and threatened species. These unintended consequences often occur when the direct benefits and costs of the policy are accounted for but the indirect/secondary benefits are ignored. Thus, a fifth way contribution that economics can make is in the construction of general equilibrium models that account for all markets simultaneously [216]. For example, the production possibilities frontier model developed in Chapter 5 is a type of general equilibrium model that recognizes the tradeoffs that must necessarily occur when more or less of one good is produced.

Some critics of the ESA argue that maintaining biodiversity comes at a cost of jobs. Debate over the northern spotted owl found in old growth forests of the Pacific Northwest focused on the loss of jobs in the timber industry that would be a consequence of maintaining habitat for owls. However, Kevin Duffy-Deno [66] examined the effects of the ESA on employment in 333 western rural counties and found no adverse employment effects of the ESA on county employment growth from 1980 to 1990.

This nonexhaustive list of ways economics can contribute to the debate regarding endangered species can be summarized into several key points. Scarcity necessitates tradeoffs and choice between alternative competing wants. Endangered species' preservation is only one item in a long and never-ending list of wants. Cost/benefit analysis is one way of prioritizing and making choices between items on the list. In making choices people respond to incentives. Prices, when they exist, provide information about relative scarcities and influence the incentives people have to produce and consume. Working to make prices reflect the marginal social benefits and costs of an action will help prevent habitat destruction and promote survivability. Furthermore, accounting for the existence of diminishing returns can help allocate resources efficiently. Finally, beware unintended consequences. These unintended consequences are not necessarily unpredictable if suitable general equilibrium models are used.

10.4 Noah's Ark Problem

If cost–benefit analysis and economics are to play a role in furthering protection of endangered species then it would be useful to have a meaningful way of prioritizing. One possible way of thinking about the problem is suggested by "The Noah's Ark Problem" [249]. In this problem Noah has to choose which animals to take with him on the ark recognizing that not all species can be saved (the ark is of limited size) and not all species are equally valuable—wild beasts were on the bottom level of the ark, birds on the second level, and humans on top. In this problem a ranking scheme is proposed that depends on four fundamental factors: distinctiveness, utility, effectiveness, and cost. We can think of the ranking scheme as determining which species get to board the ark first. The ranking formula takes the form:

$$R_i = \frac{(D_i + U_i) \times \triangle pr_i}{C_i}. \tag{10.1}$$

In Equation (10.1) R_i is the priority rank of species i, D_i is the distinctiveness of species i, U_i is the amount of utility people receive from species i, $\triangle pr_i$ measures how effective a policy can be in changing the probability that species i will survive, and C_i is the cost of the policy.

In this formulation, a species is thought of as a library. Each library is composed of a building and books in the building. The building itself yields utility—society probably values "pandas more than mosquitoes" [249,

p. 1280]—and the books in the building are a metaphor for the genetic information comprising the species. Consider two libraries that have J books in common, but library 1 holds an additional E_1 books that are not found in library 2 and library 2 holds an additional E_2 books that are not found in library 1. Library 1 has $J + E_1 = M_1$ books and library 2 has $J + E_2 = M_2$ books. If library 1 goes extinct the loss of diversity is E_1 and if library 2 goes extinct the loss of diversity is E_2. If both libraries go extinct the loss of diversity is $E_1 + E_2 + J$. Library 1 has probability pr_1 of survival and $1 - pr_1$ of extinction. Similarly, library 2 has probability pr_2 of survival and $1 - pr_2$ of extinction.

A library's distinctiveness equals its expected marginal contribution to the diversity of the total library system.[1] The expected marginal contribution of a library depends on the probability that the other library will survive and on the number of books that the library has that are different from other libraries. The distinctiveness of library 1 is $D_1 = M_1 - pr_2 J$ and the distinctiveness of library 2 is $D_2 = M_2 - pr_1 J$. For example, suppose that library 1 holds 150 books and library 2 holds 165 books. Each library holds $J = 100$ books that are also found in the other library. Library 1 holds $E_1 = 50$ additional books not found in library 2 and library 2 holds an additional $E_2 = 75$ book not found in library 1. Each library faces the risk of a fire (extinction) and has corresponding probability of survival. If $pr_1 = 0.7$ and $pr_2 = 0.4$ then the distinctiveness of library 1 is $D_1 = 150 - 0.4 \times 100 = 110$ and the distinctiveness of library 2 is $D_2 = 175 - 0.7 \times 100 = 105$. Therefore, everything else constant, library 1 would have a higher rank based on Equation (10.1) than library 2, even though it holds fewer books. In this case library 1 is more distinct because its higher probability of surviving the fire makes the books that are held in common by library 2 less valuable.

Continuing with the library metaphor suppose that a water sprinkling system can be installed at each library which reduces the risk of fire and increases the probability that the library will survive. This water sprinkling system can be thought of as a conservation program that enhances habitat. The probability of survival at library 1 increases to $pr_1' = 0.75$ and the probability of survival at library 2 increases to $pr_2' = 0.6$. Therefore, $\triangle pr_1 = pr_1' - pr_1 = 0.75 - 0.7 = 0.05$ at library 1 and $\triangle pr_2 = pr_2' - pr_2 = 0.6 - 0.4 = 0.2$ at library 2. Policies that generate larger changes in the probability of survival of a species increase its rank.

Finally, increases in the cost, C_i, of the particular conservation policy reduce the rank given to the particular species. Although it might not be easy to quantify each of the components of the ranking system given by Equation (10.1), it provides a starting point for policy discussions regarding choices between species.

There are at least two components of species' distinctiveness. First, higher

[1] The expected diversity of the two libraries is $E(Diversity) = pr_1 \times (J + E_1) + pr_2 \times (J + E_2) - pr_1 pr_2 J$.

plants produce new drugs that can then be synthetically produced. These new drugs have certain economic values as they help humans fight disease. However, such genetic information can be lost via extinction if that information cannot be substituted with other plants. Second, there are various niche species that help the functioning of the entire ecosystem. For instance, when wolves disappeared from the Greater Yellowstone ecosystem, the elk herd grew in number, contributing to overgrazing. In ocean and other fisheries, overfishing might cause a decline in adult fishes which reduces juveniles and stresses predators as well as prey.

Some researchers [59, 176] have attempted to operationalize the ranking scheme given in (10.1) by assuming that the amount of money spent by the federal government is an indicator of the species' priority ranking. Specifically, the natural logarithm of spending ($\ln Spend$) was used as a proxy for R_i. To account for the utility of a species, U_i, the researchers included a set of variables that described whether a species was a mammal, reptile, bird, amphibian, or fish. They also hypothesized that people are fond of "megafauna": i.e., large plants and animals. As proxies for distinctiveness, D_i, the researchers constructed a variable to account for whether or not the species is unique; i.e. is the only species in its genus.

To measure changes in the probability of survival, $\triangle pr_i$, the researchers used the Nature Conservancy's 1993 Global Endangerment rank to proxy the probability of improving the survivability of the species. In addition, the Fish and Wildlife Service's priority ranking for the species measures both distinctiveness and the probability of improving survivability. Finally, to proxy the cost of improving species i, C_i, the researchers examined whether or not a species is in direct conflict with development. Species in direct conflict with development have a higher opportunity cost of improving survivability. Are these variables important determinants of the amount of money allocated for saving the endangered species? If species are not valued by cost-benefit analysis then variables describing costs or utility should not be related to spending. On the other hand, if costs and benefits are playing an implicit role in the allocation of funds those costs and benefits should be correlated with spending. The research findings are left as an end of chapter problem.

10.5 Habitat Protection

One criticism of the Endangered Species Act is that it focuses on protecting species already in trouble. However, it is also important to protect species from becoming threatened or endangered. Ensuring viable habitat is one way of helping to prevent species from becoming listed. Habitat loss occurs because of habitat change, invasive species, pollution, overharvesting, and climate change [238]. Prime habitats for endangered species include old growth temperate forests, rain forests, coral reefs, and rangeland. Coral reefs are threatened by climate change and overfishing. Rain forests, temperate forests, and rangeland are threatened by expansions in agriculture and industrial and residential developments.

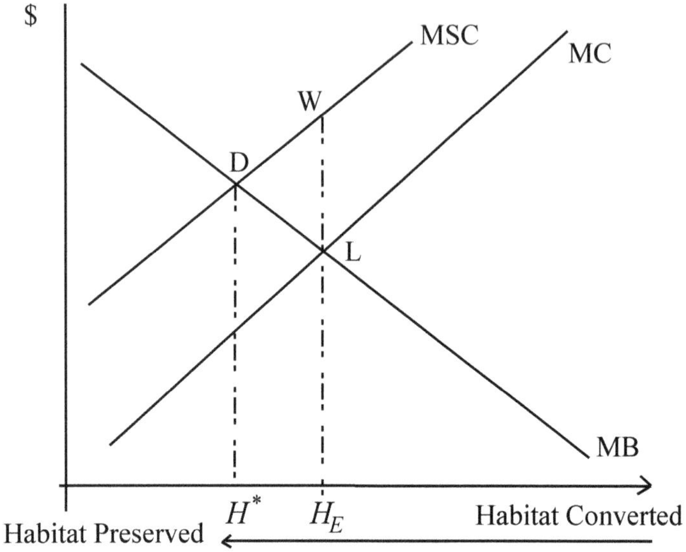

Figure 10.1: Private and Socially Optimal Habitat Preservation.

Figure 10.1 models the private benefits and costs to landowners and developers in converting wildlife habitat to other uses. Developers and landowners receive payments for converting their land to row crops and residential subdivisions, but must also bear private costs in making the conversion. For agricultural uses, trees may have to be felled, swamps drained, and irrigation ditches dug. Residential subdivisions require sewers, electric utility services, and streets and these costs increase the further the land is from already established services. The private landowner will balance the private marginal benefits and marginal costs and choose to convert H_E units of habitat. If habitat conversion is subsidized or if habitat conversion generates external costs in the form of lost ecosystem services or lost potential drugs from plants, the optimal level of habitat preservation is H^*. In contrast, the private landowner will choose to convert too many acres for agricultural/development purposes, H_E, and simultaneously preserve too few acres resulting in a deadweight loss given by the triangle DWL. However, preservationists also sometimes receive subsidies via tax credits which would result in too few acres converted for development, although those subsidies are likely less than the subsidies received by developers. Whenever subsidies are made available, individuals try to achieve the rents provided by those subsidies even if their actions are inefficient.

Public policies that make developers pay the full costs of installing streets, sewers, and utilities can help discourage habitat conversion. In addition, wildlife habitat preservation can be encouraged by private groups paying private landowners for conservation easements and development restrictions. Conservation banking provides an incentive to private landowners to maintain habitat and manage their land for species' improvement. Landowners who agree to preserve, enhance, and/or restore habitat can gain credits that can be sold to developers to offset the detrimental effects of development. For instance, if a potential development were to result in a loss of environmentally important wetlands, the developer could be required to purchase or augment existing wetlands in a different area. As greater development occurs, the price per credit could be expected to rise giving other landowners an increased incentive to bank some of their land for credit and profit. In Montana, the Nature Conservancy operates a grass bank at the Matador Ranch. Nearby ranchers who agree to "not bust sod" on their own ranches can graze their cattle at the Matador Ranch for reduced fees. The sod-buster restriction helps preserve habitat for sage grouse and other species[171]. Policies that work to reduce or end subsidies for agricultural commodities can also help discourage the conversion of forest and rangelands to agricultural uses and limit habitat loss.

10.6 CITES

The Convention on International Trade in Endangered Species of Wild Fauna and Flora (CITES) is an international agreement between governments to ensure that international trade in specimens of wild animals and plants does not threaten their survival. Drafted in 1963, CITES was signed by the United States in 1973 and came into force in 1975. Today, more than 171 countries have agreed to adhere to its provisions. In 1989, CITES banned trade in the ivory tusks of African elephants after herds declined from 1.3 million in 1979 to 600,000 in 1989 to 543,000 in 1994. One interesting piece of research [173] explaining the size of elephant populations found that countries with property rights' systems or community wildlife programs had more rapid elephant population growth rates than other countries. Political instability and the absence of representative governments lowered elephant growth rates. Countries with more riots and purges had slower elephant growth rates; the newly installed dictators have a high discount rate and want the economic benefits from elephants sooner rather than later. In contrast, countries with national property rights that gave citizens some ownership of wildlife had higher elephant population growth rates. Finally, countries that protected more area for wildlife had higher elephant population growth rates.

The basic tenet of the property rights approach is that individuals respond to incentives that are influenced by the prevailing structure of property rights. The primary problem facing common property resources is the "use it or lose it" incentive that results in the prisoner's dilemma. To improve the survivability of elephants, the social benefits from improving survivability need to be greater than the costs and the members of society bearing the costs must be appropriately compensated. Many African leaders ruthlessly seized power and their leadership positions are frequently susceptible to a sudden loss of that power. Although the dictators might adhere to a ban on ivory sales to make their governments look better, the more uncertain is their future reign, the higher is their discount rate and the less they care about the future. Citizens faced with great uncertainty also have high discount rates. Subsistence farmers commonly view elephants as pests since they receive little or no benefit from elephants, but must bear the costs of their existence. Certainly, these subsistence farmers have little or no reason to impede the activities of poachers.

With many countries in Africa having per capita incomes of $2000 or less and with ivory tusks selling for more than $6000 a pair, it is not surprising that poaching is a significant problem. Zimbabwe helped define property rights through its CAMPFIRE (Communal Area Programme for Indigenous Resources) program, helping to ensure that local inhabitants receive the benefits from wildlife [177]. The CAMPFIRE program sold permits to hunt fourteen elephants, eighty-two Cape buffalo, and twenty-six lions and leopards and provided $120,000 in revenues. Since the revenues of the program went to local villages, those villages had greater incentive to engage in conservation programs and to allocate resources to prevent poaching. In addition, farmers were compensated for the damage that wildlife did to crops and livestock.

A more recent example of paying for wildlife was provided in 2013 when the Dallas Safari Club decided to auction off the right to kill an endangered black rhino in Namibia. Defenders of the auction argued that it would raise between $250,000 and $1 million that would be given to the Conservation Trust for Namibia's Black Rhino. These defenders also argued that killing an adult male past its mating prime would allow younger male rhinos to mate, potentially increasing the herd size. However, black rhinos are endangered, and many preservationists think that it is unethical, especially for rich people from foreign lands, to take the life of an endangered species. Again, the property rights approach would be to assign property rights to the rhino, probably locally for equity purposes, and then let the owners decide the highest and best use of the rhino. For rich environmentalists to tell local indigenous people what they should do might be just as objectionable as auctioning off an endangered species.

In Chapter 1 we cautioned that economics usually takes individual preferences as given. However, if preferences can be molded, then various market transactions that might have occurred under one set of preferences might not occur under a different set of preferences. To examine the possibility for preferences to be endogenous, consider the effects of trade bans in things like ivory and other endangered species. Will these trade bans help various species to recover? Two lines of thought emerge here. Those who oppose a ban argue that the ban reduces supply and increases prices, making poaching more profitable. In contrast, those in favor of a ban argue that a trade ban can help decrease demand as ethical buyers choose to collect other nonbanned items. In addition, if potential poachers expect the ban to be successful in increasing herd size, then the larger future herd size increases supply and lowers the price of the banned items and reduces incentives to poach [133]. Governments can make credible commitments to tough anti-poaching laws by acquiring stockpiles of storable goods that are derived from elephants (ivory) and exotic birds (feathers). By stockpiling, government can sell the stockpile when prices are high, driving prices down and discouraging poaching.

On November 14, 2013 the US Fish and Wildlife Agency destroyed more than six tons of confiscated ivory. The destruction of this ivory goes against the theory that stockpiles can be used to discourage poaching. What rationale fueled this action? Apparently government officials hope to send a message to collectors that owning ivory has become taboo. In addition, bounties will be paid for the apprehension of poachers. If preferences turn against ivory, prices should fall and, with it, the illegal profits made from poaching ivory tusks will fall. However, destroying the ivory certainly reduced available supplies. If the demand for ivory does not subsequently fall, the higher priced ivory will likely increase poaching and further threaten endangered rhinos and elephants.

10.7 Choosing Distinctive Sites

Given limited resources to protect habitat and endangered species it is important to carefully choose which sites and species to protect. In Chapter 5 we examined the production possibility frontier and the output distance function as a way of measuring technical efficiency. The output distance function can also provide a way of ranking new sites for protection. Table 10.4 reports hypothetical numbers of endangered or threatened species of type 1 (y_1) and type 2 (y_2) at four existing sites and at three proposed sites. We will assume that type 1 species are insects and type 2 species are flowering plants. So that we can provide a meaningful two dimensional graph of the numerical quantities of insects and flowering plants we assume that all seven sites have the same input characteristics: for instance, all seven sites comprise the same number of acres, get the same amount of precipitation, and cost the same amount. These are just simplifying assumptions and as shown in Chapter 6 the output distance function can account for producers who use different amounts of inputs to produce multiple outputs. The four existing sites form a frontier technology for endangered and threatened species and this technology is graphed in Figure 10.2.

Table 10.4: Efficiency of New Sites

					Existing Sites and Proposed		
					Site E	Site F	Site G
Site	Type	y_1	y_2	$D_o(x,y)$	$D_o(x,y)$	$D_o(x,y)$	$D_o(x,y)$
A	Existing	1	7	1	1	1	1
B	Existing	4	5	1	1	0.96	0.92
C	Existing	6	3	1	1	0.85	1
D	Existing	7	1	1	1	0.88	1
E	Proposed	4	2	...	0.67	...	...
F	Proposed	8	3	...	...	1	...
G	Proposed	4	6	...	...	...	1

Sites A through D are efficient and so $D_O(x, y) = 1$ for each site. In addition to the four existing sites, the quantities of y_1 and y_2 at the three proposed sites are also graphed in the upper left of Figure 10.2. When proposed site E is evaluated relative to the four existing sites, the frontier technology remains the same because site E is inefficient with $D_O(x, y) = 0.67$. Since site E is inside the frontier defined by the existing sites, adding it would not enhance diversity. When proposed site G is compared to the four existing sites the frontier shifts to the northeast and is now comprised of sites A, G, C, and D as depicted in the top right corner of Figure 10.2. Site F also enhances the diversity of the four existing sites and, now, site B is inside the frontier. When site F is evaluated relative to the four existing sites the new frontier depicted in the bottom of Figure 10.2 is comprised of sites F and A. With site F included, sites B, C, and D are now inside the new frontier.

If funding limits the number of sites that can be acquired, which site, F or G, would enhance diversity the most? Comparing F and G is difficult since site F has more of y_1 and less of y_2 than site G. Without prices we also have no way to distinguish between the value generated by site F with the value generated by site G. One possible way to evaluate which site enhances diversity the most would be to examine the average efficiency of the four existing sites when the new proposed site is added. With site F, the average efficiency of sites A, B, C, and D is $\overline{D_o(x, y)} = 0.92$. With site G added, the average efficiency of the four existing sites is $\overline{D_o(x, y)} = 0.98$. Therefore, on average, the frontier shifts out further with site F added to the four existing sites than when site G is added to the four existing sites.

Even though prices are not available for species, scientists might weight the two species differently. If so, the weights can be used as pseudoprices and

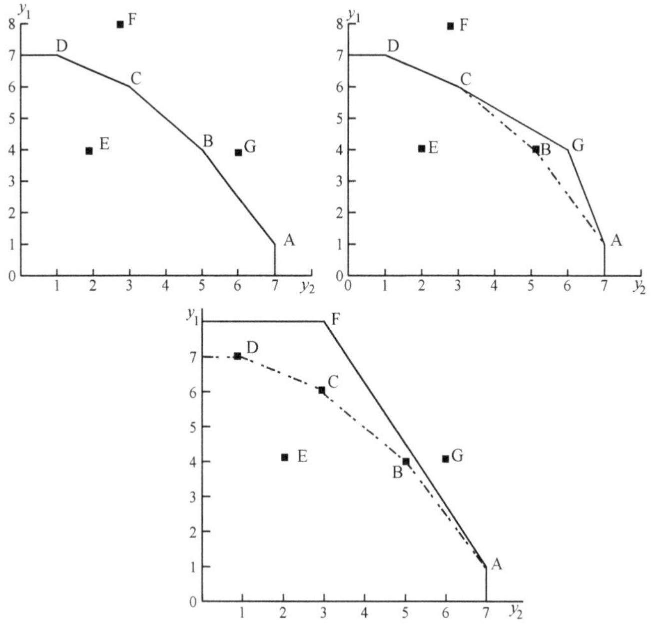

Figure 10.2: Endangered and Threatened Species Technology

a revenue maximization problem as described graphically in Chapter 5 or estimated using DEA as in Chapter 6 could be solved to determine which site adds the most value to the existing sites.

10.8 Problems and Questions to Ponder

1. The precautionary principle was used by members of the Clinton administration to oppose prescribed burns in Pacific Northwest forests. Those prescribed burns would have increased short-term risks to Northern Spotted Owl populations. Describe the unintended consequences of the precautionary principle in this case.

2. Suppose that the Greater Yellowstone Library and the Everglades Library both contain the complete works of William Shakespeare. Each library also contains some books that are not held by the other library. The Greater Yellowstone Library (library 1) holds all of the books authored by Jane Austen and the Everglades Library (library 2) holds all of the books authored by J.K. Rowling. Assume the works of Shakespeare comprise 200 books, the works of Jane Austen comprise 50 books, and the works of J.K. Rowling

comprise 80 books. Neither library is equipped with smoke alarms or fire extinguishers and so both libraries are at risk of burning. The Greater Yellowstone Library is located close to a fire station and so if a fire occurs there is a probability, $pr_1 = 0.50$, that it can be put out without damaging the books. If smoke alarms and fire extinguishers are installed at a cost of $C_1 = \$100$ the probability that a fire can be extinguished without damaging the books rises to $pr'_1 = 0.75$. The Everglades Library is far from the closest fire station and so if a fire occurs the probability that it can be extinguished without damaging the books is $pr_2 = 0.25$. Installing smoke alarms and fire extinguishers costs $C_2 = \$200$ and raises the probability of avoiding a damaging fire to $pr'_2 = 0.70$.

(a) Calculate the distinctiveness of the two libraries before smoke alarms are installed.

(b) Assume that $D_1 + U_1 = D_2 + U_2$. If there is only \$100 available to install fire extinguishers, at which library should they be installed?

(c) We previously assumed that $D_1 + U_1 = D_2 + U_2$. If only \$100 is available to install fire extinguishers, how much larger should $D_1 + U_1$ be relative to $D_2 + U_2$ to justify installing the fire extinguishers at library 1?

3. Andrew Metrick and Martin Weitzman proposed a model to operationalize the ranking scheme given in Equation (10.1). Their estimated equation follows:

$$\begin{aligned} lnSPEND = 9.39 + 0.42 \times MAMMAL + 0.59 \times BIRD \\ - 1.71 \times REPTILE - 0.78 \times AMPHIBIAN \\ + 0.85 \times ENDANGERMENT + 0.66 \times SIZE + .06 \times UNIQUE \\ - 0.10 \times PRIORITY + 1.19 \times CONFLICT \end{aligned}$$

where the variable $MAMMAL = 1$ if the species is a mammal, 0 otherwise; $BIRD = 1$ if the species is a bird, 0 otherwise; $REPTILE = 1$ if the species is a reptile, 0 otherwise; $AMPHIBIAN = 1$ if species is an amphibian, 0 otherwise. Note that the variable $FISH$ is not included in the equation since we know that if $MAMMAL = BIRD = REPTILE = AMPHIBIAN = 0$, then the species is a fish. Therefore, the coefficients for $MAMMAL$, $BIRD$, $REPTILE$, and $AMPHIBIAN$ tell us whether the species has more money spent on it (positive coefficient) or less money spent on it than a fish. The variable $ENDANGERMENT$ corresponds to the Nature Conservancy's Global Endangerment index with 1=most endangered and 5=least endangered. The variable $SIZE$ equals the natural logarithm of the average physical length of the species. The variable $UNIQUE$

takes a value of 1 if the species is the only one in its genus and 0 otherwise. The variable *PRIORITY* equals the Fish and Wildlife Service's priority ranking for the species and ranges from 1 (lowest priority) to 18 (highest priority). Finally, the variable *CONFLICT* takes a value of 1 if the species is in direct conflict with development (i.e., there is a high opportunity cost of improving survivability) and 0 if the species is not in direct conflict with development.

(a) Identify which of the variables correspond to distinctiveness (D_i), which variables correspond to direct utility (U_i), which variables measure the change in the probability of survival ($\triangle pr_i$), and which variable(s) control for costs (C_i).

(b) Explain the meaning of each of the coefficients in the estimated model. Do the coefficients have the expected sign for consistency with the Noah's Ark ranking system given by Equation (10.1)?

4. Refer back to Tables 10.2 and 10.3. Do there appear to be any patterns or regularities in which species or taxa give the highest utility in the ranking scheme given by Equation (10.1). Do these same species or taxa have a high degree of distinctiveness?

5. In a research paper that proposed a multi-attribute theory of diversity the authors [181] built on the following idea from Thomas Aquinas: "An angel is more valuable than a stone. It does not follow, however, that two angels are more valuable than one angel and one stone." In developing their theory of diversity the authors considered the following three animals: rhinoceros, whale, and shark. The rhinoceros and whale are mammals and the shark is a fish. The rhinoceros lives on land and the whale and shark swim in the ocean. Thus, there are multiple attributes to consider: is the animal a mammal or fish? Does the animal swim in the sea or live on land? If you can only save two out of the three species, which two would you choose to get the greatest diversity?

6. You are a member of the Species and Habitat Protection Committee for the private foundation WildBeasts4Ever. The table below reports the numbers of mammals (y_1) and birds (y_2) that have been protected at seven existing sites your foundation owns. A rich benefactor has just provided WildBeasts4Ever with a \$1 million dollar donation to acquire new sites. Three new sites have been proposed and the committee has been charged with evaluating the three sites and giving a recommendation on which sites should be acquired. Unfortunately, all three proposed sites cost \$1 million each so the foundation will only be able to acquire a single site.

(a) Graph the seven existing sites using EXCEL and sketch in the production possibility frontier for the existing sites.

WildBeasts4Ever Portfolio			
Site	Type	y_1	y_2
1	Existing	4	35
2	Existing	8	22
3	Existing	12	32
4	Existing	14	24
5	Existing	20	20
6	Existing	24	10
7	Existing	26	1
8	Proposed	20	29
9	Proposed	22	15
10	Proposed	27	6

(b) Add each proposed site, one by one, to the existing seven sites and sketch the new production possibility frontier. Can you eliminate any of the three proposed sites as not being diverse enough to merit addition to WildBeasts4Ever portfolio of sites?

(c) Explain the method you could employ to help choose between the two remaining proposed sites.

(d) In the past, your foundation has sometimes sold some of its existing sites and used the funds to acquire different sites that were thought to be more diverse and worthy of protection. Identify which, if any, of the seven existing sites might be a candidate for sale.

11

Interest Rates, Cost–Benefit Analysis, and Nonrenewable Resources

On what principle is it that, when we see nothing but improvement behind us, we are to expect nothing but deterioration before us?

Lord Thomas Babington Macaulay in "Southey's Colloquies on Society" [153]

11.1 Rate of Time Preference

Many decisions that society, firms, and individuals make have impacts on costs and benefits in the future as well as the present. Decisions made today to reduce greenhouse gases will have immediate costs with benefits not accruing for many years. Other kinds of actions such as clear-cutting a forest might give society immediate benefits in the form of lumber products, but impose future costs to society due to habitat destruction and erosion. To properly evaluate actions where benefits and costs occur over many periods we need account for the time value of money.

In general, individuals are impatient and prefer to receive goods and services today rather than at some point in the future. To induce those individuals to forego consumption today in exchange for consumption tomorrow they have to be compensated. For instance, the individual might require $1 in future consumption to give up $0.91 today. The more impatient the individual is, the less they are willing to give up today for an extra dollar in the future. On the producer's side of the market, goods and services can be produced today or those resources can be saved and used tomorrow. As more resources are saved for tomorrow, the opportunity cost of using those resources in terms of foregone current production tends to increase. Another way of thinking about it is that as more resources are saved for tomorrow, the return on investment goes down. Therefore, producers have to decide whether to produce today or to invest in capital to produce goods and services tomorrow.

To think about time preference we assume a simple model where a single producer and a single consumer can produce and consume goods today, $t = 0$, or tomorrow, $t = 1$. Figure 11.1 shows the producer's production possibility

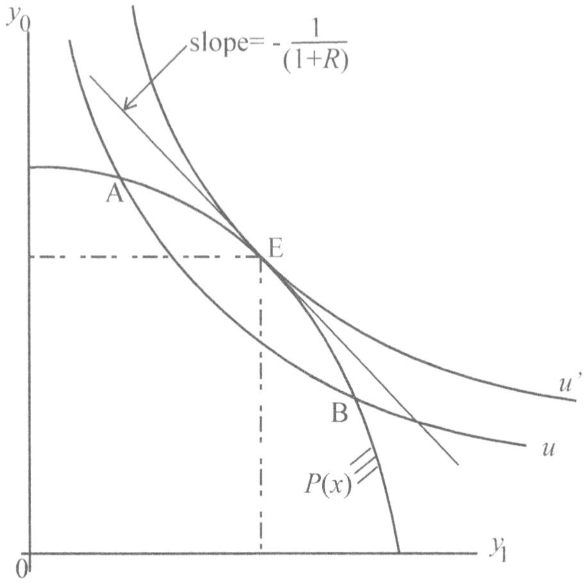

Figure 11.1: Illustration of Time Preference.

frontier for the amount of goods produced today (y_0) and goods produced tomorrow (y_1). The production possibility frontier is concave to the origin which captures the law of increasing opportunity cost. The slope of the frontier is the producer's marginal rate of transformation: $MRT = \frac{\triangle y_0}{\triangle y_1}$. Figure 11.1 also depicts two indifference curves for the single consumer. The indifference curves are convex which coincides with a diminishing marginal rate of substitution: the individual is willing to forego less and less current consumption for an extra dollar in future consumption. The slope of the indifference curve is the consumer's marginal rate of substitution (MRS) which equals the amount of present consumption an individual is willing to forego for \$1 more of future consumption. Because individuals tend to be impatient, the amount of present consumption they are willing to forego for an extra \$1 of future consumption is usually less than \$1.

At a point like A in Figure 11.1 the consumer's MRS is greater than the MRT. For instance, a consumer might be willing to give up \$0.95 today for an extra dollar tomorrow and the producer only requires \$0.91 today for an extra dollar of future production. Therefore, a trade can take place with the consumer foregoing some consumption today (saving) and the producer borrowing to invest in capital so that more goods can be produced tomorrow. At a point like B, the consumer's MRS is less than the producer's MRT and the consumer can increase their utility by borrowing to finance more current consumption. Point E corresponds to the equilibrium where the $MRS = MRT$. At the equilibrium, we also have $MRS = \frac{1}{1+R} = MRT$, where R equals the extra amount that an individual has to be compensated in future consumption to induce them to give up \$1 today. For instance, if $\frac{1}{1+R} = 0.909$ then $R = 0.10$. That is, to induce the individual to forego \$1 today they have to be compensated with \$1.10 in future consumption. Alternatively, R is the extra amount in revenues that a producer has to be compensated with in the future for foregoing an extra \$1 in revenues today. We will think of R as the interest rate recognizing that accounting for risk there are many different interest rates with R increasing as the amount of risk increases.

11.2 Time Value of Money

Because consumers and producers are impatient a dollar received today tends to be worth more than a dollar received at some point in the future. To properly account for future costs or benefits we need to discount those future cash flows back to the present. To understand discounting we first begin with a more familiar concept: compounding, which is just the inverse of discounting. If one saves \$100 today and earns an interest rate of $R = 10\%$ for $t = 1$ year, at the end of one year the \$100 will have grown to $\$110 = \$100 \times (1 + R)$. In this case the person earns \$10 in interest, which, when added to the \$100 in principal, gives \$110. If instead, the person saves \$100 today and earns $R = 10\%$ for $t = 2$ years the \$100 will have grown to $\$121 = \$100 \times (1 + R)^2$ by the end of the second year. Here, the principal grows to \$110 at the end of the first year which then yields additional interest of $\$11 = (10\% of \$110)$ in the second year giving a total of \$121. The simple compounding formula for calculating the future value (FV) of some present value (PV) amount is given as $FV = PV \times (1 + R)^t$, where R is the interest rate and t equals the number of years the principal (PV) is invested. Note that as either the interest rate or the number of years the investment is made increase, the future value of the amount will increase.

To discount, we solve the simple compounding formula for the present value of the future amount. That is, if the interest rate or discount rate is 10% and someone offers you \$110 one year from now, how much is that \$110 worth today? The present value amount is \$100, which can be obtained from the formula $PV = \frac{FV}{(1+R)^t} = \frac{\$110}{(1+0.1)^1} = \$100$. Similarly, \$121 received two years from today discounted at $R = 10\%$ is worth \$100. With discounting, the further in the future the dollar amount is received the less is its present value. Furthermore, increases in the discount rate, R, reduce the present value amount.

Some renewable natural resources such as the services provided by clean air and water or a fishery provide a stream of services in perpetuity. Each year those services provide value equivalent to a cash flow of CF. The present value of those cash flows can be written as

$$PV = \frac{CF}{(1+R)^1} + \frac{CF}{(1+R)^2} + \frac{CF}{(1+R)^3} + \ldots. \tag{11.1}$$

Equation (11.1) is the sum of an infinite geometric series with solution $PV = \frac{CF}{R}$. For instance, consider a fishery from which can be harvested a certain amount of fish that yield \$1 million in profits annually. If the discount rate is $R = 5\%$ the present value of the fishery is $\frac{\$1M}{0.05} = \$20$ million. If interest rates fall, the present value for all future harvests increases. As another example suppose that the annual costs of global warming are expected to be \$100 billion dollars in perpetuity. At a discount rate of $R = 2\%$ the present value of those perpetual costs equal \$5 trillion.

To enhance efficiency we want to compare the benefits and costs of any action or potential policy change. Therefore, we need to add up all the present and future discounted benefits and subtract the sum of the present and future discounted costs to obtain the net present value (NPV) of the policy change. Let B_t represent the benefits received in period t and let C_t represent the costs incurred in period t. The net present value of a project that lasts T years is

$$NPV = \sum_{t=0}^{T} \frac{B_t - C_t}{(1+R)^t}. \tag{11.2}$$

If a project or policy action results in a positive net present value then it is efficient to use society's resources to implement the policy. Alternatively, if the policy action results in a negative net present value, the resources that would be needed to implement the policy have more value to society in some alternative use.

Perhaps at this point, an example can help fix the idea of discounting and evaluating a project through the use of the net present value method. Suppose policy makers or regulators are interested in evaluating the effects of implementing new clear-air regulations. The regulations will require electric utilities to install filters that reduce particulate matter that is an undesirable by-product from burning coal. The reduction in particulate matter will provide health benefits to society in the future by enhancing longevity and reducing health care costs. The project is expected to have a useful life of $T = 25$ years. The costs of the project will occur in the present period, $t = 0$, and continue for three additional years. The benefits (reduced health costs) will start in period $t = 2$ and continue for the life of the project. Table 11.1 lists the expected benefits and costs of the project during the life of the project.

As is clear from the table the project passes the cost–benefit test at discount rates of 3% and 5%, but not 7%. When the costs are incurred in the early periods and benefits begin later, higher discount rates cause the project to be rejected. On the other hand, if the benefits occur early in the project, but costs are not incurred until later, higher discount rates make the project more favorable from a cost–benefit standpoint.

Table 11.1: Illustration of Net Present Value

t	B_t	C_t	PV of $(B_t - C_t)$ $R = 3\%$	PV of $(B_t - C_t)$ $R = 5\%$	PV of $(B_t - C_t)$ $R = 7\%$
0	0	−1,000	−1,000.00	−1,000.00	−1,000.00
1	0	−1,000	−970.87	−952.38	−934.58
2	300	−1,000	−659.82	−634.92	−611.41
3	300	−1,000	−640.60	−604.69	−571.41
4	300	0	266.55	246.81	228.87
5	300	0	258.78	235.06	213.90
6	300	0	251.25	223.86	199.90
7	300	0	243.94	213.20	186.82
8	300	0	236.82	203.05	174.60
9	300	0	229.93	193.38	163.18
10	300	0	223.23	184.17	152.50
11	300	0	216.73	175.40	142.53
12	300	0	210.41	167.05	133.20
13	300	0	204.29	159.10	124.49
14	300	0	198.34	151.52	116.35
15	300	0	192.56	144.31	108.73
16	300	0	186.95	137.43	101.62
17	300	0	181.50	130.89	94.97
18	300	0	176.22	124.66	88.76
19	300	0	171.09	118.72	82.95
20	300	0	166.10	113.07	77.53
21	300	0	161.26	107.68	72.45
22	300	0	156.58	102.56	67.71
23	300	0	152.01	97.67	63.28
24	300	0	147.58	93.02	59.14
25	300	0	143.28	88.59	55.27
		NPV=	\$1,104.07	\$219.22	−\$408.62

11.3 Social Rate of Discount

Clearly, the choice of discount rate is important. Choosing one that is too high will cause too many projects to fail the cost benefit test. Choosing too low a discount rate will cause too many projects to pass the cost–benefit test and prematurely exhaust available resources. In addition, there are ethical issues in discounting costs that will be imposed on future generations. Implicit in our discussion on discounting is the assumption that we are measuring the value of future consumption relative to present consumption for an individual or for people of the present generation. When projects and policies yield benefits and costs across generations, the discount rate has to be modified for how the present generation wants to value the welfare of future generations.

The social rate of discount, R, is the sum of two components [250]. First is the rate of individual time preference which we represent by δ. The parameter δ tells us how much compensation an individual requires in the future to induce them to give up \$1 today and is a measure of impatience. For instance, if $\delta = 3\%$, then an individual must receive \$1.03 one year from now to induce them to give up \$1 today. The higher δ is, the more compensation they require. The second component is the product of the growth rate of per capita consumption, g, and the elasticity of the marginal utility of consumption, η. The parameter η tells us how much future generations should be discounted since they will be richer than us. For instance, if $\eta = 2$, then a 1% increase in income causes a 2% decline in the marginal utility of income. Therefore, to the extent that future generations are richer from positive rates of growth ($g > 0$), we discount future streams of income that they would receive under any policy change. The social rate of discount is

$$R = \delta + \eta g. \tag{11.3}$$

A utility function that has a constant marginal utility elasticity of consumption (c) is given as

$$u = \frac{c^{1-\eta}}{1-\eta}, \eta \neq 1. \tag{11.4}$$

To think further about the role of η in the discounting formula 11.3 consider the following example. There are two individuals Rich and Petra. Rich consumes $c_R = \$36,500$ each year while Petra lives on the equivalent of \$2 per day or $c_P = \$730$ annually. Suppose the annual costs of global warming are \$5,000. What would be an ethical way of allocating the \$5,000 between Rich and Petra? One possibility is to take amounts from each individual so that the decline in utility is the same for both Rich and Petra. That is, we want to choose an amount to take from Rich, x_R, and amount to take from Petra, x_P, so that the change in utility is the same for both Rich and Petra. Using the utility function given by (11.4) and the constraint that $x_R + x_P = 5000$ we would want to solve the following:

$$\begin{gathered}
\text{change in Rich's utility} = \text{change in Petra's utility} \\
\frac{c_R^{1-\eta}}{1-\eta} - \frac{(c_R - x_R)^{1-\eta}}{1-\eta} = \frac{c_P^{1-\eta}}{1-\eta} - \frac{(c_P - x_P)^{1-\eta}}{1-\eta} \\
c_R^{1-\eta} - (c_R - x_r)^{1-\eta} = c_P^{1-\eta} - (c_P - x_P)^{1-\eta}.
\end{gathered} \tag{11.5}$$

One can solve for x_R and x_P using the EXCEL solver given a value of η. Table 11.2 reports the distribution of the \$5000 to be paid by Rich and Petra for alternative values of η. At first glance it appears that the choice of η makes

Table 11.2: Ethical Allocation of the Costs of Global Warming

	$\eta = 1.25$	$\eta = 1.50$	$\eta = 2$	$\eta = 3$
x_R	4,960.55	4,984.52	4,997.70	4,999.66
x_P	39.45	15.48	2.30	0.34

little difference as Rich pays almost all of the costs. However, Rich consumes 50 times more than Petra who lives on just \$2 per day. If $\eta = 1.25$ then Petra will have to pay approximately \$40 per year, the equivalent of living on only \$1 per day for almost six weeks and then \$2 per day for the other 46 weeks. If $\eta = 3$, then Petra pays very little and the burden of offsetting the costs of global warming falls almost completely on Rich.

Since the Industrial Revolution the growth rate of per capita income has averaged somewhere around 1.6%–2%. Suppose that the individual rate of time preference is $\delta = 0.02$. The elasticity of marginal utility with respect to consumption is also known as the measure of relative risk aversion. If society becomes more risk averse, the value of η increases. Most estimates of the elasticity of marginal utility range from 1 to 4 with a mean around 2. Therefore, let $\eta = 2$. If we expect a future growth rate of $g = 0.018$, the social rate of discount is $0.056 = 0.02 + 2 \times 0.018$. A slower growth rate reduces the discount rate and a higher rate of risk aversion increases the discount rate.

As our example from Table 11.2 shows, it seems ethical for people in the developed world to shoulder the burden of the costs of global warming. However, the point of incorporating the elasticity of marginal utility of income into the social rate of discount is to think about how much of the costs of global warming should be paid by people today versus people 100 years or more from now. If the average person in the future is richer than the average person today it seems ethical to make future generations pay more. However, if global warming causes economic growth to become negative, then the current generation should pay more than the future generation.

Differences in choices regarding δ, η, and g can make a big difference in the social rate of time preferences, R, and can have an effect on other policy variables. For instance, Nicholas Stern [227] in his report to the United Kingdom government, which has become known as *The Stern Review*, used values of $\delta = 0.001$, $\eta = 1$, and $g = 0.013$ resulting in a social rate of discount of $R = 0.014$. *The Stern Review* used that discount rate to calculate a carbon tax equal to \$360 per ton. In contrast, in a review of *The Stern Review*, Martin Weitzman [250] suggests that $\delta = 0.02$, $\eta = 2$, and $g = 0.02$ resulting in a social discount rate of $R = 0.06$. William Nordhaus [186] makes a similar estimate of $R = 0.055$. The higher social discount rates result in a carbon tax of only \$35/ton in 2015 rising to \$85 in 2050 to \$206 in 2100.

11.4 Do Biofuels Pass the Cost–Benefit Test?

The WHO (World Health Organization) defines a DALY (disability adjusted life year) as one year lost to premature death from disease plus the number of years an individual spends in a disabled condition weighted by the severity of the disease. The WHO estimates an annual toll of 141,000 deaths and 5.4 million lost DALYs attributable to global warming. Concern about global warming has caused countries to search for alternative renewable energy sources such as solar power, wind power, and biofuels.

Do biofuels pass a cost–benefit test? Biofuels—corn ethanol and biodiesel made from soybean oil—could increase hunger and poverty in developing countries by raising food prices. Indur Goklany [88] estimated the increase in death and disease due to biofuel policies that increase absolute poverty, defined as an individual living on less than $1.25 per day in 2005 international dollars. According to Goklany [88, p.10], "Poverty is itself a factor that could exacerbate risk factors for death and disease." Extrapolating the effects of higher grain prices that increase the number of people living in poverty Goklany estimated that biofuels policies that increase poverty "May have resulted in at least 192,000 excess deaths and 6.7 million additional lost DALYs in 2010" [88, p.9].' Given the WHOs estimates of 141,000 deaths and 5.4 million lost DALYs, cost–benefit analysis indicates that biofuels policies increase net deaths by 51,000 and result in an extra 1.3 million lost DALYs annually.

11.5 Hyperbolic Discounting

Behavioral economists argue that consumers are not rational utility maximizers, but, instead, make predictable mistakes in their choices. One of these mistakes has to do with the time value of money. In various experiments researchers have found that when subjects are offered the choice between $50 today and $100 one year from now, those subjects prefer the $50 today. However, when offered the choice between $50 five years from now and $100 six years from now, those same subjects prefer the $100 in six years. This behavior represents a preference reversal, since the latter choice is the same as the former choice except delayed by five years. In terms of discounting, the subjects have a high rate of time preference (high discount rate) when the choices are near and a low rate of time preference (low discount rate) when the choices are far in the future. Such behavior is consistent with individuals having a preference for reducing the costs of global warming which will accrue in the future, but not being willing to pull the trigger and adopt the necessary policies today.

11.6 Resource Extraction

Concern over increasing natural resource scarcity has occupied observers since Thomas Malthus set forth his theory of population and Stanley Jevons worried whether Britain would exhaust its coal deposits. In 1956, M. King Hubbert[113], a geophysicist for Shell Oil, predicted that given current oil reserves and rates of extraction, US production of oil would peak somewhere between 1965 and 1970 and world production of oil would peak in the early 2000s.

When is the optimal time to extract a nonrenewable resource such as oil, coal, or copper? To answer this question we note that when markets are competitive the law of one price says that all sellers should charge the same price, because if they do not, someone could profit by buying at a low price and reselling the item at the higher price [228]. The same idea should hold across time. That is, the discounted prices of the resource should be the same across all periods. Consider a two period problem where a seller can extract some amount today (Q_0) at price P_0 or some amount tomorrow (Q_1) at price P_1. Since the amounts extracted in the two periods must be no greater than the total fixed stock, Q, the quantities extracted must satisfy $Q_0 + Q_1 = Q$. The law of one price indicates that the discounted prices should be the same across both periods:

$$P_0 = \frac{P_1}{(1+R)}. \tag{11.6}$$

In competitive markets firms compare the price of the resource with the marginal cost of extraction. Today's marginal cost of extraction is the foregone revenue that could be earned by waiting until next period to extract: $\frac{P_1}{(1+R)}$. Consider what happens when the quantities to extract in the two periods are such that $P_0 > \frac{P_1}{(1+R)}$; that is, today's price is greater than marginal cost. Under competitive conditions the law of one price indicates that extraction should shift from the future to the present driving P_0 down and $\frac{P_1}{(1+R)}$ up until the discounted prices in the two periods are equal. If $P_0 < \frac{P_1}{(1+R)}$, it pays the firm to shift production to the future which drives P_0 up and $\frac{P_1}{(1+R)}$ down.

In the competitive equilibrium price equals marginal cost as shown in (11.6). One can rearrange (11.6) to show that in equilibrium

$$\begin{aligned} 1 + R &= \frac{P_1}{P_0} \\ R &= \frac{P_1}{P_0} - 1 \\ R &= \%\triangle P. \end{aligned} \tag{11.7}$$

That is, extraction should occur up to the point where the return from extracting today and investing the proceeds at rate R just equal the percentage change in price from today until tomorrow. Equation (11.7) is known as

Hotelling's rule for resource extraction. If $R > \%\triangle P$, more should be extracted today, which drives P_0 down and P_1 up. By extracting more today, the proceeds from the sale of the resource can be invested and earn R, which is more than can be earned by waiting an additional period before extraction. For instance, suppose the interest rate is $R = 6\%$. If the price of a barrel oil is \$100 today and is expected to rise to \$105 one year from now, it is more efficient to extract the barrel today and sell it for \$100 and then use the proceeds to invest at $R = 6\%$. One year from now the investor would have accumulated \$106 from their \$100 investment as opposed to the \$105 earned from waiting one year to extract the barrel of oil. Similarly, if $R < \%\triangle P$, less should be extracted today, which pushes P_0 up and drives P_1 down until $R = \%\triangle P$.

Consider the following example where the market demand for some hypothetical resource, call it mythonium, is given as $P = 75 - \frac{1}{2}Q$, the total resource stock is $Q = 20$, and the interest rate is $R = 6\%$. Table 11.3 lists the alternative quantities and prices of mythonium in the two periods.

Table 11.3: Competitive Extraction of a Nonrenewable Resource

Q_0	P_0	Q_1	P_1	$\frac{P_1}{1+0.06}$	$\frac{P_1}{1+0.03}$
0	75.0	20	65.0	61.32	63.11
1	74.5	19	65.5	61.79	63.59
2	74.0	18	66.0	62.26	64.08
3	73.5	17	66.5	62.74	64.56
4	73.0	16	67.0	63.21	65.05
5	72.5	15	67.5	63.68	65.53
6	72.0	14	68.0	64.15	66.02
7	71.5	13	68.5	64.62	66.50
8	71.0	12	69.0	65.09	66.99
9	70.5	11	69.5	65.57	67.48
10	70.0	10	70.0	66.04	67.96
11	69.5	9	70.5	66.51	68.45
12	69.0	8	71.0	66.98	68.93
13	68.5	7	71.5	67.45	69.42
14	68.0	6	72.0	67.92	69.90
15	67.5	5	72.5	68.40	70.39
16	67.0	4	73.0	68.87	70.87
17	66.5	3	73.5	69.34	71.36
18	66.0	2	74.0	69.81	71.84
19	65.5	1	74.5	70.28	72.33
20	65.0	0	75.0	70.75	72.82

Price falls as more mythonium is extracted in a given period. To begin, assume that $Q_0 = 1$ unit of mythonium is extracted in $t = 0$ and that $Q_1 = 19$ units are extracted in $t = 1$. To extract a second unit of mythonium in $t = 0$

would drive price down to \$74 but the marginal cost of extracting the second unit is only \$62.26 which equals the discounted market price the firm can earn from the 18th unit of mythonium in $t = 1$. In this case it pays to extract more mythonium in $t = 0$ and less in $t = 1$. It would also make sense to extract the 3rd, 4th, ..., and 14th units. However, if 15 units of mythonium are extracted in $t = 0$, the market price of the 15th unit equals \$67.50. This price is less than the discounted market price of \$68.40 from extracting $Q_1 = 5$ in $t = 1$. Therefore, it is not efficient to extract the 15th unit in $t = 0$.

The discounted prices in the two periods are not exactly equal when $Q_0 = 14$ and $Q_1 = 6$. In fact the equilibrium quantity of oil to extract is somewhere between 14 and 15 in $t = 0$ and simultaneously somewhere between 6 and 5 in $t = 1$. If fractional quantities of oil can be extracted the equilibrium quantities of oil to extract in the two periods can be found by substituting the demand function ($P = 75 - 0.5Q$) and the resource constraint ($Q_0 + Q_1 = 20$) into the equilibrium condition as follows:

$$\begin{aligned}
P_0 &= \frac{P_1}{(1+R)} \\
75 - 0.5Q_0 &= \frac{75 - 0.5Q_1}{(1+.06)} \\
75 - 0.5Q_0 &= \frac{75 - 0.5(20 - Q_0)}{(1.06)} \\
75 - \frac{65}{1.06} &= 0.5Q_0 + \frac{0.5Q_0}{1.06} \\
13.68 &= 0.97Q_0 \\
Q_0 = 14.1 \text{ and } Q_1 &= 20 - 14.1 = 5.9. \qquad (11.8)
\end{aligned}$$

We can also see in Table 11.3 that as the interest rate falls to $R = 3\%$ the optimal amount to extract in $t = 0$ falls as it become more efficient to wait until $t = 1$. This adjustment makes sense because lower interest rates means people have become more patient and willing to defer production until a later period. Figure 11.2 graphs the demand functions for the two periods. Note that the demand function in $t = 1$ equals the marginal cost function in period $t = 0$. A lower interest rate increases the marginal cost of extraction in $t = 0$ resulting in less extraction in $t = 0$.

When the market for the nonrenewable resource is controlled by a firm with monopoly power we have to modify the equilbrium condition as presented in (11.7). Instead of comparing price in the present period and the discounted future price, a firm with monopoly power compares marginal revenue in the present period with the discounted marginal revenue from the future period. Marginal revenue equals the extra revenue a firm earns when they sell one additional unit of output: $MR = \frac{\Delta TR}{\Delta Q}$, where total revenue is $TR = pQ$. Table 11.4 reports total revenue and marginal revenue for periods 0 and 1. Figure 11.3 graphs the demand and marginal revenue schedules for the two periods. When the monopolist has maximized profits their choice of Q_0 and

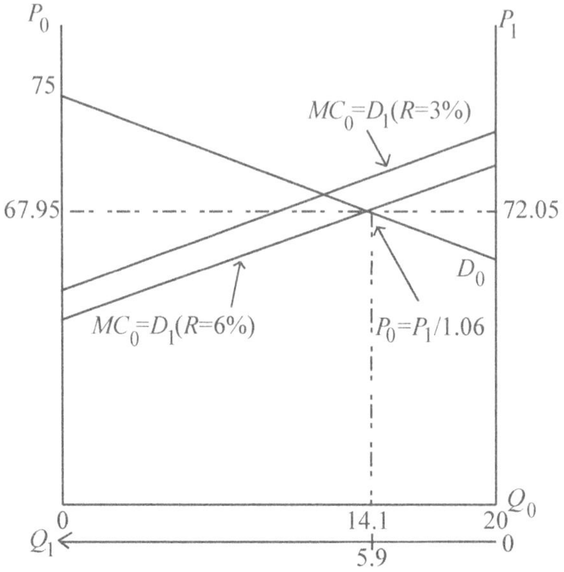

Figure 11.2: Optimal Extraction in Two Periods.

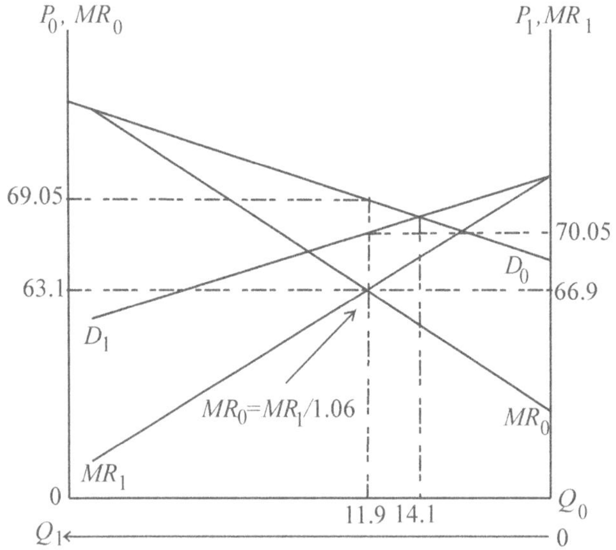

Figure 11.3: Monopoly Extraction in Two Periods.

Q_1 satisifies

$$MR_0 = \frac{MR_1}{1+R}. \tag{11.9}$$

As seen in Table 11.4 the monopolist maximizes the sum of the discounted revenues in the two periods by choosing $Q_0 = 12$ and $Q_1 = 8$. The monopolist stops at $Q_0 = 12$ since the marginal revenue for the 13th unit is $MR_0 = 62.5$ is less than the discounted marginal revenue in period 1 of 63.7. Note that the sum of the discounted revenues in the two periods at $Q_0 = 12$ and $Q_1 = 8$ is greater than the sum of discounted revenues under competition corresponding to $Q_0 = 14$ and $Q_1 = 6$. Thus, it appears that the monopolist is more conservation-minded than would occur in a competitive market. However, Joseph Stiglitz [228] points out that with an infinite time horizon all of the resource is extracted whether markets are competitive or monopolistic.

Table 11.4: Resource Extraction with Monopoly Power

Q_0	P_0	Q_1	P_1	TR_0	MR_0	TR_1	MR_1	$\frac{MR_1}{(1+R)}$	$TR_0 + \frac{TR_1}{(1+R)}$
0	75.0	20	65.0	0	x	1,300.0	x	x	1,226.4
1	74.5	19	65.5	74.5	74.5	1,244.5	55.5	52.4	1,248.6
2	74.0	18	66.0	148.0	73.5	1,188.0	56.5	53.3	1,268.8
3	73.5	17	66.5	220.5	72.5	1,130.5	57.5	54.2	1,287.0
4	73.0	16	67.0	292.0	71.5	1,072.0	58.5	55.2	1,303.3
5	72.5	15	67.5	362.5	70.5	1,012.5	59.5	56.1	1,317.7
6	72.0	14	68.0	432.0	69.5	952.0	60.5	57.1	1,330.1
7	71.5	13	68.5	500.5	68.5	890.5	61.5	58.0	1,340.6
8	71.0	12	69.0	568.0	67.5	828.0	62.5	59.0	1,355.7
9	70.5	11	69.5	634.5	66.5	764.5	63.5	59.9	1,355.7
10	70.0	10	70.0	700.0	65.5	700.0	64.5	60.8	1,360.4
11	69.5	9	70.5	764.5	64.5	634.5	65.5	61.8	1,363.1
12	69.0	8	71.0	828.0	63.5	568.0	66.5	62.7	1,363.8
13	68.5	7	71.5	890.5	62.5	500.5	67.5	63.7	1,362.7
14	68.0	6	72.0	952.0	61.5	432.0	68.5	64.6	1,359.5
15	67.5	5	72.5	1,012.5	60.5	362.5	69.5	65.6	1,354.5
16	67.0	4	73.0	1,072.0	59.5	292.0	70.5	66.5	1,347.5
17	66.5	3	73.5	1,130.5	58.5	220.5	71.5	67.5	1,338.5
18	66.0	2	74.0	1,188.0	57.5	148.0	72.5	68.4	1,327.6
19	65.5	1	74.5	1,244.5	56.5	74.5	73.5	69.3	1,314.8
20	65.0	0	75.0	1,300.0	55.5	0	74.5	74.3	1,300.0

When there are extraction costs the Hotelling condition must be modified. Let MC_i represent the marginal cost of extracting the ith unit of the nonrenewable resource. When markets are competitive, marginal profit is $M\pi = p_i - MC_i$ and when markets are monopolistic marginal profit is $M\pi = MR_i - MC_i$. Accounting for the marginal costs of resource extraction the Hotelling condition takes the form

$$M\pi_0 = \frac{M\pi_1}{(1+R)} \quad \text{or}$$
$$R = \%M\pi. \tag{11.10}$$

11.7 Trends in Nonrenewable Resource Prices

In 1980, ecologist Paul Ehrlich and economist Julian Simon made a bet about the price of five metals: chromium, copper, nickel, tin, and tungsten. They bought \$200 of each metal, \$1000 in total. If prices rose over the decade Simon would pay Ehrlich the difference between the five metals' prices and \$1000. If the prices of the metals fell, Ehrlich would pay Simon the difference between \$1000 and the five prices. Concern about overpopulation and extravagent American lifestyles caused Ehrlich to think that prices would rise. In contrast, Simon thought that prices would fall because any increasing scarcity of a particular metal would be met by human ingenuity seeking substitutes. As mentioned in Chapter 1 Simon won the bet and received a check from Ehrlich for \$576.07. The two men's professional lives are chronicled in an appropriately named book called *The Bet* by Paul Sabin.

Figure 11.4 plots an index of the inflation adjusted prices of the five metals that Ehrlich and Simon bet on. The year 1980 serves as a base year when the index takes a value of 1, or \$1000. For the 112 year span from 1900 to 2011, the index rises and falls, but mostly moves sideways. However, at the end of 2011 the five metals were 21% above their 1980 level, and were approximately equal to the peak that occurred in 1916.

Which direction are the five metals' prices headed? Increasing scarcity of the metals along with the Hotelling rule suggest that the index should continue to increase at the interest rate that nonrenewable resource owners use to discount the future. The effect of China's demand on the five metals' prices cannot be underestimated. From 1952 to 1980, China's real GDP per capita grew by 112% while US real GDP per capita grew by 87%. However, from 1980 to 2011, China grew by 422% while the United States grew by only 65%. Moreover, from 2000 to 2011, China grew by 136% compared to only 5.5% in the United States [165]. With a population over three times the size of the United States, China's demand for commodities is likely putting significant upward pressure on prices. If Paul Ehrlich is correct, mineral prices should continue to rise reflecting increased resource scarcity. On the other hand, in 2002 the five mineral prices reached their lowest level since 1900. If Julian Simon is correct, the upward trend in mineral prices that has occurred since 2002 will only be a blip on the screen as users react to increased prices by searching for more efficient means of using the minerals and for substitutes that can be had at a lower cost.

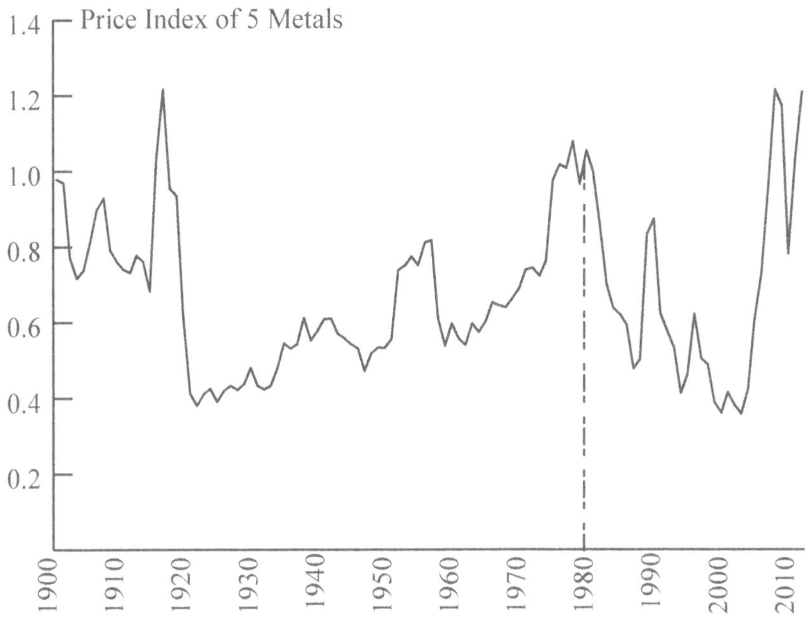

Figure 11.4: Price Index of Chromium, Copper, Nickel, Tin, and Tungsten, 1980=1.0 (Source: Historical Statistics for Mineral and Material Commodities in the United States, US Geological Survey, http://minerals.usgs.gov/ds/2005/140/).

11.8 Problems

1. Big State University is considering replacing its coal fired generating plant with a new power plant that uses natural gas as the fuel. The new plant will be more energy efficient (a benefit) and will help reduce carbon dioxide emissions (a benefit). The cost of the plant is \$1 million dollars. The expected life of the plant is $T = 5$ years. The university expects the benefits of the plant to increase during each year of operation due to an expected tax on carbon dioxide emissions. The costs and benefits of the plant are listed in the table below. Complete the columns for the present value of the net benefits $(B_t - C_t)$ and calculate the net present value of the project at an interest rate of $R = 5\%$ and $R = 8\%$.

Time t	Benefits B_t	Costs C_t	PV of $(B_t - C_t)$ $R = 5\%$	PV of $(B_t - C_t)$ $R = 8\%$
0	0	\$1,000,000		
1	\$150,000	0		
2	\$200,000	0		
3	\$250,000	0		
4	\$300,000	0		
5	\$350,000	0		

2. What problem(s) might arise for society in designing policies to combat global warming if the costs of reducing CO_2 are near and the benefits of reducing CO_2 are far in the future and citizens implicitly use hyperbolic discounting to evaluate projects?

3. Consider an economy with two periods: $t = 0$ (today) and $t = 1$ (tomorrow). The demand for some exhaustible natural resource, like oil, is given as $P = 500 - 0.5Q_d$ in both periods. There are $Q = 100$ units of oil in the ground, so that $Q_0 + Q_1 = 100$. Assume a zero marginal cost of extraction in both periods.

   (a) Let $R = 5\% = 0.05$. Set up a spreadsheet to determine the various combinations of Q_0, P_0, and the simultaneous combinations of Q_1 and P_1. Now, find the competitive equilibrium quantity of extraction in period 0 (which is simultaneously the competitive quantity of extraction in period 1) by implementing the condition that if $R > \frac{P_1 - P_0}{P_0}$ more should be extracted in t_0 and if $R < \frac{P_1 - P_0}{P_0}$ less should be extracted in t_0. Repeat for an interest rate of $R = 10\% = 0.10$.

   (b) Graph the demand and supply schedules for period $t = 0$ putting P on the vertical axis and Q on the horizontal axis. The supply curve in period $t = 0$ equals the discounted prices that can be earned in $t = 1$.

   (c) Instead of a competitive resource market assume that all of the oil is controlled by a monopolist. Continue your spreadsheet by determining the monopolist's total revenue and marginal revenue in each period: TR_0, MR_0, and the simultaneous combinations of TR_1 and MR_1. Discount the marginal revenue in $t = 1$ when $R = 0.05$ and when $R = 0.10$.

   (d) Determine the amount of oil that should be extracted in period 0 and in period 1 when the discount rate is $R = 0.05$ and when the discount rate is $R = 0.10$ under conditions of monopoly.

   (e) Compare the competitive equilibrium quantity with the

monopoly quantity of the exhaustible resource to extract in the two periods.

(f) Produce a graph of the marginal revenue schedule in period 0 and the marginal revenue schedule in period 1 when $R = 0.05$ and when $R = 0.10$. Use the XY scatter diagram and put Q_0 on the horizontal axis, and on the vertical axis graph MR_0, MR_1 when $R = 0.05$, and MR_1 when $R = 0.10$.

(g) As R increases, what happens to the optimal amount to extract in periods 0 and 1?

12

Renewable Resources

> But after forty days without a fish the boy's parents had told him that the old man was now definitely and finally *salao*, which is the worst form of unlucky, and the boy had gone at their orders in another boat which caught three good fish the first week.
>
> Ernest Hemingway, *The Old Man and the Sea* [104, p. 1]

12.1 Not Enough Fish in the Sea

The tragedy of the commons is on display in much of the world's commercial ocean fisheries. Trawlers use nets that stretch for over a mile in length and are weighted with a roller. As the nets are dragged along the bottom of the ocean everything in the way is flattened and fish are indiscriminantly harvested with much waste [42]. Figure 12.1 depicts tons of harvested fish for several commercially harvested stocks. The collapse of bluefin tuna, cod, and Alaskan king crab are clearly seen. The use of global positioning systems, sonar, and bigger vessels contribute to overfishing. As fish stocks dwindle, fishing seasons shorten, and fishers engage in what the popular television show calls the "Deadliest Catch." Overharvesting of one species, such as bluefin tuna, has led commercial fishers to pursue other species, such as skipjack tuna.

Although bad luck might have have once contributed to a fishless day at sea, today it is more likely that bad incentives are the culprit. In January of 2013, a 489-pound bluefin tuna sold for $1.76 million at the Tokyo fish auction. High prices encourage fishers to "mine" the fish stock in a kind of marine gold rush. Without property rights in future fish today's fishers have no reason to limit their harvest to allow fish to grow larger and reach reproductive stage since more and larger fish only serve to encourage new boats to enter the industry.

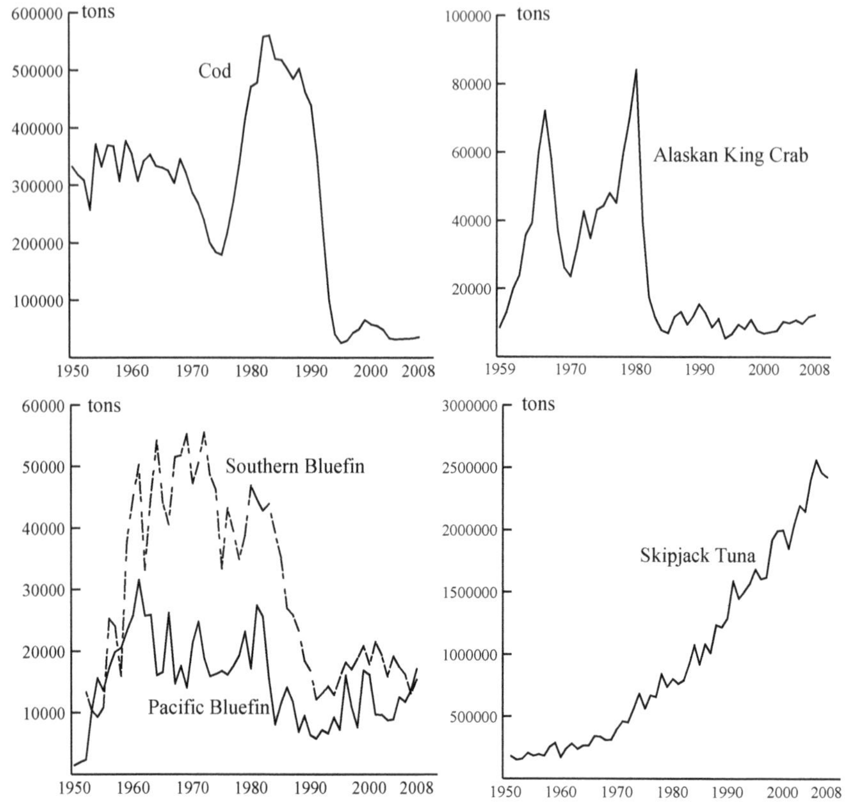

Figure 12.1: Harvested Fish (tons) 1950-2008 (Source: FAO FISHSTAT Plus).

12.2 Bioeconomic Fishery Model

12.2.1 Biological Growth Functions

To examine the effects of fishing effort on the fishery biomass we develop both a biological and economic model of the fishery [102]. Let $S(t)$ represent the fish stock or biomass at time t. Growth of the biomass depends on its size at the beginning of the period and we represent this relation as $dS(t)/dt = G(S)$. Biological growth of the biomass depends on the number of fish that are hatched during the period (recruitment) plus the physical growth of existing fish minus the natural mortality of the existing biomass.

Several functions can be used to model the growth of the biomass such as the depensated function and the logistic growth function. The logistic function takes the form

$$G(S) = \lambda S(1 - \frac{S}{K}), \tag{12.1}$$

where λ is the growth rate of the biomass when S is small and K is the natural carrying capacity that can be supported. Table 12.1 and Figure 12.2 represent a logistic growth function with $K = 300$ and two different λ parameters: $\lambda = 1$ and $\lambda = 0.5$. Clearly, when there is no biomass there can be no growth. When the biomass is small, but positive, the growth in the biomass is relatively rapid as there is little competition between fish. Given $\lambda = 1$ the biomass approximately doubles when S is small. For instance, when $S = 1$, the biomass grows by 0.997. Similarly, when $S = 2$ the biomass grows by 1.987. For the logistic growth function maximum growth in the stock occurs at a biomass of $S^* = \frac{1}{2}K$. After reaching its maximum, the growth in the fish stock declines as the biomass reaches the habitat's carrying capacity where the birth rate and mortality rate are in equilibrium. At this point, growth in the stock is again zero. Slower growing species have smaller growth at every level of the stock. For instance, when $\lambda = 0.5$, the growth function shifts downward and is half what it was when $\lambda = 1$.

Table 12.1: Logistic Growth Function

	$\lambda = 1$	$\lambda = 0.5$
Stock $= S$	Growth $= G(S)$	Growth $= G(S)$
0	0.000	x
0.01	0.010	0.050
1	0.997	0.498
2	1.987	0.993
10	9.667	4.833
25	22.917	11.458
50	41.667	20.833
75	56.250	28.125
100	66.667	33.333
125	72.917	36.458
150	75.000	37.500
175	72.917	36.458
200	66.667	33.333
225	56.250	28.125
250	41.667	20.833
275	22.917	11.458
290	9.667	4.833
298	1.987	0.993
299	0.997	0.498
300	0.000	0.000

As seen in Table 12.1 and Figure 12.2 the same amount of growth can be

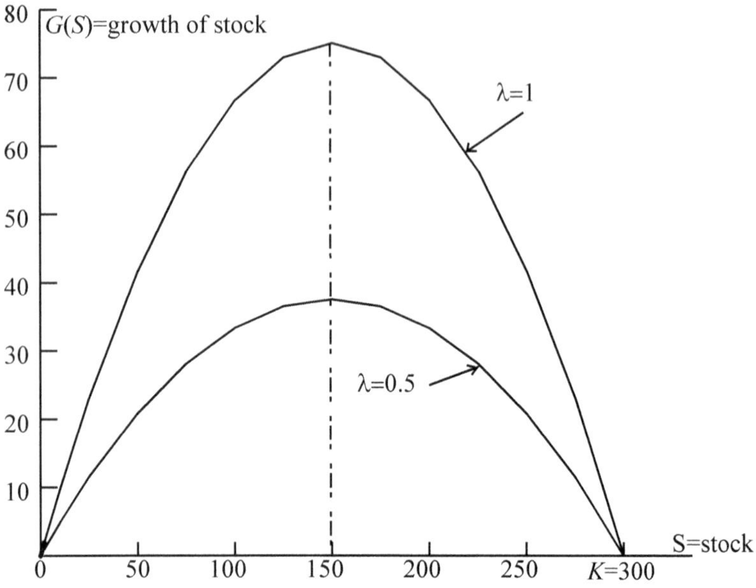

Figure 12.2: Logistic Growth Function.

generated from two different stock levels. Fish stocks of $S = 25$ and $S = 275$ both generate growth of $G(S) = 22.917$. Similarly, fish stocks of $S = 50$ and $S = 250$ both generate growth of $G(S) = 41.667$. The maximum growth in the stock, called the maximum sustained yield, is $MSY = 75$ and occurs at $S^* = 150$.

Using the data in Table 12.1 we know that when a nonexploited fish has reached its maximum carrying capacity, K, there is no growth in the stock: $G(S) = 0$. Suppose that fishers arrive and begin harvesting some amount of fish, say $y = 22.917$. Since $G(S) = 0$ at $K = 300$ and assuming a growth parameter of $\lambda = 1$, the stock will begin to shrink and will decline to $S = 275$. If fishers harvest $y = 22.917$ when $S = 275$, small changes in the stock will return the stock to $S = 275$. For instance, if the stock experiences a positive shock and begins the period at $S = 290$, the growth of the stock will be $G(S) = 9.667$ which is less than the harvest, causing the stock to fall back to $S = 275$. On the other hand, if the stock suffers a negative exogenous shock due to a poor spawn and falls to $S = 250$, the stock will grow by $G(S) = 41.667$ which is greater than the harvest and the stock will return to $S = 275$. In fact, given a harvest of $y = 22.917$, any stock of fish greater than $S = 25$ but less than $S = 275$ will yield growth greater than the harvest and return the stock to $S = 275$. Thus, the pair $S = 275$ and $y = 22.917$ is a stable equilibrium.

Now, suppose the harvest is $y = 22.917$, but the stock starts at $S = 25$. This situation is an unstable equilibrium since any small change in the stock will cause a movement away from $S = 25$. For instance, an oil spill that pollutes the fishery and causes the stock to fall below $S = 25$ will yield growth less than the harvest and the stock will go extinct. Positive shocks to the stock will cause growth in the stock to be greater than the harvest resulting in the stock growing to $S = 275$.

For each stock of fish to the right of S^* there is a harvest that results in a stable equilibrium. For stocks less than S^* the associated harvest is unstable. Finally, if the stock is at S^*, any harvest above the maximum sustained yield (MSY) will be unsustainable and will eventually result in extinction.

12.2.2 Static Bioeconomic Equilibrium

In this section we combine the biological growth function with the incentive of fishers to maximize profits. We first assume that the fishery is a perfectly competitive industry so that new fishers can freely enter or exit the industry in response to profit opportunities. Such a situation is often called an open-access fishery. Fishers are assumed to use variable inputs (x) that include labor, fuel, and capital such as nets, sonar, and boats. For instance, we might define one unit of input as one boat, five workers, and 500 gallons of fuel. Two units of input would then equal two boats, ten workers, and 1000 gallons of fuel. In addition, a fixed input—the stock of fish (S)—determines the amount of fish that can be harvested in a given period. We let y represent the number of fish harvested. The production function takes the form $y = f(x, S)$ and is depicted in Figure 12.3 for two stocks of fish, S_1 and S_2 where $S_2 > S_1$. Diminishing returns are evident in that output (harvest) increases at a decreasing rate as more inputs are used along with a given stock of fish. As the fish stock increases a given level of fishing effort yields a larger harvest and is represented by an upward shift in the production function. Given inputs, x, when the stock of fish is S_1 the harvest is y_1 and when the stock of fish is S_2 the harvest is y_2.

For a given level of variable input, an equilibrium or sustainable harvest occurs when the harvest equals the growth of the fish stock. We can write this condition as $y = G(S)$. This condition is depicted graphically in Figure 12.4 where production depends on the fish stock for a given level of input. When the level of fishing effort is fixed at x_2 the fish production function is given by $f(x_2, S)$, the equilibrium harvest is $G(S)$, and the equilibrium fish stock is S_2. As shown in Figure 12.3 when the amount of input used increases, more fish can be caught. In Figure 12.4 more input is represented by the fish production function rotating to the northwest to $f(x_1, S)$ where $x_1 > x_2$. Note that the same amount of fish can be caught for two different levels of effort. However, the increased input use rotates the production function to the northwest so that the fish production function intersects the biological growth function to the left of the stock associated with maximum sustained yield. This higher level of input use is inefficient since a reduced level of effort could yield either more fish or the same amount of fish.

Starting at maximum carrying capacity, K, Figure 12.4 shows that as more inputs are employed the total number of fish harvested increases until the stock

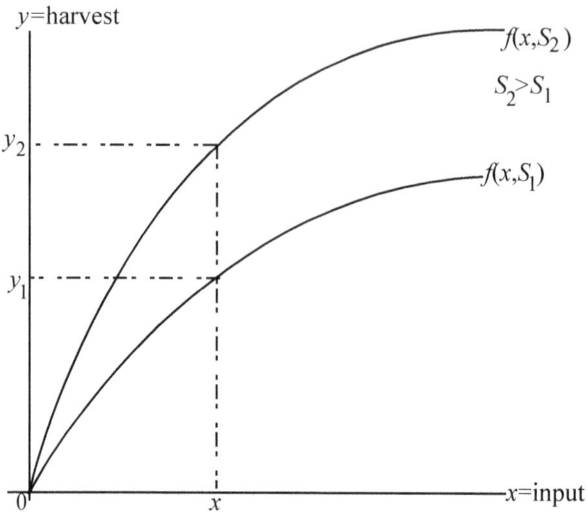

Figure 12.3: Fishery Production Function.

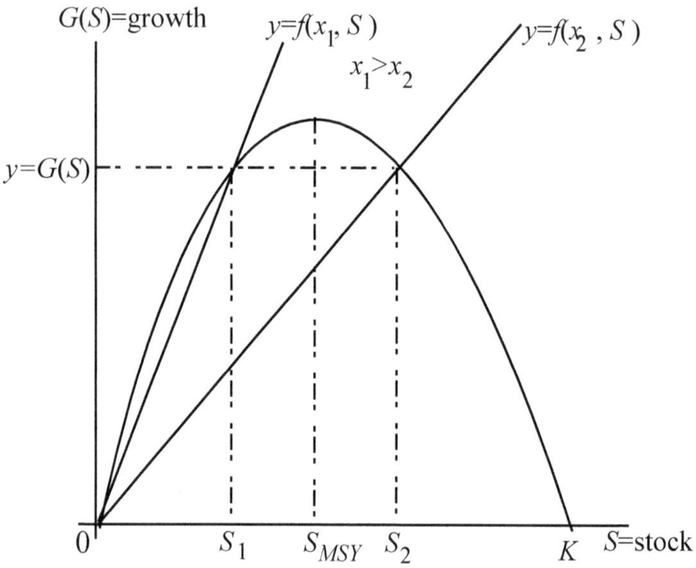

Figure 12.4: Equilibrium Fish Harvest.

has been depleted to equal that associated with maximum sustained yield, S_{MSY}. Similarly, as more inputs are employed total revenues, $TR = p \times y$, will first increase and then decrease. However, as more inputs are employed, total costs given by $TC = w \times x$ will always increase. Allocative efficiency occurs when the fishers choose a harvest that maximizes profits. This harvest occurs where marginal revenue per extra unit of input equals the marginal cost of an extra input: $MR = MC$. Marginal revenues are derived as $MR = \frac{\triangle TR}{\triangle x}$ and marginal costs are derived as $MC = \frac{\triangle TC}{\triangle x}$. Table 12.2 uses the logistic growth function from Table 12.1 to calculate total revenues and total costs assuming that the price of fish is $p = \$3$, the price of the input is $w = \$18.75$, and the biological growth parameter is $\lambda = 1$. Starting from maximum carrying capacity of $K = 300$, we assume that as the stock of fish decreases by 25, the amount of input required to catch fish equal to the growth of the stock increases by 1. For instance, when the stock is $S = 275$ it takes $x = 1$ to catch $y = 22.917$ fish. When the stock is $S = 250$ it takes $x = 2$ to catch $y = 41.667$ fish, and so forth.

Table 12.2: Fishery Revenue and Cost Functions

Input x	Harvest y	$TR =$ $\$3 \times y$	$TC =$ $\$18.75 \times x$	Profits= $TR - TC$	$AR =$ $\frac{TR}{x}$	$MR =$ $\frac{\triangle TR}{\triangle x}$	$MC =$ $\frac{\triangle TC}{\triangle x}$
0	0.000	0.00	0.00	0.00	x	x	x
1	22.917	68.75	18.75	50.00	68.75	68.75	18.75
2	41.667	125.00	37.50	87.50	62.50	56.25	18.75
3	56.250	168.75	56.25	112.50	56.25	43.75	18.75
4	66.667	200.00	75.00	125.00	50.00	31.25	18.75
5	72.917	218.45	93.75	124.70	43.69	18.45	18.75
6	75.000	225.00	112.50	112.50	37.50	6.55	18.75
7	72.917	218.75	131.25	87.50	31.25	−6.25	18.75
8	66.667	200.00	150.00	50.00	25.00	−18.75	18.75
9	56.250	168.75	168.75	0.00	18.75	−31.25	18.75
10	41.667	125.00	187.50	−62.50	12.50	−43.75	18.75
11	22.917	68.75	206.25	−137.50	6.25	−56.25	18.75
12	0.000	0.00	225.00	−225.00	0.00	−68.75	18.75

Profits are maximized when $x = 4$ units of input are employed and $y = 66.667$ fish are harvested. From Table 12.1 we know that such a harvest is associated with a fish stock equal to $S = 200$. However, we assumed that the fishery is perfectly competitive with free entry. Since profits of $\pi = 125$ are earned by the incumbent fishers, more firms and inputs will be attracted into the industry. Profits are driven to zero when $x = 9$ units of input are employed. In this case the harvest is $y = 56.25$ and the stock of fish falls to $S = 75$. Thus, the tragedy of the commons results in too many boats chasing too few fish.

The top panel of Figure 12.5 depicts the total revenue and total cost functions. The TR function has the same shape as the logistic growth function. The bottom panel depicts the average revenue, marginal revenue, and marginal cost functions. Average revenue equals total revenue per unit of input: $AR = \frac{TR}{x}$.

If each unit of input earns average revenue then additional inputs (boats and fishers) are attracted into the industry when average revenues exceed the marginal cost of hiring the inputs. The competitive equilibrium occurs where $AR = MC$ at x_c. In contrast, it is economically efficient for boats to enter (exit) the industry when the marginal revenue earned by another boat, MR, is greater (less) than the marginal costs. The optimal level of input use that maximizes profits is x^* and occurs where $MR = MC$.

Now, imagine there is either an advance in technology or a decrease in input prices. In these cases the total cost of production line rotates down to TC' and the marginal cost line shifts down to MC' as depicted in Figure 12.5. Lower costs cause profits to increase and under conditions of perfect competition more inputs are attracted into the industry. Given that the slope of the MR line is greater in absolute value than the slope of the AR line, the optimal amount of fishing effort will also increase, but by less than the amount that would occur under conditions of free entry.

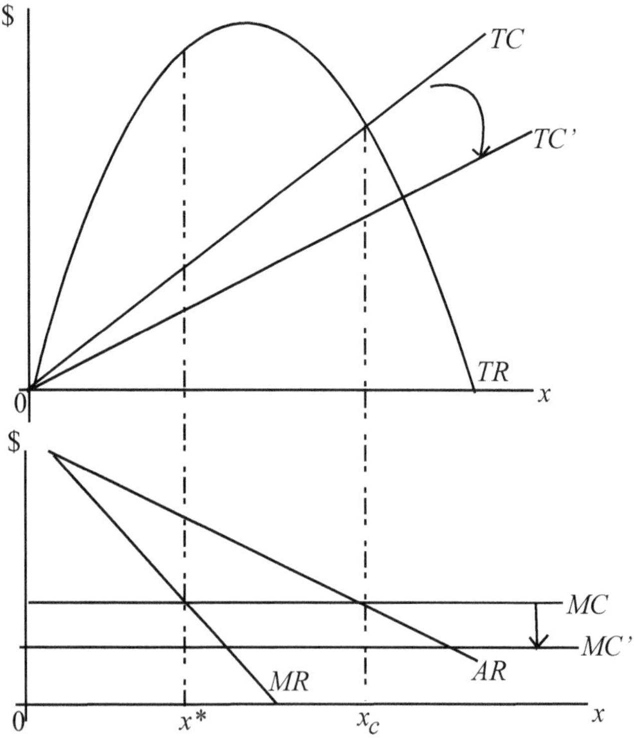

Figure 12.5: Competitive Equilibrium vs. Optimal Harvest Equilibrium.

Instead of changes in costs, changes in demand which affect prices and total revenues can also change fishing effort. If consumer preferences turn in favor of fish, the increased demand causes prices to rise. In this case the total revenue curve shown in Figure 12.5 shifts upward causing higher profits for the incumbents and attracting new entrants into the industry.

12.2.3 Aspects of Dynamic Bioeconomic Equilibrium

In the previous section we derived the static single period equilibrium. When harvests occur over multiple years the value of future harvests have to be discounted back to the present. If the goal of the fishery manager is to maximize the value of the fishery, what is the dynamic equilibrium condition? The value of the fishery equals the sum of the discounted profits over the life of the fishery. The condition for a bioeconomic equilibrium requires that the marginal benefits of delaying the harvest must be offset by the marginal costs. In this case, the marginal cost of delaying the harvest equals the lost interest income that could have been earned by reinvesting the marginal profits at the interest rate R. In Chapter 11 we learned that the Hotelling equilibrium condition for a nonrenewable resource is that the amount to extract corresponds to where the percent change in marginal profits just equals the interest rate. Let $M\pi$ represent marginal profit where $M\pi = MR - MC$. The Hotelling equilibrium for a nonrenewable resource is $\frac{M\pi_{t+1} - M\pi_t}{M\pi_t} = R$.

With a renewable resource the equilibrium condition has to be modified to account for the growth of the stock and for the effects that changes in the stock have on the marginal cost of harvesting. In Chapter 11 we learned that the present value of a perpetuity is equal to the perpetual cash flow (CF) divided by the interest rate (R): $PV = \frac{CF}{R}$. Suppose we have a fishery with capacity $K = 300$ and $\lambda = 1$ such as that depicted in Table 12.1 with $MSY = 75$. Assume that the stock of fish is in its pristine state with $K = 300$. If the price of fish is \$3 per unit and the interest rate is $R = 5\%$, what size of stock maximizes the value of the fishery?

To address this question we start by assuming that there is no cost to harvesting and then, later, relax that assumption. Clearly, if $K = 300$, no fish are harvested and the present value of the fishery is \$0. If 150 fish are harvested, the stock falls to that associated with maximum sustained yield: $MSY = 75$. In this case, the owner of the fishery nets $\$3 \times 150$ from the initial harvest of 150 fish and then earns $\$3 \times 75$ for ever after. The present value of the fishery is $PV = \$3 \times 150 + \frac{\$3 \times 75}{0.05} = \$4,950$.

Could harvesting more than 150 fish and then settling for a lower yield result in even greater value? When there are no harvesting costs the equilibrium condition requires that the marginal benefits of delaying the harvest equal the marginal cost of delaying the harvest. If marginal profits of $M\pi$ can be earned by harvesting, then the marginal cost of delaying the harvest is $MC = R \times M\pi$ which equals the lost interest income from not earning the marginal profits. The marginal benefit of delaying the harvest is that the stock of fish increases which changes the growth of the stock $G(S)$.[1] We call the change in $G(S)$ due to a one unit change in the stock the marginal growth of the stock $G'(S)$ which equals the slope of the growth function. An approximation to the marginal growth of the stock is $G'(S) = \frac{\triangle G(S)}{\triangle S}$. Thus, the marginal benefits of delaying the harvest are $MB = M\pi \times G'(S)$.

Equating MB and MC yields $M\pi \times R = M\pi \times G'(S)$, which can be simplified to $R = G'(S)$. Given a positive interest rate the marginal growth of the stock must also be positive. A positive growth rate means that the stock is less than that associated with maximum sustained yield. For the logistic growth function given by Equation (12.1) the marginal growth of the stock is $G'(S) = \lambda - 2\lambda\frac{S}{K}$. With $R = 5\%$, $\lambda = 1$, and $K = 300$, the equilibrium stock is found by solving $0.05 = 1 - 2\frac{S}{300}$ for S: rearrangement yields $S = 142.5$. One can easily check that $G(S) = 74.815$ when $S = 142.5$. The sum of discounted profits from harvesting 157.5 in the first period and then harvesting 74.815 fish forever after results in a fishery with present value of $PV = \$3 \times 157.5 + \frac{\$3 \times 74.815}{0.05} = \$4,961.4$ which is greater than that associated with maximum sustained yield. In this formulation as the interest rate increases the equilibrium stock falls. One can easily check that if $R = 0.10$, the equilibrium stock falls to $S = 135$. In this example, any interest rate above 100% causes extinction in that the equilibrium stock falls to $S = 0$. With slower growing stocks, extinction occurs at lower interest rates.

When there are harvesting costs the equilibrium condition has to be modified. Since a smaller stock of fish increases the amount of input needed to harvest a given amount of fish, harvesting today increases the future marginal cost of production. This effect is called the marginal external cost of harvesting. Recall from Table 12.2 that increases in the number of fishers causes the harvest to increase but at a decreasing rate. Since each fisher is assumed to catch an average amount of fish, new fishers reduce the average caught by each of the incumbent fishers. In contrast, delaying the harvest increases the stock of fish and reduces the marginal external cost of harvesting.

Accounting for the external costs of harvest the equilibrium condition is modified to

$$M\pi \times R = M\pi \times G'(S) - MEC, \text{ which can be rearranged as}$$

$$R = G'(S) - \frac{MEC}{M\pi}. \tag{12.2}$$

Since a larger stock reduces the MEC, we have $MEC < 0$, which means that $-\frac{MEC}{M\pi} > 0$. If the reduction in the marginal external cost from delaying the harvest is large enough, it is possible for the marginal growth of the stock to be negative in equilibrium. Negative marginal growth in the stock is given as $G'(S) < 0$ and implies an equilibrium stock above that associated with maximum sustained yield. For example, let $R = 5\%$ and let the decline in

[1] The marginal growth of the stock is found by taking the derivative of $G(S)$ with respect to S: $G'(S) = \frac{\partial G(S)}{\partial S}$.

the marginal external costs as a percent of marginal profits be 7% when the stock increases by 1. Then, the dynamic equilibrium requires that $0.05 = G'(S) + 0.07$ which yields $G'(S) = -2\%$. However, the higher the interest rate, the greater the chance that the equilibrium stock will be less than that associated with maximum sustained yield. At an interest rate of $R = 10\%$ and a decline in marginal external costs of 7%, the equilibrium growth rate of the stock is $G'(S) = 3\%$, implying that the stock is less than that associated with maximum sustained yield. In an open access fishery incumbent fishers have a high discount rate since delaying the harvest attracts new entrants which causes marginal profits to fall toward zero. An infinite interest rate would imply fishers have no preference for future fish and harvest would occur until the marginal profits of the last fish caught are zero. This state of affairs is also the static equilibrium condition offered in the previous section.

12.3 Policies to Reduce Overfishing

Various policies have been employed to reduce overfishing such as taxes, gear restrictions, restrictions on the total allowable catch, seasons, and individual transferable quotas (ITQs).

As we learned in the previous section an open access fishery attracts new entrants until profits are dissipated. In industries where new entrants do not impose external costs on incumbent firms the increased competition is a good thing as it encourages innovation and lowers prices for consumers. However, in fisheries, new entrants who catch some of the stock of fish impose a negative externality on incumbents who now catch fewer fish. We examined negative externalities in Chapter 8 and one available policy option is to impose a tax equal to the marginal external cost of the action.

Consider the fishery that was first depicted in Table 12.2 with the table partially recreated in Table 12.3. Each boat catches its average share of the total harvest and earns an average amount of revenues from that catch. However, beyond the first boat, each additional boat reduces the average catch of the incumbent boats and imposes an external cost equal to the decline in average revenue multiplied by the total number of boats that were already fishing before the new boat arrived. For instance, when the second boat arrives the average catch of the two boats is $\frac{41.667}{2} = 20.83$ and average revenue is $AR = 20.83 \times \$3 = \$62.50$. Since the first boat had been earning \$68.75 before the second boat arrived, the marginal external cost is $MEC = (68.75 - 62.50) \times 1 = 6.25$. When the third boat arrives, the three boats' average catch is $\frac{56.25}{3} = 18.75$ fish, yielding average revenues of $AR = \$3 \times 18.75 = \$56.25$. The third boat reduces the average catch and average revenue of the first two boats and its marginal external cost is $MEC = (62.50 - 56.25) \times 2 = 12.50$. Earlier we showed that $x = 4$ maximized profits. Similarly, comparing average revenue with marginal social costs (MSC) shows that it is socially optimal for $x = 4$ boats to enter the fishery, but not the fifth boat since $AR = 43.69$ is less than $MSC = 43.75$. Imposing a tax equal to \$25 per boat would give boats beyond the fourth boat the incentive to seek opportunities elsewhere, since for these boats $AR < MC + tax$.

Table 12.3: Externalities in the Fishery

Input x	Harvest y	$TR = \$3 \times y$	$AR = \frac{TR}{x}$	$MEC = \triangle AR \times \Delta x$	$MC = \frac{\Delta TC}{\Delta x}$	$MSC = MEC + MC$
0	0.000	0.00	x	x	x	x
1	22.917	68.75	68.75	0.00	18.75	18.75
2	41.667	125.00	62.50	6.25	18.75	25.00
3	56.250	168.75	56.25	12.50	18.75	31.25
4	66.667	200.00	50.00	18.75	18.75	37.50
5	72.917	218.45	43.69	25.00	18.75	43.75
6	75.000	225.00	37.50	31.25	18.75	50.00
7	72.917	218.75	31.25	37.50	18.75	56.25
8	66.667	200.00	25.00	43.75	18.75	62.50
9	56.250	168.75	18.75	50.00	18.75	68.75
10	41.667	125.00	12.50	56.25	18.75	75.00
11	22.917	68.75	6.25	62.50	18.75	81.25
12	0.000	0.00	0.00	68.75	18.75	87.50

An example of a tax would be requiring fishers to pay a license fee before entry is granted into the fishery. In practice, determining the optimal size tax on fishing inputs can be difficult because fishers can substitute various kinds of nontaxed inputs for the taxed input. For instance, if each boat must have a license but the tax is independent of the size of the boat, a rational outcome would be to substitute large boats for small ones. If boats of different sizes have to pay different taxes, boat owners might switch to the use of other forms of capital such as global positioning systems, larger nets, and more labor. Taxing all of the various kinds of inputs can become unnecessarily complex and administratively costly. Alternatively, an excise tax per unit of fish landed could be imposed. Although this kind of tax might be simpler to impose, fishers would still have the incentive to search for cost-reducing innovations which would require the tax to be continuously updated.

A second policy alternative would be to impose gear restrictions. For instance, to limit the catch, boats might be restricted in size, the number of hooks per line might be limited, nets might be restricted to some maximum length, and minimal mesh sizes in nets might be required to allow smaller fish to escape. These gear restrictions tend to increase costs of production and would be expected to reduce fishing effort. However, when one type of gear is restricted, fishers can be expected to switch to other nonrestricted types of gear to increase their catch. Furthermore, if a single boat fishes for several different species with different types of gear required for each species, difficulties in monitoring fishing methods can limit the effectiveness of gear restrictions [116].

Third, fishery managers might impose a season or restrict the fishery to some total allowable catch. With seasons, fishers have the incentive to fish intensively while the season is open to maximize production during the window of opportunity. With a total allowable catch, a phenomenon known as **derby fishing** occurs with fishers racing to catch as many fish as possible before the total allowable catch is reached. Both seasons and catch limits result in large quantities of fish reaching the market in a short span of time. When the market is flooded with fish, prices and profits fall. Although fresh fish typically command higher prices than frozen fish, much of the fish supply ends up frozen or, worse, wasted. In addition, seasons and catch limits tend to increase danger and risk in an already dangerous occupation. Fishers have the incentive to go out in inclement weather to catch as many fish as possible before the season closes. Furthermore, seasons and catch limits give fishers the incentive to make inefficient investments in bigger boats equipped with the latest technology.

Individual transferable quotas (ITQs) combine private property rights with a total allowable catch. Each fisher is either assigned or bids for some catch share with the total catch shares summing to one. At the beginning of the season fishery biologists determine the total allowable catch or quota to be allowed. Those without a share cannot fish. Those with a share can either engage in fishing or sell their catch share to another fisher. Those fishers who own catch shares but have high costs have the incentive to sell (transfer) their shares to fishers with lower costs. If there are economies of scale in catching a particular species the ITQs will be acquired by fishers who can take advantage of those scale economies. On the other hand, if there are diseconomies of scale, then numerous small fishers will each acquire small shares in the total allowable catch. Since fishers acquire property rights to a share of the fish, those same fishers have the incentive to report violators since those violators raise costs and reduce profits for those who own shares. In addition, ITQs allow the harvest to be spread throughout the year resulting in both higher profits and greater consumer surplus as more fresh fish are available in place of frozen fish.

New Zealand was an early adopter of ITQs and their use spread to other countries in the 1980s. However, only about 2% of global fish stocks are governed by ITQs. Recent research examining more than 11,500 fish species found that when ITQs were adapted, the probability of overexploitation fell by half[54, 53]. One fishery that has recovered is the Pacific Halibut depicted in Figure 12.6. In 1991, Canada adopted ITQs for halibut and in 1995, the United States adopted ITQs. From 1975 to 1995 the halibut fishery was subject to seasonal closings. In 1975, the season lasted 125 days. By 1980 the season had fallen to 25 days and by 1994 the season lasted only 2 days. During that twenty-year span the total allowable catch was exceeded in two out of three years. With ITQs, the number of fishing vessels decreased by 10%, by-catch of other species decreased by 80%, there was a 77% decrease in ghost-fishing where abandoned or lost nets take fish, and the value of the harvest increased by 34% [226].

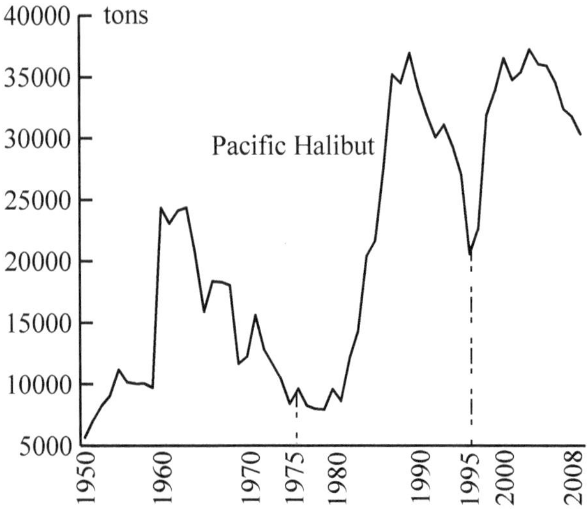

Figure 12.6: Landings of Pacific Halibut (tons), 1950–2008 (Source: FAO FISHSTAT Plus.)

12.4 Two Examples of Overexploitation: Whales and Buffalo

The growth function and bioeconomic equilibrium conditions for an open access fishery can also be applied to other renewable resources. In this section we describe the economic conditions that contributed to the overexploitation of two mammals: whales and buffalo.

12.4.1 Whales

Robert Allen and Ian Keay [7] performed a simulation study examining the plight of whales from the seventeenth century to 1911. In 1596 William Barrents discovered a whaling grounds near the Greenland island named by Barrents as Spitzbergen. During the seventeenth century the Eastern Arctic bowhead whale (*Balaena Mysticetus*) was hunted from the shores and then at sea near Spitzbergen and later the Davis Straits. In 1602 the Dutch granted the Noordsche Compaigne a monopoly charter to hunt whales, but later revoked the charter and encouraged entry into the industry. To encourage their own whaling fleet they imposed a 4% import duty and outlawed the use of foreign sailors to increase employment opportunities for their own citizens. Although initially monopolistic, the industrial organization soon approximated perfect competition and attracted British, German, and Basque whaling fleets.

In the early spring whaling ships from the Dutch ports of Rotterdam, Ryp, and Amsterdam set out for Greenland. Initially, whaling operations were land-based. Spotters took positions in tall, land-based towers and when whales were spotted, longboat crews were sent out to harpoon the whales. The harpoons were attached to ropes that could reach depths of 300 fathoms. After the whale was killed, the tail was cut off and it was hauled to shore for processing. Two primary products—baleen and blubber—were obtained from the whale. The blubber was processed and turned into whale oil which was used as a lubricant and for luminescence. The baleen came from the cartilage and bone in the head of the whale and was used to make springs [7].

The open-access nature of the industry and initially high profits increased the number of whalers and caused whale stocks to fall. In addition, both the British and the Dutch governments were impressed by the sailing skills of the whalers and began to offer a subsidy to whaling ships to train sailors. This subsidy caused costs to fall and contributed to declining whale stocks. As whaling stocks declined, shore-based operations were abandoned in favor of whaling vessels from which whales were spotted from tall masts and longboats were launched. At first, blubber was stored in barrels and taken to land for processing. However, various innovations allowed the blubber to be processed at sea, reducing travel costs.

When profits fall due to free entry, the competitive model assumes that resources will seek other opportunities that offer a higher expected return. However, whaling ships were outfitted with special protection against sea ice and the crew tended to have knowledge that was specific to whaling. Such industry-specific capital and knowledge made the capital and labor less mobile and caused the size of the fleet to remain large even in the face of declining profits. However, the whalers appeared to respond to economic incentives because as the Spitzbergen stock of whales fell, the whaling ships diversified and began to hunt whales in the Davis Straits.

The conditions of open-access, subsidies, technological progress, and industry-specific knowledge and capital all contributed to declining whale stocks. "Today there is no identifiable Greenland-Spitzbergen stock of Bow-headwhales, and only a small number of individuals remain in the Davis Strait stock [7, p. 401]."

12.4.2 Buffalo

Prior to 1870, the size of the buffalo herd on the high plains of the United States was estimated to be between eight and ten million [100]. Initially, the westward advance of settlers and their hunting of buffalo exterminated the relatively sparse herds of Kentucky. Later, habitat destruction caused by agricultural expansion pushed herds out of the tall-grass prairies further west. As pioneers pushed further west, meat hunting helped reduce herds, but most meat was consumed locally. If buffalo could have been shipped live, its demand as meat would have helped to reduce herd size. However, the obvious difficulties in capturing and then shipping live buffalo kept that from happening. Furthermore, slaughtering buffalo and then shipping the meat was prohibitively expensive and a large market for buffalo meat never developed. Instead, most buffalo were hunted for their winter robes, locally consumed meat, and tongues. Most observers thought that the rate of natural increase would sufficiently offset hunting pressures so that the herd would decline gradually. It is remarkable then, that the buffalo was largely exterminated during the short span of 1871-1883.

What caused the buffalo's demise? Some researchers have argued that the high cost of fencing large areas made it difficult to impossible to establish private property rights, resulting in a tragedy of the commons. An alternative theory argues that an innovation in tanning the hides of buffalo created a new market for buffalo hides. At first, tanners in the United States were unable to turn the buffalo hides into a useful product. However, when several bales of hides were shipped to Europe, successful experimentation by English and German tanners allowed the hides to be turned into useful leather. This innovation increased the demand for "flint" hides which are the hairless hides taken from buffalo in nonwinter months. According to Taylor [233] a good hunter could earn between \$2 and \$3 per hide and the hunter could kill several thousand per season. Compared with cowboy wages of \$50 per month there were huge profits to be earned in buffalo hunting. Furthermore, cow hides were a close substitute for buffalo hides resulting in a very elastic demand for hides. When demand is elastic, larger increases in supply cause only small decreases in price, so that the profit opportunities from hunting buffalo did not appreciably decline as more hunters entered the market. Finally, the penetration of railroads into the high plains allowed cheap transportation of the hides to market.

Westward expansion and habitat destruction would probably have eventually caused the buffalo's extermination. However, the tanning innovation increased the demand for buffalo hides and railroad penetration into the high plains helped reduce transportation costs. Cattle hides were a close substitute for buffalo hides and the increased supply of buffalo hides resulted in only a small downward effect on hide prices. Given the open access conditions in hunting buffalo and the lack of regulations and license requirements to take buffalo, the herd was decimated. By the late 1880s there were fewer than 100 animals remaining [233].

12.5 Forest Resources

From 1630 to 1900 the amount of forest land in the United States declined from 1.05 billion acres to 748 million acres. Much of the decline was caused by the conversion of forest lands to agriculture, primarily in the east. However, from 1900 to 2002 the amount of forested land in the United States has remained within plus or minus 5% of 745 million acres [223]. Forests provide important ecosystem services; forests serve as a carbon sink, help prevent soil erosion, and provide habitat for diverse species. In addition, forests provide timber for houses, furniture, firewood, and a host of other products. In closed economies, where the demand for forest products is local, forest cover tends to increase as income increases [81]. In addition, conservation-based policies that try to reduce the demand for forest products like recycling paper might actually reduce forest cover. Since 99% of US timber production comes from private forests, any reduction in the demand for paper might cause forest owners to cut down the trees and plant corn or build a subdivision instead [86].

Forest managers face several problems in deciding when and how often to harvest a stand of trees. In addition, social efficiency requires that we account for any jointly produced by-products of harvesting such as greater soil erosion, reduced recreational benefits, and fewer ecosystem services provided by the forest. Like fisheries, the harvest decision depends on the interest rate and the growth rate of the stand of timber. In turn, the growth rate of the timber depends on its age.

We begin with a simple problem of when to harvest a stand of trees all of the same age. These trees might have been replanted after a clear-cutting operation or they might have been established after a forest fire that killed all existing trees. In general, the volume of timber within a stand will grow at an increasing rate during its early years and then at some point grow more slowly. As the forest reaches maturity it is possible that the volume of timber will decline as some trees succumb to disease, lightning strikes, and insects.

Table 12.4 depicts a hypothetical relation between the volume of trees (y) in a stand and the age of the stand (t). The first two columns of Table 12.4 represent the production function for tree volume over time, where soil quality, sunlight, and water are the fixed inputs. The third column is called the annual increment which is the marginal product from allowing the trees to grow another year: $MP = \frac{\triangle y}{\triangle t}$. The fourth column is the mean annual increment (MAI) which equals total volume per age: MAI=$\frac{y}{t}$. If the forester is only going to harvest the trees once then the volume to be harvested is maximized by harvesting at $t = 110$. If instead, the forester is going to harvest and replant

Table 12.4: Forest Volume and Profits

Age t	Volume y	MP $\frac{\triangle y}{\triangle t}$	MAI $\frac{y}{t}$	growth rate $\frac{MP}{(y_t+y_{t-10})/2}$	Net Rev. $\$2 \times y$	PV of π $R=1\%$	PV of π $R=3\%$
0	0	x	x	x	0	-1,000	-1,000
10	1,000	100	100	0.200	2,000	811	488
20	5,000	400	250	0.133	10,000	7,195	4,537
30	12,000	700	400	0.082	24,000	16,806	8,888
40	21,000	900	525	0.055	42,000	27,209	11,875
50	32,000	1,100	640	0.042	64,000	37,914	13,599
60	42,000	1,000	700	0.027	85,000	45,238	13,258
70	50,000	800	714	0.017	100,000	48,831	11,630
80	56,000	600	700	0.011	112,000	49,525	9,525
90	60,000	400	667	0.007	120,000	48,007	7,391
100	63,000	300	630	0.005	126,000	45,584	5,556
110	64,000	100	582	0.002	128,000	41,841	3,956
120	64,000	0	533	0.000	128,000	37,783	2,688

continuously and we are interested in maximizing biological volume, then it would make sense to maximize MAI which occurs at $t = 70$. Note, if the trees were allowed to grow for $t = 110$ years the volume of trees harvested after two rotations (220 years) would be $y = 128,000$, but if harvesting and replanting occurs every 70 years, the volume harvested after only 210 years is $y = 150,000$. The fifth column reports the growth rate of timber: $g = \frac{MP}{(y_t+y_{t-10})/2}$, where we use the average volume between two periods in the denominator of the growth rate.

While maximizing MAI is intuitive it ignores the profits that can be earned by harvesting and the time value of money. The sixth column of Table 12.4 reports net harvest revenues. Net harvest revenues equal the difference between price (p) and unit cost (c) from harvesting multiplied by the volume harvested, y. Let the price/cost margin for timber be $(p - c) = \$2$ with net revenues $= (p - c)y$. To obtain profits we have to subtract the planting costs from net revenues. Planting costs are incurred at $t = 0$ and equal \$1000. The last two columns report the present value of profits from harvesting the timber:

$$PV \text{ of } \pi = \frac{\text{net revenues}}{(1+R)^t} - \text{ planting costs.} \tag{12.3}$$

When $R = 1\%$ harvesting the trees at $t = 80$ maximizes profits. When $R = 3\%$ harvesting the trees at $t = 50$ years maximizes profits. We can see that harvest occurs earlier as the interest rate rises and the harvester becomes more impatient.

The economics behind when to harvest arises from recognizing the costs and benefits of delaying the harvest. If the harvester delays the harvest they gain profits from the extra tree growth. These marginal profits are $M\pi = (p - c) \times MP$, where $MP = \frac{\triangle y}{\triangle t}$ is the extra tree volume from allowing the trees to grow another year. However, those extra profits from tree growth have to be balanced against the lost interest income that the forester could have earned if they harvested the trees and invested the proceeds at R. This lost interest income is the marginal cost of waiting and equals $MC = R(p - c)y$. The optimal time to harvest occurs when the marginal benefits from delaying the harvest just offset the marginal costs:

$$\begin{aligned} M\pi &= MC \\ (p - c) \times MP &= R(p - c) \times y \\ \frac{MP}{y} &= R. \end{aligned} \tag{12.4}$$

Equation (12.4) tells us that if the growth rate of the trees is greater than the interest rate the forester can increase profits by delaying the harvest. Referring back to Table 12.4 and comparing the interest rate with the tree growth rate (column 5) we can see that when $R = 1\%$ the time to harvest is in $t = 80$ and that corresponds to maximum profits. When $R = 3\%$ the optimal time to harvest is at $t = 50$ years.

Now, consider the case where the forester has just replanted and is going allow the trees to grow and then harvest and replant forever. What is the optimal rotation length? At the end of each rotation the forester would receive profits of $\pi = (p - c)y - F$, where F is the cost of replanting at the beginning of the next rotation. Again, the forester wants to choose when to harvest and replant by balancing the marginal benefits of delaying the harvest and allowing the trees to grow another period with the marginal costs of waiting. Before, when the trees were to be harvested only once, the only waiting cost was the lost interest income. Now, however, when the harvest is delayed the forester loses the interest income from the current volume of trees in the stand and and also loses the interest income from all future harvests. The present value of the lost interest income from all future harvests equals the price of the land. Thus, the opportunity cost of the land equals RL, where L is the price of the land. For example, if $R = 5\%$ and the piece of land sells for $L = \$100,000$, the forester loses an additional \$5,000 in interest income from delaying the harvest. Since the marginal cost of delaying the harvest increases, the optimal rotation will tend to be shorter than when the trees are harvested only a single time.

In addition to the value of the timber in the forest, there are other non-timber values that should also be accounted for. Old growth forests provide wildlife habitat for birds which nest in tree cavities, such as owls and woodpeckers. Forests also provide recreation services to hikers, campers, fishers, and hunters. In addition, forests help maintain aquatic ecosystems by slowing the runoff of snowmelt and rain. Forests can also serve as a carbon sink for greenhouse gas emissions. However, some species of wildlife such as deer prefer open areas as opposed to a closed canopy forest. Accounting for these nonmarket benefits can change the optimal rotation, but whether the rotation lengthens or shortens must be determined by the value of the nontimber uses. In general, though, if the amenity values associated with forests with younger trees are greater, the rotation will shorten. On the other hand, if the amenity values associated with older forests are greater, the rotation will lengthen [11].

12.6 Problems

1. The herring fishery off the coast of New England has a logistic biological growth function where $\lambda = 0.80$ and $K = 100,000$ tons.
   (a) What stock is consistent with maximum sustained yield? What will be the maximum sustained yield $G(S_{MSY})$?
   (b) Given a stock size of $S = 40,000$ tons calculate $G(S)$.
   (c) Assume no harvesting costs and a herring fishery that is in its pristine state. If $R = 0.12$ and the price of herring is $p =$ \$1,500/ton, what is the present value of the fishery if the initial harvest decreases the stock from K down to the stock associated with maximum sustained yield, and then future harvests equal the maximum sustained yield?
   (d) Assuming no harvesting costs and starting from a pristine stock, what is the equilibrium stock, S, and equilibrium harvest, $y = G(S)$, that maximizes the present value of the fishery? What is that present value assuming $p =$ \$1500/ton?
   (e) Assume that $R = 0.15$ and a pristine stock. Recalculate the bioeconomic equilibrium stock, equilibrium harvest, and present value of the fishery.
2. An alternative to the logistic biological growth function is the depensated growth function. With this function, growth of the biomass is negative below some critical amount called the minimum viable population which we label K_{min}. Let K represent the maximum carrying capacity and let λ represent the growth rate of the stock at the minimum viable population. The depensated growth function takes the form: $G(S) = \lambda S(\frac{S}{K_{min}} - 1)(1 - \frac{S}{K})$. Suppose we have a fishery with the following parameters: $\lambda = 0.5$, $K_{min} = 10$, and $K = 100$.
   (a) Use EXCEL to produce a spreadsheet of S and $G(S)$ for the depensated growth function. Let S range from $S = 0$ to $S = 100$ in increments of 1 and then calculate $G(S)$ for each of those stock values.

(b) What is the growth of the stock at stock values less than the minimum viable population K_{min}?

(c) What stock yields the maximum sustained yield?

(d) Approximate the marginal growth in the stock as $G'(S) \approx \frac{\triangle G(S)}{\triangle S}$. Assume no harvest costs and $R = 0.13$. What stock maximizes the value of the fishery?

3. The following table illustrates the production function for a fishery. The production process uses a boat with accompanying crew (x) to produce pounds of fish (y). The current price of fish is \$5/lb. The cost of a boat and crew is \$30.

x	y	TR	AR	MR	$\pi =$ Profits	MEC	MSC
0	0	0	x	x	0	x	x
1	17					x	
2	32						
3	45						
4	56						
5	65						
6	72						
7	77						
8	80						
9	81						
10	80						
11	77						
12	72						
13	65						
14	56						
15	45						

(a) Calculate total revenues ($TR = p \times y$), average revenue per boat ($AR = TR/x$), marginal revenue of an extra boat ($MR = \frac{\triangle TR}{\triangle x}$), and profits ($\pi = TR - TC$).

(b) If a boat captain assumes that he or she will receive the average revenue per boat if they add a boat onto the fishery, how many boats will we observe if each boat captain ignores the effects of their fishing on others' catch?

(c) A fishery can be thought of as a common property resource. In an open access fishery with no private property rights to fish each additional boat imposes a negative externality on other boats in that each additional boat reduces the average revenue of previous boats. The marginal external cost (MEC) of an additional boat equals the decline in AR multiplied by the number of boats. Complete the column for the MEC of each additional boat.

(d) The private marginal cost of another boat on the fishery is $MC = \$30$. Social efficiency requires that whenever $AR > MC + MEC$ more boats should be allowed to fish, but when $AR < MC + MEC$ fewer boats should be allowed to fish. What are the optimal number of boats and the optimal pounds of fish to harvest from the fishery? How does your answer compare to the open access equilibrium? Does the optimal number of boats maximize profits?

(e) What policies or set of rules and regulations might help the fishery achieve social efficiency?

4. The table below depicts the production relation between the volume of timber (y) in a forest and the forest's age (t). There is no usable timber in the first thirty years.

t	y	$MP = \frac{\Delta y}{\Delta t}$	MAI	$g = \frac{\Delta y}{(y_t + y_{t-10})/2}$
0	0	x	x	x
30	20			
40	50			
50	100			
60	140			
70	175			
80	205			
90	225			
100	235			
110	240			
120	243			
130	245			

(a) Complete the columns for the marginal product of allowing the timber to grow an additional year (MP), the mean average increment (MAI), and the growth rate (g) of timber volume.

(b) If the goal of the forester is to choose a rotation that maximizes timber volume, when should the timber be harvested?

(c) If the interest rate the forester uses to discount the future is $R = 10\%$, when should the forest be harvested?

(d) If the interest rate is $R = 4.5\%$, when should the forest be harvested?

5. Paul B. harvests the fast-growing Wispruce tree with his partner Babe. The forest production function is given in the table below.

t	y	$MP = \frac{\Delta y}{\Delta t}$	$g = \frac{\Delta y}{(y_t + y_{t-1})/2}$	PV of π
0	0	x	x	x
1	100			
2	500			
3	1,500			
4	2,900			
5	4,200			
6	5,000			
7	5,500			
8	5,800			
9	6,050			
10	6,200			
11	6,300			
12	6,300			

(a) Complete the columns for MP and the growth rate of the forest, g.

(b) If the discount rate that Paul and Babe use is $R = 3\%$, at what t should they harvest the trees?

(c) Paul and Babe sell their trees at a price of $p = \$4$ and they can harvest the trees at a constant marginal cost of $c = \$1.5$. If the discount rate is $R = 3\%$ calculate discounted profits (PV of π) from the harvest at every t and determine when to harvest to maximize discounted profits.

(d) Compare your answers in parts (b) and (c).

6. Will the following changes increase or decrease fish stocks?

(a) Fishing boats acquire new technology, say geographic positioning systems.

(b) Lower fish prices.

(c) The owners of fishing boats receive an investment tax credit when they make new capital investments in their boats.

(d) Climate change causes the logistic growth parameter, λ, to fall.

13

Transaction Costs and Institutional Choice

The enforcement of property rights through private-order institutions reflects something deep and supremely important about British institutions in the eighteenth century. The culture of respectability and gentility helped solve the standard collective action problems that bedevil the production of public goods. The emergence of a plethora of networks, clubs, friendly societies, academies, and associations created a civil society, in which the private provision of public goods became a reality and created what might be called a *civil economy*.

Joel Mokyr [178, p. 381]

13.1 Introduction

Suppose a society has some available technology that allows inputs (x) to be transformed into jointly produced desirable and undesirable outputs (y, b). We represent the technology as

$$T = \text{the set of inputs } x \text{ that can produce the outputs } (y, b). \tag{13.1}$$

The desirable outputs include both private goods and public goods. Likewise, the undesirable outputs include both private bads and public bads. Assume that each individual has identical preferences over the desirable and undesirable outputs which can be represented by a utility function $u(y, b)$. What is the best way to solve the following problem?

$$\text{maximize } u(y, b) \text{ by choosing } (x, y, b) \text{ subject to } (x, y, b) \in T. \tag{13.2}$$

Various institutions can be chosen to allocate resources so as to maximize individual and societal well-being. One possibility is to assign private property rights and then let private market exchanges spontaneously arise. Another possibility is to allow a central planner to solve the problem. Such a central planner might be a dictator or might be elected by citizen voters. Hybrid institutional choices are also possible with individuals making some decisions privately while other decisions are made collectively. Although it is relatively easy to write down the equilibrium conditions for the problem above, all institutions have varying degrees of inefficiency. Depending on individual and societal preferences one institutional choice might be more appropriate than another at maximizing well-being. In this chapter we examine the transaction costs and inefficiencies of various institutions used to allocate resources.

13.2 Transaction Costs

Transaction costs are costs associated with defining and maintaining property rights and fostering an exchange system [5]. These costs are associated with the choice of institution used to allocate resources. Using either government or the market will entail transaction costs. In the absence of public goods or externalities, private markets tend to efficiently allocate resources when markets are competitive and all participants have perfect information about prices, quality, availability, etc. Since these conditions are seldom met the market mechanism results in less than complete efficiency. However, using government to allocate resources also entails transaction costs. To implement the conditions required to allocate resources efficiently when there are public goods or externalities, government officials need information about individual preferences and such information is seldom known. Furthermore, government officials are likely to be self-interested just like other individuals and the pursuit of their own self-interest might not always be in society's interest.

13.2.1 Transaction Costs of Using the Market

The transaction costs of using the market include the costs of defining private property rights, obtaining a medium of exchange (money), assuring perfect information, and maintaining competition. These duties are sometimes assigned to government. For instance, the Federal Reserve System tries to control the money supply to promote full employment and price level stability or at least a level of inflation that can be anticipated correctly by all economic agents. Government also helps define and protect private property rights through our court system and through government services such as police protection, fire protection, and national defense. The Department of Justice oversees mergers and acquisitions between large corporations to try to promote and maintain competition. Other government agencies such as the Food and Drug Administration and the Department of Agriculture are involved in the production of information by testing new drugs and inspecting food supplies to assure minimum standards of quality. These transaction costs are part of the costs of using the market to allocate goods and services.

Private individuals and firms are also involved in producing information and defining and protecting private property rights. Ranchers in the American

West invested in brands for cattle and horses to define property rights to their livestock. Barbed wire was an innovation that allowed ranchers to privatize the open range. Today, television broadcasters make their programs available via satellite but scramble their signals. Movies that are sold as DVDs are encrypted to prevent copying. Information is also privately produced as firms such as Underwriters Laboratory, Good Housekeeping, and Consumer Reports test products, give seals of approval, and provide information to consumers about the quality of products. Buyers on eBay leave feedback for potential customers regarding the quality of service they received. Industry trade associations provide information to firms about how to treat waste, and, more importantly, how to reduce waste [127] which adds to profitability.

Although the Hollywood caricature of business people is one of mistrust and greed, private owners have a lot to lose (at least in competitive markets) when they engage in price gouging or offer shoddy or dangerous merchandise. Moreover, firms that pollute see their stock prices decline when that information becomes public [127, 130]. For instance, the first Toxics Release Inventory became publicly available in 1989. For each additional chemical that a firm reported as being released, the firm's stock value dropped by $236 thousand [97]. As another example, on April 20, 2010 British Petroleum's stock price was $60.48. That day, the Deepwater Horizon platform exploded in the Gulf of Mexico resulting in a large oil spill. On June 25, 2010 BP's stock price had fallen to $27.02. When the well was capped on July 15, 2010 BP's stock price had recovered to $38.92, but with 3.131 billion shares outstanding the aggregate loss to the firm's value totaled more than $67.5 billion. Even without civil and criminal penalties most firms take actions to prevent those kinds of losses.

Part of the transaction costs of using the market are derived from the fact that most trades take place between strangers. Private owners frequently undertake activities to foster trust and send signals to consumers about their reputation by investing in physical capital that has little value other than to convey to consumers that they have a lot to lose by misleading the public or offering shoddy merchandise. In essence, the firm's reputation is in the hands of consumers and dishonest firms lose potential profits when they lose the business of repeat customers. Lee and McKenzie [139] argue that firms invest in things like logos, advertising campaigns touting "In business for three generations," and installing permanent fixtures to send a signal to consumers that they are in business for the long run. Such signals are usually lacking from a seller touting a Rolex watch from under their coat or from out of the back of a van.

13.2.2 Transaction Costs of Using Government

Now lets think about the government as the means of allocating resources for public goods, correcting externalities, and providing information. The transaction costs of using government to allocate resources for public goods include determining each individual's marginal willingness to pay so that the aggregate willingness to pay can be compared with marginal cost. On a similar note, when there are externalities government authorities have to determine the appropriate external costs and benefits so that social efficiency can be achieved.

In private competitive markets individuals "reveal" their willingness to pay for private goods by their actions. The price system aggregates the preferences of many disparate consumers and the production technologies of many different producers to reveal the relative scarcities of private goods and services. However, the existence of public goods or externalities means that prices frequently do not exist. To implement efficient policies government officials have to somehow elicit privately held information about inviduals' preferences for public goods and then find ways of financing the public expenditures. With the exception of the rare lump-sum tax or transfer scheme, the financing of public expenditures via taxes results in efficiency or deadweight losses when taxes on labor, capital, and consumption distort decisions to work, save, and consume. Furthermore, the administration of the tax system uses scarce resources as government has to monitor taxpayers and as taxpayers employ lawyers and accountants to reduce their tax burden. These are part of the transaction costs of using government to allocate resources.

Sometimes governments use democratic voting to determine the amount of resources to allocate for public goods or the amount of regulation or taxes to impose on private market participants to correct for externalities. Using the ballot box as a way of eliciting citizen preferences also entails transaction costs. Some voters might choose to be "rationally ignorant" since the private costs of acquiring information to vote for appropriate policies is high while the probability of influencing an election is small. Other voters might be well informed, but biased in the sense that they only accept as valid information that which confirms their prior stance on an issue. Furthermore, there is no guarantee that democratic voting, even among well-informed voters, will choose an allocation of resources consistent with Pareto efficiency.

13.3 Private Market Failure

Private markets can fail to efficiently allocate resources when public goods or externalities exist, when there is imperfect or asymmetric information between potential traders, or when increasing returns to scale leads to a monopoly. Increasing returns to scale exist when it is cheaper to build one large hydroelectric dam to generate electricity and run one set of electric lines to all consumers via a monopoly than to have competition among firms resulting in hydroelectric dams on numerous small rivers and streams and multiple sets of electric lines. Market exchanges can also fail to realize Pareto efficiency when economic agents have asymmetric information. If used car sellers know more about car quality than buyers, only "lemons" might end up being offered for sale. Similarly, if borrowers know more about potential investment projects than lenders, high interest rates can drive the borrowers for safer investment projects out of the market resulting in only "risky" borrowers seeking fund. Buyers and sellers can address these kinds of information problems by resorting to what is called two-part pricing of the product, as when used car sellers offer some cars at low prices but with no warranty and offer other cars at higher prices but with a warranty, but those solutions are part of the transaction costs of using the market.

The private competitive model generally assumes that when firms become inefficient with rising costs and declining quality, its customers will quit patronizing the firm (exit) and seek a better option from another firm. However, Albert Hirschman [109] in his book *Exit, Voice, and Loyalty* argues that "voice" is also an important mechanism for providing feedback to firms. Consumers use "voice" when they file complaints with the firm about product quality. Such consumers typically display at least some degree of loyalty to the firm as they choose to protest rather than exit. The use of exit and voice might be thought of as part of the transaction costs of using the market. The costs of using exit will tend to be low when there are a large number of competing firms and the costs of using voice will tend to be low when firms have monopoly power and exit is not a viable option.

Recent research in behavioral economics indicates that consumers and producers do not always exhibit rational behavior. One kind of behavioral regularity that we addressed in Chapter 11 was hyperbolic discounting. Another behavioral result shows that how a particular choice is framed can affect the decision maker's choice even when alternative choices are the same. For example, consider two pricing options for people who stay multiple nights at a resort: under option A, the people are told that their accomodations will cost \$82/night and if they are willing to forego fresh towels and linens each day they will receive a \$2 discount. Under option B, people are told that their accomodations will cost \$80/night and fresh towels and linen cost \$2 per day. The two choices are identical, but can elicit different responses when framed as a gain as in option A versus a loss as in option B. Thus, part of the transaction costs of using the market arises from decisionmakers who are less than fully rational resulting in unrealized gains from trade.

Finally, society might desire a more equitable distribution of income than is generated by private market transactions. Government can step in and theoretically generate an "equitable" distribution of resources via lump-sum taxes and transfer and then let private markets allocate resources. According to the second fundamental welfare theorem introduced in Chapter 1 the lump-sum redistribution and then private market allocation can generate a Pareto efficient outcome. Unfortunately, lump-sum taxes and subsidies are generally not available as policy tools and government must rely on taxes and subsidies that distort an individual's behavior to consume, save, invest, and work.

13.4 Nonmarket Failure

The use of government to allocate resources and promote equity can also entail various kinds of nonmarket failure in addition to the transaction costs associated with voting. Can specific supply/demand characteristics of nonmarket activities that give rise to nonmarket failure be identified [257]? First, it is oftentimes difficult to distinguish the outputs of nonmarket activities from the inputs used to produce those activities. Certainly the number of acres in a national park like Yellowstone is an input in creating recreational services, but, the number of acres in the park is also an output when part of a person's experience is being surrounded by several million acres of undeveloped land. Second, it is hard to monitor and distinguish output quality because individuals' feedback is harder to discern and slower to materialize in nonmarket situations. Since the public sector often is less competitive, the citizen/voters use of exit will be less and the use of voice will be greater. However, voicing complaints and marshalling a critical mass of dissatisfied voters takes time and its outcome is often less certain. In fact, small special interest groups can more easily avoid the free-rider problem than larger groups. When special interest groups are successful, the political process results in inefficiency.

Third, a lack of competition makes it hard to evaluate nonmarket production. To the extent that a government agency like the National Forest Service or Environmental Protection Agency has monopoly power it becomes more difficult to evaluate their performance without a comparative producer. Fourth, it is hard to evaluate performance in nonmarket situations without documents like income statements, which help provide a clearer picture of producer performance. Even in cases such as some national parks, where entrance fees and money costs of operations are available, the entrance fees likely do not represent the benefits to society and various cross-subsidies often makes the costs of operations opaque.

Because of the specific supply/demand characteristics of nonmarket activities, Charles Wolf [257] identifies four types of nonmarket failures, which are the counterpart of private market failures. These nonmarket failures include internalities and private goals, redundant and rising costs, derived externalities, and distributional inequities.

Internalities are "the goals that apply within nonmarket organizations to guide, regulate, and evaluate agency performance and the performance of agency personnel [258, p. 66]." Regulators are self-interested and frequently pursue their interests in ways which might conflict with society's goals. As a consequence, some bureaucrats might pursue larger and larger budgets or more personnel or the most sophisticated technology if those things enhance their own utility even when they do not enhance societal welfare. Thus, if the Forest Service receives part of their budget from timber sales or miles of road built we should probably expect the Forest Service to pursue the sale of timber and the building of roads. Other bureaucrats might pursue the acquisition and control of information so as to increase their influence and power. Although China has collected information on pollution, that information has not been released to the public for scrutiny. These private goals of bureaucrats are internalities. Furthermore, influencing the demand side of nonmarket activities is the increased public awareness of market type failures, which frequently rewards legislators who publicize those shortcomings and propose solutions that have long-run consequences that are different and sometimes counter to the short-term consequences. Such short-sighted behavior is endemic when legislators must face voters every two, or four, or six years.

Wolf also argues that a consequence of regulation and direct production of goods by nonmarket entities is that costs tend to rise over time with some costs redundant. These **redundant and rising costs** might arise from the difficulty of measuring output quality, or distinguishing outputs from inputs, or from infeasible social objectives, or from the lack of incentives to terminate inefficient plans. As an example, municipal recycling programs often require households to separate their waste stream into paper, plastic, aluminum, and glass even though machines can identify and separate materials more cheaply. Costs might also rise over time because of Baumol's disease [21]. Whereas many activities in the private sector experience rising labor productivity because of capital/labor substitution, public sector activities, such as the performing arts or fire-fighting crews in national forests, tend to be less amenable to such capital/labor substitution. Baumol's disease arises because public sector workers who can work at high wages in the private sector have to be compensated similarly to private sector workers even if they experience no gains in productivity. Although fire crews who battle fires in national forests can use bulldozers to help build fire lines, environmental concerns often curtail such capital use so that fighting fires is still a very labor intensive activity.

Derived externalities are the unintended side effects of the policy. For instance, the series of hydroelectric dams built on the Columbia and Snake Rivers to generate the electricity intensive aluminum used in planes during WWII caused Pacific salmon and steelhead runs to decline precipitously. As another example, President Eisenhower's Interstate Highway Act helped contribute to urban sprawl and increased demand for gasoline. Another example would be the Endangered Species Act which helped contribute to a "shoot, shovel, and shut-up" strategy among some private landowners.

Finally, **distributional inequities** can arise in nonmarket activities. Public enterprises tend to have monopoly power and often seek to expand their monopoly power via the political process. Those rent-seeking activities result in resources used to redistribute income from those without political power to those who have political power. Although redistributing income is not necessarily bad, it tends to cause resources to be shifted into rent-seeking activities which reduces the amount of resources available to produce goods and services.

13.5 Private Property Rights

Private property rights are part of the foundation of a capitalist/market economy. Richard Stroup [229] writes that private property rights must have the three Ds. First, rights to property must be defined. Second, property rights have to be defended and. third, property rights have to divestible. The use of barbed wire, fences, livestock brands, locks, encrypted software, and scrambled television signals are examples of how the private sector has defined property rights. Property rights are also defined by government via copyrights, property titles, and patents. Property rights also have to be enforced and defended against those who would take the property for their own use. Here, government plays an important role by enforcing and defending property rights via police and fire protection, national defense, and the court system. However, property rights are also enforced and defended in the private sector through neighborhood watch programs, gated communities, and private detectives. Elinor Ostrom [191] recognized the importance of users monitoring each other in their use of common property resources such as irrigation ditches, communal pastures, and fishing areas. Finally, for resources to have maximum value property rights have to be divestible; property owners must have the right to exchange, transfer, and mortgage their property. If marketable emissions' permits or ITQs are to have maximum value, trades have to be allowed between individuals and across time.

Private property rights provide several incentives that help promote efficient resource use. First, private property rights facilitate trade between individuals so that resources can be allocated toward their highest and best use. Even without production taking place, trade and exchange allow resources to be moved from those who place a low value on the resource to those who place a high value on the resource, resulting in positive gains to both parties. In this manner, private property rights give owners the incentive to use their property in ways that benefit others or else bear the costs of their actions.

Voluntary trade brings the interests of disparate individuals into line in a way that benefits both parties. Government allocation of resources often creates a source of conflict between winners and losers. As an example consider a policy proposal to create a park. The proposal passes at the ballot box 60% to 40% and the park is created with taxpayer money. Those who do not use the park or do not think they are getting their money's worth from the park must still pay. Now, consider a private developer who creates a park and charges admission. Those individuals who value the park pay the owner for the privilege of using the park and consider themselves better off. Those who do not use the park do not have to pay and are no worse off from its creation.

Second, private property rights provide owners with the incentive to conserve and protect resources. In Missouri, roads and highways are publicly owned and frequently littered with trash. When something is publicly owned, everyone shares the responsibility for its upkeep. If everyone is responsible, it might very well be that no one takes responsibility as people attempt to free ride on other's contributions. In contrast, individuals typically do not trash their own property since those actions would lower its value.

Third, private property rights provide owners the incentive to innovate since they can capture the profits or rents associated with their innovation through higher revenues or lower costs. For instance, the owners of mineral rights or oil pools undertake innovation to enhance recovery and increase profits. Furthermore, if there are well-defined property rights to clean air, then polluters, such as electric utilities or car drivers, must pay for the damage they cause in burning fossil fuels. Clearly, those bearing such costs will search for ways to reduce pollution so that they can lower their costs.

Fourth, the owners of private property also have the incentive to think about the future. Since secure property rights can be traded from one generation to another, current owners have the incentive to maximize the future stream of income that is generated by the property, not just the current income from the property.

North American Indian tribes chose the institution of property rights in a variety of instances to promote incentives. Charles Mann [163, p. 42] writes about the importance of property rights for Indians of the Massachusetts Alliance, circa 1600, in the settlement of Patuxet, the home of Tisquantum.

> Patuxet, like its neighboring settlements, was governed by a schem, who upheld the law, negotiated treaties, controlled foreign contacts, collected tribute, declared war, provided for widows and orphans, and allocated farmland whenever there were disputes over it. (Dawnlanders lived in a loose scatter, but they knew which family could use which land—"very exact and punctuall," Roger Williams, founder of Rhode Island colony, called Indian care for property lines.)

Mann further argues that as sixteenth century New England grew to more than 100,000 people, boundaries between groups of people became more formal. Brief and mild armed conflict between Indian groups became more frequent as a means of defending property rights.

The institutions of private property rights and markets have been useful in helping avoid societal collapse in the face of rapid climate change. Weiss and Bradley [247] and deMenocal [62] describe the effects of long-term drought on various societies. In particular, the Classic Maya collapse in the nineth century AD coincided with a period of long-term drought in Mesoamerica. Interestingly, Mann [163, p. 278] argues that the Mayan states had been "run by all-powerful monarchs who embodied the religion and monopolized trade." Although the wetter southern cities had the greatest endowment of natural resources they were the first to fall and fell the hardest. In contrast, the drier and less endowed northern cities actually prospered. Why? In reaction to the drought the people of Chichen Itza (a northern city) instituted "sweeping economic" changes. Mann writes

> In the new regime, economic power passed to a new class of people: merchants who exchanged salt, chocolate, and cotton from Chichen Itza for a host of goods from elsewhere in Mesoamerica. In previous centuries trade focused on symbolic goods that directly engaged the king, such as jewelry for the royal family. During the drought, something like markets emerged. Dahlins (a researcher) calculated that the evaporation pans outside a coastal satellite of Chichen Itza would have produced at least three thousand tons of salt for export every year; in return, the Maya acquired tons of obsidian for blades, semiprecious stones for jewelry, volcanic ash for tempering pottery, and, most important, maize. ... Chichen Itza apparently traded its way through drought.

Rather than a romantic notion of an American Indian environmental ethic Terry Anderson [8] argues that Indians responded to incentives given the presence or absence of private property rights. Fruit and nut trees, horses, and garden plots tended to be privately owned, usually by a family or clan rather than an individual, and were efficiently used. Although buffalo hunts were communally staged, hunters marked their arrows which provided them a claim in a portion of the meat.

Indians were also capable of overexploitation in the presence of a common property resource, such as the beaver in the Hudson Bay drainage. During the late 1600s the Hudson Bay Company established a series of trading posts in James Bay and competed with the French who were in the market around the Great Lakes Region. Indians, who had previously viewed beaver as a source of food and clothes, began to view beaver as a way of obtaining European goods. Ann Carlos and Frank Lewis [34] examined beaver prices, skins traded, and beaver populations for the period 1700–1763. In the far northwest part of Canada around the Fort Churchill post the Hudson Bay Company was able to exercise monopsony power in their purchases of skins. Being the only buyer of skins, the Hudson Bay Company paid below market prices and purchased fewer skins than would have been the case under competitive conditions. As a consequence of the low prices, beaver populations remained at levels near maximum sustained yield at Fort Churchill throughout the period 1700–1763. However, at the posts of York Factory and Fort Albany, which were closer to the French trading posts, the Hudson Bay Company responded to increased French competition by raising the price of beaver skins around 1740. Indians responded to the higher prices by increasing supply which caused beaver populations to fall. In addition, Carlos and Lewis [34] present evidence that shows that as beaver became more scarce due to overharvesting, Indians switched to trapping marten, which also experienced declines in population.

13.6 Water Rights

Water rights are an important exception to property rights schemes that satisfy the three Ds. Water rights became increasingly important with the opening of arid land to the west of the one hundredth meridian in the United States. Riparian rights typically assign a right to landowners whose property adjoins water, such as a lake or river. Riparian rights give owners access, the right to fish, the right to wharf out to deeper water, the right to alter or protect shorelines, the right to recreate, and the right to claim title to the bed on nonnavigable waterways. The riparian right to use the water cannot be transferred to another party without also transferring the land [89]. If demand for water by the group of landowners exceeds supply, the amount of water that can be used typically depends on the frontage owned by the landowner. Riparian rights tend to be inefficient since the people who hold land titles adjacent to the water (the riparian owner) might not have the most beneficial use of water.

In some states, such as Missouri, landowners maintain riparian rights, but if the waterway is navigable, other people can use the waterway and land up to the high water mark. In other states, such as Wyoming, landowners cannot impede navigable uses of the stream by the public, but they can limit access to the land surrounding and underneath the water so that users may not wade in the river.

Water rights can also be assigned by prior appropriation. That is, the first user of the water can claim the right to a given amount and becomes the senior appropriator based on the date of the original claim. Those who follow are junior appropriators. A water right can be both junior to those who acquire a prior appropriation and senior to those who acquire a later appropriation. The doctrine of prior appropriation became increasingly used due to the discovery of mineral resources such as gold and silver and the development of irrigation districts[192]. Gold mining typically used sluice boxes to separate gold from the mining residue. In California, water could be appropriated for beneficial uses and such uses allowed the transfer of water to gold mines via canals and ditches with the user claiming the right to divert a given quantity of water for a particular beneficial use. Irrigation districts allowed water to be transferred to arid farmland via irrigation ditches and canals. Users who can beneficially use all unappropriated water in a stream can claim an appropriation for that amount unless a state has a minimum flow restriction.

As initially defined, the doctrine of prior appropriation usually meant that the water had to be used or the right would be lost. Generally, the beneficial uses of water were for mining, agriculture, navigation, and power generation. Maintaining in-stream flows for recreational or wildlife purposes was not counted as a beneficial use and so when stream flows dropped, fishers and boaters found their use of water curtailed. Recently, states have recognized recreation as a beneficial use.

With water rights assigned by prior appropriation, all users do not have an equal share in the risk associated with natural events such as droughts. Senior rights' holders will generally be subject to less risk than junior users as junior appropriators are closed down in reverse order of priority [89].

Figure 13.1 illustrates a hypothetical river that can be diverted at various headgates labeled A, B, C, D, and E. Let the senior user be at A, followed by B, C, D, and, finally, the most junior user located at E. In a normal year, the flow at A's gate is 1000 cubic feet per second. Senior user A has an appropriation for 300 cubic feet per second. User B who is junior to A, but senior to users C, D, and E, has an appropriation for 200 cubic feet per second. User C with an appropriation of 200 cubic feet per second is junior to both A and B but senior to D and E. Users D and E have appropriations of 100 cubic per second and 50 cubic feet per second. In a normal year, all appropriations are met. If a drought occurs so that only 500 cubic feet of water are available, users E, then D, and then C lose their appropriation and the river runs dry after B's appropriation.

If the senior rights' holder is located at E, followed by junior users at D, C, B, and then A, in the drought year when only 500 cubic feet of water are available, users A and B will have to close their gates and let water pass to the downstream senior users.

The appropriation system has tended to foster waste since senior users may pad their appropriation for more water than they need for the specific beneficial use. For instance, senior user A might claim 300 cubic feet even though they only need 200 cubic feet. This excessive claim impacts the junior user at C in years when only 500 cubic feet are available, since user C loses their claim.

The doctrine of prior appropriation allows a market for water to develop. However, Libecap [144] notes that the water market is inefficient since agricultural to urban trades frequently take place at higher prices than agricultural to agricultural trades. For instance, in the Reno/Truckee basin of Nevada, the median price of an agricultural to urban trade was $17,685 per acre foot while the median price of an agricultural to agricultural sale had a median price of only $1500 per acre foot.

The trading process is further complicated by return flows. In Figure 13.1, suppose user A has the senior rights to 300 cubic feet per second and diverts that amount. When plants do not absorb all available water, some water appropriated by a user may return to the stream. Junior users often depend on the return flow, especially in drought years. Suppose the return flow is 150 cubic feet per second by user A and 50 cubic feet per second by user B. Then, in a drought year, with only 500 cubic feet per second flow, user C can receive their entire appropriation and if they have any return flow, user D might receive part of their appropriation. Now, consider a transfer of 300 cubic feet per second by senior user A to a user located downstream of C's

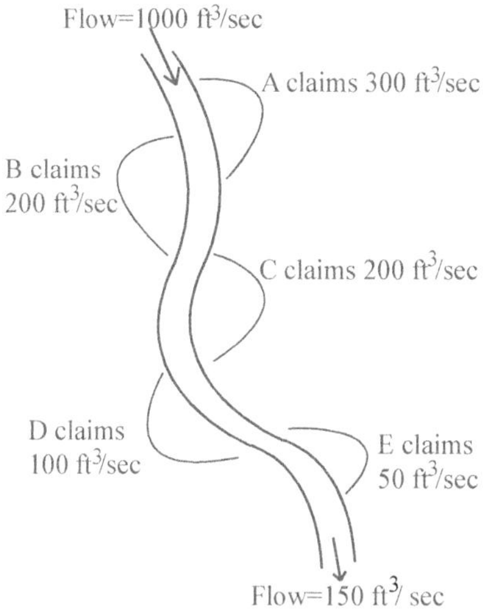

Figure 13.1: Water Diversions and Junior/Senior Water Rights.

gate. Such a transfer reduces the return flow that C had depended on during drought years and creates a negative externality that can be regulated by the state to ensure "no injury" due to the water transfer [89].

13.7 Institutional Choice

Because there are transaction costs of using either markets or government to allocate resources the efficient institutional choice will depend on those transaction costs and society's preferences between private and public goods. Private markets with competitive profit maximizing firms tend to be efficient at producing private goods, but less effective at producing public goods or when using resources for which private property rights are not well defined. However, private firms and other private organizations do provide some public goods through charitable donations of time, resources, and money. Sometimes, government will be more effective at producing public goods or at correcting externalities when private property rights are not well defined. In addition, government will sometimes be more effective than private firms when production entails economies of scale which give rise to a natural monopoly or when information is asymmetric between buyers and sellers. Other nonmarket institutions such as private clubs and associations, churches and charitable organizations, cooperatives, tribes, and families can also sometimes solve the free-rider problem and be effective at providing both private and public goods.

Figure 13.2 represents three alternative institutional forms for allocating resources: private markets, private nonmarket groups, and government. There are transaction costs associated with each institutional form. As well as the polar extremes located at each vertex, hybrid institutional forms can also exist. For instance, point S corresponds to a mix of private markets and government, say 40% market allocation and 60% government allocation. Point T corresponds to a mix of private markets and private nonmarket institutions with no reliance on government. Point U corresponds to a mix of government and private nonmarket institutions with no reliance on markets. Point R represents a hybrid of all three institutions, say 50% markets, 40% government, and 10% private nonmarket associations. Let I represent the triple corresponding to society's institutional choice, where $I =$ (private, government, nonmarket) are the shares of resources allocated by private markets, government, and nonmarket institutions. For instance, $I = (0.3, 0.5, 0.2)$ means 30% market allocation, 50% government allocation, and 20% private nonmarket allocation.

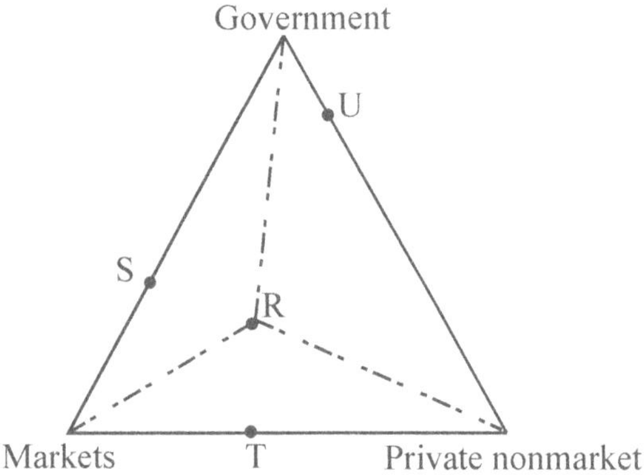

Figure 13.2: Institutional Forms.

Figure 13.3 depicts three production possibilities frontiers for three different institutional choices: $I_1 = (0.7, 0.2, 0.1)$, $I_2 = (0.4, 0.5, 0.1)$, and $I_3 = (0.2, 0.7, 0.1)$. Differences in institutional comparative advantage mean that some institutional choices will be better at producing one good than the other. If the citizens of a country have a high relative preference for private goods with the most preferred mix of private and public goods at point E, then I_1 would be the most appropriate mix of institutions, with some reliance on government institutions but most resources allocated by the market mechanism and private nonmarket associations. If instead, citizens have a high relative preference for public goods with their most preferred point at D then I_3 would be the most appropriate mix of institutions.

Suppose a country is currently at a point like C in Figure 13.3, but its most preferred level of public goods and private goods corresponds to a point such as D. In this case, the institutions have been chosen correctly but there

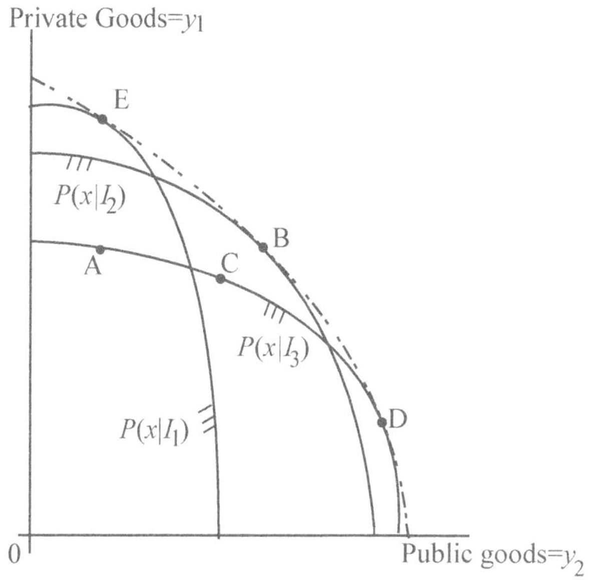

Figure 13.3: PPF for Different Institutional Choices.

is allocative inefficiency in that society prefers fewer private goods and more public goods than the current mix at C. Suppose instead the country is at point C but that the most preferred outcome corresponds to point B. At C the country has chosen not only the wrong mix of outputs to produce, but also the wrong mix of institutions. To get to a point such as B the country should reduce its reliance on government and increase its reliance on market and private nonmarket processes in order to increase the amounts of private goods and public goods. Finally, suppose the country has chosen I_3 and is at a point such as A. If the country's most preferred mix is at E, the country could expand private goods while holding public goods constant via less reliance on government allocation and more reliance on markets and private nonmarket institutions. For each possible combination of institutional choice there will be a separate production possibility set and frontier. The envelope of all the frontiers allowing institutional choice to vary is given by the dashed curve. One might think of the distance between a given institutional frontier and the most preferred point on the dashed frontier as part of the transaction cost inefficiency associated with a particular institutional choice.

Figure 13.4 depicts the production possibility sets for two institutional choices when undesirable outputs are jointly produced by-products of desirable output production. The set $P(x|I_1)$ might correspond to an institutional choice of $I_1 = (0.2, 0.65, 0.15)$ and results in very little desirable output production but also very little undesirable output production. The set $P(x|I_2)$ corresponds to a different institutional choice of $I_2 = (0.6, 0.3, 0.1)$. When the institutional choice corresponds to I_1, well-being is maximized at point A and when the insitutional choice corresponds to I_2 well-being is maximized at point B. Since point A lies on a higher indifference curve than point B the appropriate institutional choice is I_1. In this case society prefers a bundle with less desirable goods and less undesirable bads to a bundle with more desirable goods and more undesirable bads. However, a different set of preferences could easily mean that society prefers both more undesirable by-products and desirable outputs to a bundle with less of both outputs. We save this possibility for an end of chapter problem.

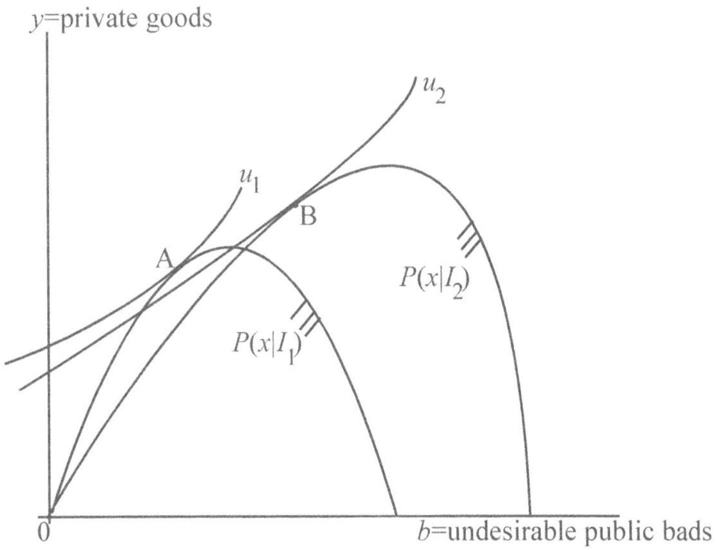

Figure 13.4: Institutional Choices with Desirable and Undesirable Outputs.

13.8 Summing Up

Neither private markets nor government are perfectly efficient at allocating resources. Private markets frequently fail to allocate resources efficiently, but that failure is no guarantee that government allocation will be better. If citizens can choose institutions then as preferences change and new problems arise citizens will make marginal adjustments in institutions. Even if individuals prefer a relatively large mix of private goods to public goods, the case of water rights showed that private property rights are not always designed in ways to promote efficiency. In addition, the break-up of the former Soviet Union revealed that the tragedy of the commons was all too common even in a centrally planned command economy. Today, China is experiencing urban air pollution that endangers the health of their citizens and the central government seems powerless, or at least unwilling, to address that problem.

13.9 Questions

1. Citizens in the state of Webtopia have a given amount of inputs (x) that include labor, land and natural resources, and capital. The mix of institutions the citizens use to allocate resources include private markets, government, and private nonmarket associations. The figure below depicts the production possibility sets for two different mixes of institutional choice: I_1 corresponds to 20% markets, 65% government, and 15% nonmarket associations and I_2 corresponds to 60% markets, 30% government, and 10% nonmarket associations. Citizens have a set of preferences such that if I_1 is chosen point X is their most preferred bundle and if I_2 is chosen point Y is their most preferred bundle. Citizens also prefer bundle Y to bundle X. Draw indifference curves onto the diagram that correspond to the preferences of the citizens.

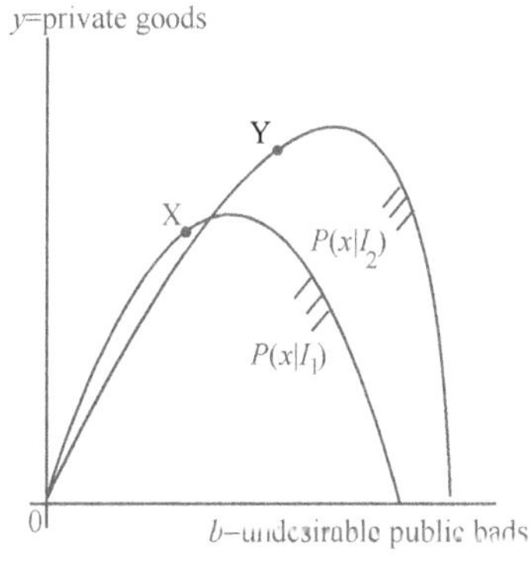

2. A standard view among economists is that the allocation of resources via private markets will result in too few resources allocated for public goods and too many resources allocated to the production of desirable outputs that yield jointly produced undesirable public bads. However, the quote by Joel Mokyr at the beginning of the chapter suggests that this reasoning might not always be true. What are examples of environmental amenities that private individuals or groups provide or disamenities that private individuals or groups help to reduce?

14

Retrospect and Future Prospects

The elite's environmental deterioration is often the common man's improved standard of living.

Anthony Downs [65, p. 44]

14.1 Retrospect

The Industrial Revolution ended thousands of years of very slow economic growth. The past 250 years have seen per capita incomes in the developed world grow by a factor of 15 to 23, and possibly more when the improved quality of goods and services is accounted for [184]. Furthermore, life expectancy increased from around 36 years to 76 years today. In spite of the phenomenal growth in the developed world, 2 billion people still live on less than $2 per day. Researchers have put forth many hypotheses for the spark that lit the Industrial Revolution including the scientific revolution, the enlightenment, greater competition between various religions and countries, and increased liberty and dignity for the bourgeoisie.

An important factor that is at least correlated, if not the causal effect of the Industrial Revolution is the switching of renewable energy sources—water, draught animals, wind, wood, and human power—for nonrenewable energy of coal and oil. As a consequence of the first two laws of thermodynamics the production of desirable goods always results in jointly produced undesirable polluting by-products. Burning fossil fuels to power economic growth increased the amount of unavailable energy (entropy) and waste in an irreversible process. The use of other resources, such as chorofluorocarbons as refrigerants and propellants, opened a hole in the ozone layer and increased the risk of skin cancer from ultraviolet radiation. The agreement to rapidly phase out CFCs was possible with the use of side-payments between countries.

When poor, the benefits that people gain from desirable outputs such as food, clothing, housing, and heat typically more than offset the costs of pollution. However, as individuals and societies become wealthier, undesirable byproducts garnish greater attention. The environmental Kuznets curve (EKC) hypothesis theorizes that pollution will first increase as incomes increase followed by declines in pollution as further economic growth occurs. The cause of the decline might be due to an income elasticity of demand for environmental amenities that is low at low incomes, but then rises as incomes increase. In addition, the mix of outputs might change as economic growth causes a shift in demand toward services that rely on human capital and away from the dirtier physical capital and energy intensive manufacturing sector. Rising incomes might also focus regulatory efforts on reducing pollution. Moreover, as firms are forced to comply with regulatory statutes they might realize economies of scale in abating pollution which allows them to abate increasing quantities at lower average cost.

Technological progress and resource substitutability have helped mitigate resource scarcity. At the start of the iron age, forests were denuded for fuel to make charcoal for smelting ore. The rapid decline in forests gave producers the incentive to search for subsititutes, such as coal. Innovation led to greater efficiency in steam engines which also opened up new uses for the steam engines in manufacturing and transportation-steam ships and locomotives. Paradoxically, greater efficiency in steam engines increased the demand for energy to power the engines in a phenomenon known as Jevons's Paradox.

Economic growth in the absence of private property rights has led to irreversible extinctions of certain kinds of wildlife such as the passenger pigeon and Seychelles giant tortoise. Other species, such as the American bison and beaver, recovered after experiencing near extinctions. Some species of birds and amphibians have been harmed by indiscriminate spraying of pesticides, notably DDT. However, those same pesticides also had benefits in reducing the spread of some diseases such as malaria, which still inflicts more than 200 million people worldwide.

In 1798, with the preceding thousand years as his evidence, Thomas Malthus posited that mankind's struggle for prosperity would be offset by increases in population. Today, the developed world has avoided the Malthusian trap and lives in prosperity. In addition, rather than increasing population, the trend toward greater freedom, rising incomes, and education—especially for women—have helped to reduce population growth to the replacement rate and below in developed countries. Rather than running out of food and resources, man's ingenuity helped create greater efficiencies and discover substitutes from resources that had previously gone unused.

14.2 Future Prospects

Economics is about the choices that individuals make when constrained by scarcity: limited resources that have alternative competing uses. Some of the many choices individuals and society have to make concern environmental resources: should we produce and consume more material goods and fewer wilderness areas, parks, and wildlife? Is it better to produce and consume more material goods and more jointly produced undesirable by-products such as pollution, or would we prosper by producing fewer material goods and less of the jointly produced undesirable by-products?

Money income is a biased measure of individual and societal well-being. When policy makers concentrate on growth in money income and ignore the effects that production and consumption have on the environment, present and future generations are subjected to a lower standard of living than what might have been realized. However, a goal of zero pollution, while not only unattainable, would also reduce the standard of living. By recognizing that resources have alternative competing uses, economics can help to frame environmental issues by analyzing the trade-offs facing society in the use of those resources.

Greenhouse gas emissions are increasing, but their effects on future well-being are uncertain. The EKC hypothesis posits that enough economic growth will eventually lead to declining levels of pollution. Empirical evidence has been found for the EKC hypothesis for pollutants such as sulfur dioxide, nitrogen oxide, carbon monoxide, carbon dioxide, biological oxygen demand, gasoline lead emissions, toxic waste sites, and hazardous waste and provides a case for cautious optimism regarding the future. However, although per capita carbon emissions are starting to fall in higher income countries, much of the world's population is still on the positively sloped portion of the environmental Kuznets curve and so global carbon emissions and the stock of carbon in the atmosphere will continue to rise as China, India, and other low income countries experience economic growth that more than offsets the decline in emissions in developed countries.

The fact that people in the developed world can worry about how our energy demands harm wildlife and impact the welfare of future generations is evidence of our abundant wealth. Unfortunately, the prospects of the 2 billion people who still live on $2 per day are not as bright. Living at or near the Malthusian subsistence level leaves people little inclination to care about the effects of global warming or wildlife which are more often viewed as a potential meal or pest, rather than an endangered species.

So far, attempts by countries to coordinate and reduce greenhouse gas emissions have run into the classic public goods' free-rider problem. Moreover, individual government policies designed to promote "green" alternative energy sources have encountered problems of their own. Because of subsidies, Germany, a country with solar irradiance equal to Seattle, has more installed solar energy capacity than any other country in the world. However, fluctuating solar energy generation has made it unprofitable to run base-load nuclear power plants which are costly to start and stop with solar's fluctuating daytime/nightime generation pattern. Instead of nuclear power plants, nighttime electricity demands are now met by coal-fired plants with the result that Germany's greenhouse gas emissions actually grew in 2012 [35].

In the wake of the Fukushima nuclear accident in Japan that occurred as a result of an earthquake followed by a tsunami, shutting down nuclear power plants seems to be a good idea. Although more than 18,000 people died in the tsunami, radiation leaking from the plant is expected to cause fewer than 200 deaths. In contrast, policies that reduce greenhouse gas emissions can be expected to reduce not only the effects of climate change, but also reduce premature deaths from inhalation of particulate matter. In a simulation, a policy imposing a \$50–\$380 per ton of carbon is expected to save between 300 to 700 thousand premature deaths in 2030, rising to 1.4 to 3 million by 2100 [252]. In 2012, nuclear power generated approximately 18% of total electricity while renewables generated about 5% [70]. Solar power accounted for less than 10% of total renewable electricity, so it is unlikely that solar power can substitute for nuclear power in the near future.

In the United States, subsidies for electric cars have helped increase sales, but under certain conditions electric cars and hybrid vehicles can also raise carbon emissions. One reason why this effect occurs is because the electricity used to charge the batteries can come from coal-fired electric utilities. Unless those utilities are subject to emissions taxes or must purchase permits to emit, their carbon emissions will increase offsetting declines in carbon emission from gasoline. Furthermore, by lowering the price of traveling one mile, electric cars increase the number of miles driven—the Jevons's Paradox [86]. Finally, the Corporate Average Fuel Efficiency (CAFE) program allows "manufacturers to count each all-electric vehicle they sell as two vehicles for computing the sales-weighted average emissions rate [146, p. 35]." It is an easy calculation to see that if 10,000 miles are driven, two conventional vehicles that get 40 miles per gallon will consume more gasoline than one hybrid vehicle that gets 80 miles per gallon.

In China, a government policy provided free winter heating for people living north of the Huai River. The subsequent increase in heating demand increased total suspended particulates and, according to one study, resulted in 2.5 billion lost life-years for the 500 million people affected by the policy [38]. While most countries in the developed world tax fossil fuel use, less developed countries tend to subsidize its use, often as a way to redistribute income. The subsidies are 80% of cost in Venezuela, 70% in Iran, 80% in Saudi Arabia, 18% in Russia, and 4.6% in China[115]. Phasing out such subsidies could reduce greenhouse gas emissions and have the added benefit of reducing the adverse health effects of air pollution. In addition, if consumers and producers in those countries were to face world market prices, incentives to innovate and improve energy efficiency would be enhanced. Finally, most economists think that subsidizing a particular commodity is an inefficient way of redistributing income compared to a guaranteed minimum income or cash transfer.

Once subsidies are in place it is difficult to get rid of them. Interest groups that benefit from the subsidy are vocal and insistent that the subsidy remain in place. Furthermore, when the costs of the subsidy are widely spread throughout the general populace each individual bears only a small portion of the costs, so it is often not worth the effort for an individual to lobby against a particular subsidy. Finally, government subsidies give incentives for new firms to enter the market to try and capture any rents associated with the subsidy.

A rational approach to promoting the general welfare in regard to climate change would be for government to either impose a cap on greenhouse gas emissions and then issue tradeable permits or impose a tax per unit of carbon equivalent emissions. A cap on greenhouse gas emissions coupled with tradeable permits would give producers the incentive to search for less carbon-intensive production techniques. Developing less carbon-intensive production techniques would reduce the producer's need to purchase permits or allow the producer to earn revenues from the sale of permits already owned. Consumers who wanted to purchase carbon-intensive products, such as gasoline, would be forced to pay higher prices that would be derived from the firms having to purchase permits to supply the product. Higher prices would give consumers the incentive to purchase less carbon-intensive products. Alternatively, if a tax on carbon-equivalent emissions were imposed each seller's marginal cost of production would rise to reflect the tax and would be passed on to consumers in the form of higher prices. Again, firms would have the incentive to innovate and develop production techniques that reduce carbon emissions and consumers would have the incentive to find substitutes for the carbon-intensive products.

From an economic standpoint one potential advantage of a carbon tax would be that the carbon tax revenues could be used to reduce other distorting and less equitable taxes, such as the payroll tax. In that case, the carbon tax would be revenue neutral and might help to allay fears of a government expansion. Furthermore, many taxes are already in place and the administrative costs for collecting carbon tax revenues should not be too burdensome. However, determining the correct per unit tax is problematic. Although the United States estimates the social cost of carbon to be \$36/ton, that estimate is only speculative: we simply do not know what the marginal damages of carbon will be on future generations. However, coupling an emissions tax to the tropical troposphere temperature would allow the tax to be adjusted up or down in response to changes in measured temperatures.

Capping emissions, issuing permits, and allowing permit trading is also economically feasible means of reducing greenhouse gas emissions and has some advantages over emission taxes. Given uncertainty in the marginal cost of abatement, a system of permits would have a smaller deadweight loss than taxes the steeper is the marginal benefit function from abatement. In addition, permits could be issued to firms rather than auctioned, so that firms would not experience windfall losses from having to initially purchase the permits and those firms would likely be more inclined to support the permit program. The success behind sulfur dioxide trading among electric utilities and the rapid phasing out of lead in gasoline due to permit trading offers some degree of optimism that cap and trade could be an effective program for reducing greenhouse gas emissions.

If emission taxes or permit trading are implemented as a policy to reduce greenhouse gas emissions they should be coupled with a quick phasing out of subsidies for alternative "renewable" energy sources. Subsidies for alternative energy sources are unnecessary and even detrimental in the quest for reduced energy consumption. The higher prices of carbon-based energy will be sufficient for firms to develop and consumers to purchase low-carbon alternatives. In fact, subsidies for renewable energy like solar or biofuels can actually harm the environment by keeping the price of permits low, thereby reducing the incentive for firms to develop low carbon technologies. Such a situation has happened in European countries where the subsidization of wind and solar power has caused the price of carbon permits to plummet, increased reliance on coal in Germany, and delayed the search for lower-carbon substitutes.

14.3 Conclusion

In 1970 the Environmental Protection Agency was started via an executive order by President Nixon. The later 1970s and early 1980s saw debate between environmentalists who wanted command and control type regulations for the environment on one hand and those advocating more market-based instruments of environmental policy, including Presidents Reagan and Bush. Market-based policy instruments have been effective in reducing sulfur dioxide emissions and lead emissions in gasoline via permit trading, traffic congestion in London via congestion charges, and overfishing via individual transferable quotas.

There are plenty of areas where market-based policy instruments could be successfully used. The drought in California that began in 2013 and continues into 2014 has pitted environmentalists who want to maintain in-stream flows for fishes and wildlife against agricultural and urban interests. Current water rationing schemes create conflict between the various interests. Instead, a market-based policy would raise the price of water. Similarly, if floods and hurricanes are expected to increase in frequency and magnitude it makes little sense to build higher levees. Instead, end the subsidies for homeowners and businesses that build in floodplains and coastal regions. Some environmentalists rightly bemoan the trophy houses built on large lots as "unfriendly" to the environment because the increased heating and cooling costs contribute to greenhouse gas emissions and the large lots contribute to urban sprawl. Attempts to make houses more energy efficient through the use of passive solar and Energy Star applicances ignores the even greater contribution that might occur if the federal government would end the tax deductibility of mortgage interest.

Climate change skeptics are likely wrong to think that successful opposition to a carbon tax or permit trading will result in business as usual. Instead, a rejection of cap and trade or emission taxes will likely lead to greater command

and control type regulations that are more detrimental to the economy and less effective for the environment. For instance, recent mandates that require a certain percentage of energy production to come from renewables will be more costly and less effectual than higher energy prices. The Energy Information Administration reports that US energy consumption in 2012 was approximately 100 quadrillion BTUs. It is better for 90% of our energy consumption in 2020 and beyond to come from fossil fuels if total energy consumption is only 90 quadrillion BTUs than it is for 75% of our energy consumption in 2020 to come from fossil fuels if our total energy consumption is 110 quadrillion BTUs. We are more likely to get to lower energy consumption if all energy prices, including the price of renewables, are higher.

People have long had a propensity to "truck, barter, and trade." The market process has helped enrich countries that have adopted it. Although some people decry the market for putting a price on everything, markets and prices have also reduced some of man's more baser instincts, especially violence and the subjugation of the many by the few. Private property rights are the foundation of a market economy. Environmental policies that promote property rights in air, land, water, wildlife, fish, and forests would allow markets to emerge and individual decision makers to learn about relative prices via the market system. Such knowledge would serve to enrich the current generation and enhance the environmental heritage for future generations.

Answers to Selected End of Chapter Questions

Chapter 1

5. An unintended consequence of a recycling law is that if it becomes more costly to legally throw things away, some people might engage in illegal dumping. In addition, since recyling takes time, it is inefficient for people who have a high time value to recycle.

8. Julian Simon wanted to bet on factors that directly affect human well-being. For example as sulfur dioxide increases in Asia, people are worse off, but increased economic activity that caused sulfur dioxide to increase makes people better off. The ecologists want to bet on trends in nature, but Simon is more concerned about trends in human welfare.

Chapter 2

2. $CS = (35 - 20) \times 850,000 + (30 - 20) \times (900,000 - 850,000) + (25 - 20) \times (950,000 - 900,000) + (20 - 20) \times (1,000,000 - 950,000) = \$13,500,000$.

3. At $P = \$3000$, the number of acres supplied is $Q_s = 10,000$. Producers' surplus is $PS = (3000-750) \times (1000-0) + (3000-1000) \times (2000-1000) + \ldots + (3000-2750) \times (9000-8000) + (3000-3000) \times (10000-9000) = \$10,125,000$.

4. At $P = \$1,500$, $Q_d = 1,000$ and $CS = \$500,000$.

5. (a) At $P = \$3,750$, $Q_s = 2$ tons and $PS = \$3,000$. (b) At $P = \$5,250$, $Q_s = 3$ tons and $PS = \$6,750$. (c) As P increases, PS increases.

8. When economic profits are positive, new firms will be entering the market and price will be falling. Firms operating in the range of decreasing returns to scale will be forced to build a smaller plant size in order to reduce their average costs. (b) When economic profits are negative some firms will be exiting the market and prices will be rising for the remaining firms. Firms that are operating in the range of increasing returns to scale will have the incentive to expand plant size in an effort to decrease average costs of production.

9. (b) The efficient quantity of clean air is $Q = 6$. (c) $Q = 6$. (d) Net surplus falls. At $Q = 9$, MB$= 7$ which is less than MC$= 16$ indicating that

the resources used to produce the last unit of clean air have a higher value in some other alternative.

10. (a) With method A, each firm has costs of $C = 2 \times 625^{0.9}$=\$656.63. Transport costs equal $0.05 \times \$656.63 = \$32.83$ per firm. For the 8 firms the total transport costs are \$262.65. (b) With method B, the single firm has costs of $C = 2 \times 5000^{0.9} = \$4266.81$ and transport costs of $0.05 \times \$4266.81 = 213.34$. (c) Transport costs are smaller for method B. (d) There are increasing returns to scale since ATC for each farmer using method A is $ATC = \frac{656.63}{625} = \$1.05$ and for the single farmer using method B, $ATC = \frac{4266.81}{5000} = \$0.85$.

11. (a) Average calories in 1980 equal 500 per person and in 2000 equal 1417 per person. From 1980 to 2000 marginal calories equal 1,250 per person. (b) Looking at average calories per person gives a misleading picture of whether people are better or worse off. People in the DCs receive the same amount of calories and people in the LDCs receive more calories per person. Thus, the world population is better off. (c) Since marginal calories are less than average calories in 1980, the average is pulled down.

Chapter 4

2. (a) \$17,500. (b) \$50,000. (c) \$23,333.

3. (a) −0.5%. (b) −5%. (c) −8%.

4. (a) In general, the vertical intercept is interpreted as the effect of exogenous factors not accounted for in the equation. The negative sign might be interpreted as the rate of assimilation of the pollutant, or the rate at which technological progress is allowing the pollutant to be reduced. (b) The positive sign for c_1 indicates that the higher last period's income was the greater will be the growth rate of the pollutant, and the lower will be the sustainable growth rate. (c) If $(-c_0 - c_1 \times y_{t-1})$ is positive (negative), then as c_2 becomes smaller, the sustainable growth rate of income becomes larger (smaller). (d) A \$1,000 increase in last period's income increases the growth rate of the pollutant by $0.01 = 1\%$.

5. (a) 2.35%. (b) 0.7%.

Chapter 5

1. Timber companies A, D, and F produce on the frontier and are technically efficient. Companies B, C, and E produce inside the frontier and are technically inefficient. Free (strong) disposability is seen in that for any of the points A–F, one can move either to the west (dispose of wildlife) or to the south (dispose

of timber). When $p_{timber} = \$3$ and $p_{wildlife} = \$6$ the revenue maximizing combination corresponds to point D. When $p_{timber} = \$3$ and $p_{wildlife} = \$2$ the revenue maximizing combination corresponds to point A.

2. (c and d) The paper firms 1, 2, 4, 5, and 6 produce on the frontier and are technically efficient. Firms 3 and 7 are inside the frontier and are technically inefficient. (e) For any of the seven points it is possible to move to the south (dispose of paper) which depicts strong disposability of the good output. Weak disposability is depicted in that for any of the seven points it is possible to draw a line from the point back to the origin and all points on those lines will be part of the production possiblity set. (f) Firm 6 maximizes revenues with $R = \$40$ because they produce the most paper. (g) Firm 4 maximizes net revenues with $R = \$12.5 = 1 \times 20 - 0.5 \times 15$. (h) At the higher charge for chemical waste firm 2 now maximizes revenue with $R = \$5$.

3. (b) In period $t+1$ the efficient timber companies are A, C, D, and F. All companies experienced an increase in productivity. (c) Technological progress.

4. (b) Technological progress except for firm 2 where the PPF in $t+1$ is inside the PPF for period t. (c) Firm 3 is inefficient in $t+1$. Firm 7 is inefficient since they could produce $y = 30$ but only $b = 28$ if it were to operate like firm 5. Firm 2 experienced a decline in productivity. The other six firms all experienced an increase in productivity.

8. (a) $R^t = \$30,000$. (b) A change corresponding to combination B is strongly sustainable. Changes to combinations B, C, and D satisfy weak sustainability.

9. The switch to natural gas from a coal-fired plant does not satisfy strong sustainability. However, the change in the costs (C) of the undesirable outputs is $dC = q_1 \times \Delta b_1 + q_2 \times \Delta b_2 + q_3 \times \Delta b_3$ is $dC = \$15 \times (-10000) + \$500 \times (+80) + \$200 \times (+100) = -\$90,000$. The switch to natural gas satisfies weak sustainability.

Chapter 6

1. The DEA reference technology for the four national forests takes the form

$$
\begin{aligned}
T = (x_1, x_2, y, b) \text{ such that} \\
x_1 &\geq 1000z_1 + 1500z_2 + 2000z_3 + 2500z_4, \\
x_2 &\geq 100z_1 + 200z_2 + 200z_3 + 400z_4, \\
y &\leq 25z_1 + 30z_2 + 50z_3 + 40z_4, \\
b &= 2z_1 + 5z_2 + 10z_3 + 12z_4 \\
z_1 &\geq 0, z_2 \geq 0, z_3 \geq 0, z_4 \geq 0.
\end{aligned}
$$

2. The column for actual revenues is $py - qb$ and maximum revenues is $R(x, y, b)$. The optimal quantities that solve the revenue function are y^* and b^*. Take the ratio of $py - qb$ and $R(x, p, q)$ to get revenue efficiency.

k	$\frac{1}{TE1}$	$TE2$	$py - qb$	$R(x, p, q)$	y^*	b^*
1	1.00	1.00	55	55	3.0	1
2	1.00	1.00	85	85	5.0	3
3	1.97	0.33	105	241	13.8	7
4	1.00	0.49	150	271	15.8	9
5	1.00	0.59	220	243	13.4	5
6	1.08	0.81	340	381	21.8	11
7	1.06	0.83	430	521	29.8	15
8	1.00	0.44	210	425	25.0	15
9	1.11	0.88	300	340	20.0	12
10	1.20	0.83	180	422	23.6	10

Chapter 7

2. (b) $Q = 6$. (d) $t_A = \$50$, $t_B = \$100$, $t_C = \$150$. (e) $Q_A = 8$, $Q_B = 5$, $Q_C = 5$. (f) Income is redistributed from citizens B and C to A. (g) Lindahl taxes are $t_A^* = \$65$, $t_B^* = \$95$, $t_C^* = \$100$. (h) The efficient quantity is $Q = 5$.

3. (a) 33.3 miles (b) Lindahl taxes are $t_{canoers}^* = \$1,666.66$, $t_{fishers}^* = \$2,833.34$, $t_{tubers}^* = \$500$.

4. (a) Sum up the marginal rates of substitution and set equal to the marginal rate of transformation and then solve for b: $(3 + 0.1b) \times 500 = 2,000$. Solving for b yields b=10. (b) $MRS_i = 4$. (c) The total amount of private good that can be produced at the efficient level of pollution is $10 \times 2000 = 20,000$. Since heat is a private good each resident receives $H_i = \frac{20,000}{500} = 40$. (d) Utility is $u_i = 40 - 3 \times 10 - 0.05 \times 10^2 = 5$. (e) Relative to the efficient level of pollution (b=10), if one more unit of pollution is produced then 2,000 more units of heat can be generated with each resident receiving $4 = \frac{2,000}{500}$ extra units of heat. With $H_i = 44$ and $b = 11$, $u_i = 4.95$ and the consumer is worse off. (f) $H_i = 36$, $b = 9$, yielding $u_i = 4.95$.

Chapter 8

1. (a) $P_E = \$35$, $Q_E = 50$. (b) The socially efficient quantity is $Q^* = 40$. (c) The socially efficient price is $P^* = \$40$. (d) A Pigouvian tax=10 would internalize the externality. (e) With a tax=10, the price the producer receives would fall from $P_E = \$35$ to \$30. The price the consumer pays would rise from $P_E = \$35$ to \$40. The higher price gives consumers the incentive to reduce electricity use, thereby reducing the production of the undesirable output of SO_2.

2. (a) $P_E = 70$, $Q_E = 25$. (c) $P^* = 90$, $Q^* = 10$. (d) $P^* = 85$, $Q^* = 30$. (e) $P^* = 105$, $Q^* = 15$. (f) Tax equal to the difference in MEC and MEB: tax$= 50 - 25 = 25$.

3. (b) Nick will want to smoke $Q = 10$ cigarettes because that quantity maximizes his total benefits. If Nick smokes $Q = 10$ cigarettes Hack will incur a total cost of 95. (c) Nick will choose not to smoke the tenth cigarette because Hack is willing to pay Nick 14 not to smoke the tenth cigarette and Nick only values the tenth cigarette at 7. A similar argument can be made for the nineth, eighth, and seventh cigarettes. However, Hack only bears a marginal cost of 10 from Nick smoking the sixth cigarette and Nick receives benefits from the sixth cigarette of 11, so no further voluntary reduction in smoking is possible. Nick and Hack will voluntarily agree to limit Nick's smoking to $Q = 6$ cigarettes. The equilibrium price that Hack will pay Nick for not smoking will be somewhere between 10 and 11 cents per cigarette. (d) $Q = 0$ cigarettes is not optimal because at the margin Nick receives more benefits from smoking than the costs borne by Hack. (e) Nick is willing to pay Hack up to 16 for the first cigarette and Hack only requires a payment of 5. For the second, third, fourth, fifth, and sixth cigarettes Nick is willing to pay more than Hack requires. Hack would rather have dirtier air, but money in his pocket up to $Q = 6$ cigarettes. Similarly, Nick would rather have less money in his pocket in exchange for the privilege of smoking up to $Q = 6$ cigarettes. The equilibrium price that Nick pays Hack is between 10 and 11 cents per cigarette. (f) The answers are the same.

4 (a) It is not efficient to ban waste water discharges since the marginal benefits are \$25,000 and the marginal costs are \$20,000. (b) A Pareto improving trade is possible since Bad Water is willing to pay \$25,000 and the FLD Club requires compensation of only \$20,000. (c) If Bad Water owns the property rights the FLD Club could pay \$14,000 for the installation of the waste water treatment and receive benefits of \$20,000 so a Pareto improving trade is possible.

5. (b) The total cost of the regulation is \$45 = \$30 (Ford) +\$15 (GM). (c) 6 tons of emissions remain. (d) The bid functions for each firm are the reverse of the MC schedules. (e) Ford will purchase 4 permits and GM will purchase 2 permits. (f) Ford will clean up 1 unit of pollution at a cost of \$10

Ford				GM			
Q of Pollution	Pollution Abated	MC	TC	Q of Pollution	Pollution abated	MC	TC
5	0	x	0	5	0	x	0
4	1	10	10	4	1	5	5
3	2	20	30	3	2	10	15
2	3	30	60	2	3	15	30
1	4	40	100	1	4	20	50
0	5	50	150	0	5	25	75

and GM will clean up 3 units of pollution at a cost of \$30. Together, these costs of \$10 + \$30 = \$40 are cheaper than the regulatory scheme. (g) The firm with the lowest MC of abatement cleans up a unit of pollution and can thus sell a permit to another firm and make a profit. (h) The market quantities demanded are $Q = 1$ at $P = \$50$, $Q = 2$ when $P = \$40$, $Q = 3$ at $P = \$30$, $Q = 4$ at $P = \$25$, $Q = 6$ at $P = \$20$, $Q = 7$ at $P = \$15$, $Q = 9$ at $P = \$10$, and $Q = 10$ at $P = \$5$. Given 6 permits the equilibrium price of a permit is \$20. (i) With the Nature Conservancy the new market demand schedule is $Q = 1$ at $P = \$50$, $Q = 2$ when $P = \$40$, $Q = 4$ at $P = \$30$, $Q = 6$ at $P = \$25$, $Q = 9$ at $P = \$20$, $Q = 11$ at $P = \$15$, $Q = 14$ at $P = \$10$, and $Q = 16$ at $P = \$5$. (j) $P_E = \$25$, Ford buys 3 permits, GM buys 1 permit, and The Nature Conservancy buys 2 permits.

6. (a) Greasy Brothers abates 3 units of pollution and Stale Air abates 1 unit of pollution. (b) Greasy Brothers abates 5 units of pollution and Stale Air abates 2 units of pollution. (c) Greasy Brothers abates 6 units of pollution and Stale Air abates 3 units of pollution. (d) The equal marginal cost principle.

7. (a) $EMC = \frac{1}{2}[(7 + A) + (1 + A)] = 4 + A$. (b) Set $EMC = MB$ and solve to get $A = 8$, $T = 12$. (c) $\mid MB \mid > EMC$. (d) (i) With MC_H as the true marginal cost the optimal level of abatement is $A^* = 7$. (d) (ii) When the true marginal cost is MC_H the MC_H at $A = 8$ is 14, but at $A = 8$ the $MB = 12$. The DWL$=\frac{1}{2}(14 - 12) \times (8 - 7) = 1$. (e) With a tax of $T = 12$ the firm chooses to abate up to the point where $T = MC_H$. Given $T = 12$ (from part (b)), the firm abates $A = 5$. At $A = 5$, the marginal benefit of abatement is $MB = 18$. The deadweight loss is DWL$=\frac{1}{2}(18 - 12) \times (7 - 5) = 6$. Note, the DWL of the tax is greater than the DWL of the quantity control given that $\mid MB \mid > EMC$.

8. (b) All 10 workers will work at the lake. (c) $MEC = 0$ for $N_L = 1$, $MEC = 1$ for $N_L = 2, \ldots, MEC = 9$ for $N_L = 10$. (d) Total fish production is maximized with $N_L = 5$ and N_F=5. The optimal tax should equal 3 fish. With this tax the sixth, seventh, ..., and tenth lake fishers will prefer to fish at the farm.

9. Tyler 9 to 1, Bradley 12.5 to 1, Florence 18.5 to 1, Cuyahoga 55 to 1, Hudson 89 to 1.

Chapter 9

1. When Days=300 we have $P = 6,500 - \frac{1}{2}Q$ or $Q = 13,000 - 2P$. Given $P = 1,000$ we have $Q = 11,000$ and $CS = \frac{1}{2}(6,500 - 1,000)(11,000 - 0) = \$30,250,000$. When Days=330 we have $P = 7,100 - \frac{1}{2}Q$ or $Q = 14,200 - 2P$. Given $P = 1,000$ we have $Q = 12,200$ and $CS = \frac{1}{2}(7,100 - 1,000)(12,200 - 0) = \$37,210,000$. The change in consumer surplus is $\triangle CS = \$6,960,000$.

2. House price appreciation in Cape Girardeau due to the expansion in National Forest Land is $\triangle H_{Cape} = 0.25 \times (20,500 - 20,000) = 125$. House price appreciation in Jackson is calculated similarly.

3. When Cape Girardeau starts with ACRES=20,000, house price appreciation due to the expansion in National Forest Land is $\triangle H_{Cape} = 0.25 \times (20,500 - 20,000) - 0.000001 \times (20,500^2 - 20,000^2) = 104.75$. When Cape Girardeau starts with ACRES=21,000, house price appreciation due to a 500 acre expansion in National Forest Land is $\triangle H_{Cape} = 0.25 \times (21,500 - 21,000) - 0.000001 \times (21,500^2 - 21000^2) = 103.75$. As the number of acres increases, the willingness to pay for an expansion in acres declines.

4. The decrease in wages is $\triangle wages = 3 \times (100 - 90) + 0.2 \times (100^2 - 90^2) = 410$. When exposure to hazardous waste starts at 50 and falls to 40 the decrease in wages is $\triangle wages = 3 \times (50 - 40) + 0.2 \times (50^2 - 40^2) = 210$. As exposure to hazardous waste falls, the individual's willingness to pay (in lower wages) for a further reduction in waste declines.

5. When $WQ = 1$ the inverse compensated demand function is $P_{CR} = 502 - Q_{CR}$ or $Q_{CR} = 502 - P_{CR}$. When $P_{CR} = 25$ we have $Q_{CR} = 477$ and $CS = \frac{1}{2}(502 - 25)(477 - 0) = \$113,764.50$. When $WQ = 4$ and $P_{CR} = 25$ we have $Q_{CR} = 483$ and $CS = \frac{1}{2}(508 - 25)(483 - 0) = \$116,644.50$. The change in consumer surplus is $\triangle CS = \$2,880$.

6. The lower bound estimate is $0.3 \times \$30 + 0.3 \times \$20 + 0.15 \times \$20 + 0.25 \times \$0 = \$16.50$. Assuming that no one would pay more than \$60 for the park the upper bound estimate is $0.3 \times \$60 + 0.3 \times \$29 + 0.15 \times \$19 + 0.25 \times \$9 = \$31.80$. With 20,000 in population we get a range of the benefits of the new park between \$330,000 and \$636,000.

7. Setting the expected value equal to zero we have $\frac{9,995}{10,000}(w \times 40 \times 50) + \frac{5}{10,000} \times (-\$5,000,000) = 0$. Solving for the wage premium gives $w = \$1.25$.

8. Workers who take \$3 less to live in Lander are revealing a willingness to pay of $\$3 \times 40 \times 50 = \$6,000$ for the benefits of the National Forest land. With 3,000 in the labor force the willingness to pay is $\$6,000 \times 3,000 = \$18$ million.

9. (a) The marginal rate of transformation is $MRT = \frac{22.5M - 20M}{750 - 0} = \$3,333.33$. (b) The Green account should be credited for $\$3,333.33 \times 750 = \$2,500,000$. (c) Given a $MRT = 7$ at point X, the shadow price of SO_2 is $q = \$35 \times 7 = \$245$ per ton. (d) The charge that should be deducted from national income is $\$245 \times 1000 = \$245,000$.

10. (c) Yes, at every level of fish feed (x_1), the marginal product of extra pen space declines. (d) When $x_1 = 6$ and $x_2 = 2$ the respective marginal products are $MP_1 = 12$ and $MP_2 = 20$. Given $w_1 = \$10$ the equilibrium condition is $\frac{20}{12} = \frac{w_2}{\$10}$. Solving yields $w_2 = \$16.66$. (e) When $x_1 = 6$ and $x_2 = 3$ the respective marginal products are $MP_1 = 14$ and $MP_2 = 19$. Given $w_1 = \$10$ the equilibrium condition is $\frac{19}{14} = \frac{w_2}{\$10}$. Solving yields $w_2 = \$13.57$.

11. The equilibrium condition is $\frac{MP_1}{MP_2} = \frac{w_1}{w_2}$. Substituting gives $\frac{20{,}000}{500} = \frac{w_1}{\$20}$. Solving yields a shadow value of another acre of woodland in the watershed of \$800.

Chapter 10

1. Preventing prescribed burns probably helped owls in the short run, but by raising the probability of larger crown fires it might have inadvertently contributed to a lower probability of owl survival in the long run.

2. (a) At the Yellowstone library distinctiveness is $D_1 = 250 - 0.25 \times 200 = 200$. At the Everglades library distinctiveness is $D_2 = 280 - 0.5 \times 200 = 180$. (b) Smoke alarms change the probabilities as $\triangle pr_1 = 0.5 - 0.25 = 0.25$ and $\triangle pr_2 = 0.70 - 0.25 = 0.45$. Let $D_1 + U_1 = D_2 + U_2 = constant > 0$. The ranks of the two libraries are $R_1 = constant \times \frac{0.25}{\$100}$ and $R_2 = constant \times \frac{0.45}{\$100}$ so $R_2 > R_1$. Install the smoke alarms at the Everglades library. (c) Since $R_2 = c \times \frac{0.45}{\$100}$ is 1.8 times greater than $R_1 = c \times \frac{0.25}{\$100}$, then $D_1 + U_1$ at the Yellowstone library would have to be 1.8 times greater $D_2 + U_2$ at the Everglades library.

3. (a) The variable UNIQUE corresponds to distinctiveness. Utility might depend on size and whether the animal is mammal, bird, fish, etc. The variable Conflict would correspond to the cost of saving the species. Changes in the probability of saving the species might depend on the Nature Conservancy's Endangerment ranking and the Fish and Wildlife Service's priority ranking. (b) The signs and magnitudes of the coefficients indicate that we seem to like birds the best, then mammals, then fishes, then amphibians, and finally reptiles. More money is spent on the least endangered species and the higher the priority ranking of the species the less money is spent on the species. These findings seem contrary to what we would expect unless more money has a very small probability of changing the survival of endangered species or species with a high priority rank. The findings also reveal more money to be spent on large species and more on unique species which is to be expected given Noah's ranking system. Finally, the more a species conflicts with development the more money is spent which seems counter to Noah's ranking system.

5. Since there is only one fish, but two mammals we ought to save the shark. After saving the shark we already have an animal that swims so a land mammal—the rhinoceros—should probably be saved.

6. (a) The production frontier is comprised of sites 1, 3, 5, 6, and 7. (b) Proposed site 9 is a linear combination of existing sites 5 and 6 and would not shift the frontier. (c) Calculate the average efficiency of the existing sites and proposed new site. The proposed site that gives the lowest average efficiency is the one that shifts the frontier out the most and should be chosen. When site H is included, the 7 existing sites have average efficiency of $\overline{D_o(x,y)} = 0.90$. When site J is included, the 7 existing sites have average efficiency of $\overline{D_o(x,y)} = 0.93$. Thus, site H shifts the frontier out further than site J. (d) Existing sites 2 and 4 are inside the frontier of existing sites and would be good candidates for sale.

Chapter 11

1. At $R = 5\%$, the net present value is $NPV = \$61,627$. At $R = 8\%$, the net present value is $NPV = -\$32,472$.

3. (a) When $R = 5\%$, the competitive equilibrium prices and quantities in the two periods are $P_0 = 463.4$ and $Q_0 = 73.17$ and $P_1 = 486.58$ and $Q_1 = 26.83$. At $R = 10\%$, the prices and quantities are $P_0 = 452.4$ and $Q_0 = 95.2$ and $P_1 = 497.6$ and $Q_1 = 4.8$. (d) At $R = 5\%$, under monopoly we have $P_0 = 469.5$ and $Q_0 = 61$ and $P_1 = 480.5$ and $Q_1 = 39$. At $R = 10\%$, under monopoly we have $P_0 = 464.5$ and $Q_0 = 71$ and $P_1 = 485.5$ and $Q_1 = 29$. (e) Monopolist extracts less in $t = 0$ and more in $t = 1$ than the competitive firm. (g) As R increases, the amount to extract in $t = 0$ increases.

Chapter 12

1. (a) $S_{MSY} = 50,000$ tons and $MSY = 20,000$ tons. (b) $G(S) = 12,000$ tons. (c) $PV = 50,000 \times \$1500 + \frac{20000 \times \$1500}{0.12} = \$325,000,000$. (d) $S = 42,500$, $PV = 57,500 \times \$1500 + \frac{19,550 \times \$1500}{0.12} = \$330,625,000$.

2. (b) Negative. (c) $S_{MSY} = 68$, $MSY = 63.104$. (d) $S = 66$.

3. (b) At $x = 12$ boats, profits are zero. (c) Marginal external costs are \$5 for the second boat and increase by \$5 for each additional boat. (d) The optimal number of boats is $x = 6$. $x = 6$ boats maximizes profits at $\pi = \$180$.

4. (a) $MP = 0.66$, 3, 5, 4, 3.5, 3, 2, 1, 0.5, 0.3, 0.2. $MAI = 0.66$, 1.25, 2, 2.33, 2.5, 2.56, 2.5, 2.35, 2.18, 2.03, 1.88. $g = 1.33$, 0.86, 1.11, 0.73, 0.54, 0.4, 0.24, 0.11, 0.05, 0.048, 0.026, 0.016. (b) Harvest the trees at age $t = 80$ years. (c) Harvest the trees at age $t = 100$ years. (d) Harvest the trees at age $t = 110$ years.

5. (a) $MP =$ 100, 400, 1000, 1400, 1300, 800, 500, 300, 250, 150, 100, 0.

MAI= 100, 250, 500, 725, 840, 833.3, 785.7, 725, 672.2, 620, 572.7, 525. g= 2, 1.33, 1, 0.636, 0.366, 0.174, 0.095, 0.053, 0.042, 0.024, 0.016, 0. (b) Harvest at t=9. (c) PV of π= 0 at $t = 0$, \$339.8 at $t = 1$, \$1649.5 at $t = 2$, \$4804.5 at $t = 3$, \$9018.1 at $t = 4$, \$12680.4 at $t = 5$, \$14656 at $t = 6$, \$15652 at $t = 7$, \$16025 at $t = 8$, \$16228.9 at $t = 9$, \$16146.8 at $t = 10$, \$15929.4 at t=11, and \$15465.4 at $t = 12$. (d) Profits are maximized at $t = 9$ which yields the same answer as when comparing the growth rate of the trees with the discount rate.

Bibliography

[1] Acemoglu, D., and J.A. Robinson. *Why Nations Fail: The Origins of Power, Prosperity, and Poverty.* Crown Publishers, 2012.

[2] Ackerman, F., and E.A. Stanton. *Climate Economics: The State of the Art.* Routledge, 2013.

[3] Adler, J.H. The fable of federal environmental regulation: Reconsidering the federal role in environmental protection. *Case Western Reserve Law Review*, 55(1):93–1132, 2004.

[4] Alcott, B. Jevons' paradox. *Ecological Economics*, 54:9–21, 2005.

[5] Allen, D. W. *Transaction Costs*, pages 893–926. In [25], 2000.

[6] Allen, R.C. A review of Gregory Clark's *A Farewell to Alms: A Brief Economic History of the World*. *Journal of Economic Literature*, 46(4):946–973, 2008.

[7] Allen, R.C., and I. Keay. Saving the whales: Lessons from the extinction of the Eastern Arctic Bowhead. *The Journal of Economic History*, 64(2):400–432.

[8] Anderson, T.L. *Conservation-Native American Style*, pages 1–21. In [9], 1997.

[9] Anderson, T.L., ed. *Breaking the Environmental Policy Gridlock.* Hoover Institution Press, 1997.

[10] Anomaly, J. Public goods and government action. *Politics, Philosophy, and Economics*, pages 1–20, 2013.

[11] Aronsson, T., and K-G. Löfgren. *Renewable resources: forestry*, pages 122–140. In [240], 1999.

[12] Arrow, K., B. Bolin, R. Costanza, and P. Dasgupta. Economic growth, carry capacity, and the environment. *Science*, 268(5210):520–521, 1995.

[13] Arrow, K., P. Dasgupta, L. Goulder, G. Daily, P. Ehrlich, G. Heal, S. Levin, K.G. Mäler, S. Schneider, D. Starrett, and B. Walker. Are we consuming too much? *The Journal of Economic Perspectives*, 18(3):147–172, 2004.

[14] Azomahou, T., F. Laisney, and P.N. Van. Economic development and CO2 emissions: A nonparametric panel approach. *Journal of Public Economics*, 90:1347–1363, 2006.

[15] Baarschers, W.H. *Eco-Facts and Eco-Fiction: Understanding the Environmental Debate.* Routledge, 1996.

[16] Banker, R.D., A. Charnes, and W. W. Cooper. Some models for estimating technical and scale inefficiencies in data envelopment analysis. *Management Science*, 17:35–44, 1984.

[17] Barbier, E.B. *Capitalizing on Nature: Ecosystems as Natural Assets.* Cambridge and New York: Cambridge University Press, 2011.

[18] Barnett, H., and C. Morse. *Scarcity and Growth: The Economics of Natural Resource Availability.* Baltimore, Maryland: Johns Hopkins University Press, 1963.

[19] Barro, R.J. *Getting it Right: Markets and Choices in a Free Society.* The MIT Press, 1996.

[20] Baumgärtner, S., H. Dyckhoff, M. Faber, J. Proops, and J. Schiller. The concept of joint production and ecological economics. *Ecological Economics*, 36:365–372, 2001.

[21] Baumol, W.J., and W.G. Bowen. *Performing Arts: The Economic Dilemma.* The Twentieth Century Fund, New York, 1966.

[22] Berrens, R.P., A.K. Bohara, K. Gawande, and P. Wang. Testing the inverted-U hypothesis for US hazardous waste: An application of the generalized gamma model. *Economics Letters*, 55:435–440, 1997.

[23] Boden, T.A., G. Marland, and R.J. Andres. *Global, Regional, and National Fossil-Fuel CO2 Emissions.* Carbon Dioxide Information Analysis Center, Oak Ridge National Laboratory, US Dept. of Energy, Oak Ridge, TN, 2011.

[24] Bolin, B. *Geophysical and Geochemical Aspects of Environmental Degradation.* Volume 1 of [159], 2003.

[25] Bouckaert, B., and G. De Geest, ed. *Enclyclopedia of Law and Economics: Volume 1, The History and Methodology of Law and Economics.* Edward Elgar Press, 2000.

[26] Boulding, K. *The economics of coming spaceship Earth*, pages 3–14. In [117], 1966.

[27] Bradford, D.F., and G. G. Hildebrandt. Observable preferences for public goods. *Journal of Public Economics*, 8(2):111–131, 1977.

[28] Brand, S. *Whole Earth Discipline: An Ecopragmatist Manifesto.* New York: Viking, 2009.

[29] Brännlund, R. Productivity and environmental regulations. *Umeä Economic Studies*, (728), 2008.

[30] Brodberg, T., P-O. Marklund, E. Samakovlis, and H. Hammar. Testing the Porter hypothesis: The effects of environmental investments on efficiency in Swedish industry. *Journal of Productivity Analysis*, 40:43–56, 2013.

[31] Brooks, N., and R. Sethi. The distribution of pollution: Community characteristics and exposure to air toxics. *Journal of Environmental Economics and Management*, 32:233–250, 1997.

[32] Brown, G.M., and J.F. Shogren. Economics of the Endangered Species Act. *Journal of Economic Perspectives*, 12(3):3–20, 1998.

[33] Caldari, K. Alfred Marshall's idea of progress and sustainable development. *Journal of the History of Economic Thought*, 26(4):519–536, 2004.

[34] Carlos, A.M., and F.D. Lewis. Indians, the beaver, and the bay: The economics of depletion in the lands of the Hudson's Bay Company, 1700-1763. *The Journal of Economic History*, 53(3):465–494, 1993.

[35] Carlyle, R. Should other nations follow Germany's lead on promoting solar power? *Forbes*, October, 4 2013.

[36] Carson, R. *Silent Spring.* Fawcett Publications, Inc., 1962.

[37] Charnes, A., W.W. Cooper, and E. Rhodes. Measuring efficiency of decision making units. *European Journal of Operational Research*, 2:429–444, 1978.

[38] Chen, Y., A. Ebenstein, M. Greenstone, and H. Li. Evidence on the impact of sustained exposure to air pollution on life expectancy from China's Huai River policy. *PNAS Early Edition*, http://www.pnas.org/cgi/doi/10.1073/pnas.1300018110:1–6, 2013.

[39] The World Factbook. http://www.cia.gov/library/publication/the-world-factbook/index.html, 2013.

[40] Clark, G. *A Farewell to Alms.* Princeton University Press, 2007.

[41] Clay, K., and W. Troesken. *Did Frederick Brodie Discover the World's First Environmental Kuznets Curve? Coal Smoke and the Rise and Fall of the London Fog*, pages 281–309. In [144], 2011.

[42] Clover, C. *The End of the Line: How Overfishing is Changing the World and What We Eat.* The New Press, 2006.

[43] Coase, R. The problem of social cost. *The Journal of Law and Economics*, 3:1–44, 1960.

[44] Coelli, T., L. Lauwers, and G. V. Huylenbroeck. Environmental efficiency measurement and the materials balance condition. *Journal of Productivity Analysis*, 28:3–12, 2007.

[45] Cohen, B.L. Catalog of risks extended and updated. *Health Physics*, 61(3):317–335, 1991.

[46] Cohen, B.L. Risks in perspective. *Journal of American Physicians and Surgeons*, 8(2):50–53, 2003.

[47] Cole, M.A., A.J. Rayner, and J.M. Bates. The environmental Kuznets' curve: an empirical analysis. *Environmental and Development Economics*, 2(4):401–416, 1997.

[48] Coman, K. Some unsettled problems of irrigation. *American Economic Review*, 1(1):1–19, 1911.

[49] Commission for Racial Justice. Toxic Wastes and Race in the United States: A National Report on the Racial and Socio-economic Characteristics of Communities with Hazardous Waste Sites. Technical report, United Church of Christ, 1987.

[50] Cooney, R., and B. Dickson, ed. *Biodiversity and the Precautionary Principle: Risk and Uncertainty in Conservation and Sustainable Use.* Sterling, VA: Earthscan, 2005.

[51] Copeland, B.R., and M.S. Taylor. Trade, growth and the environment. *Journal of Economic Literature*, 42:7–71, 2004.

[52] Costanza, R., J. Cumberland, H. Daly, R. Goodland, and R. Norgaard. *An Introduction to Ecological Economics.* St. Lucie Press, 1997.

[53] Costello, C., J. Lynham, S.E. Lester, and S.D. Gaines. Economic incentives and global fisheries sustainability. *Annual Review of Resource Economics*, 2:299–318, 2010.

[54] Costello, C., S.D. Gaines, and J. Lynham. Can catch shares prevent fisheries collapse? *Science*, 321(5896):1678–1681, 2008.

[55] Council on Enviromental Quality. CEQ 25th Anniversary Report. Technical report, Council on Environmental Quality: The White House, 1994.

[56] Dales, J.H. Land, water, and ownership. *Canadian Journal of Economics*, 1(4):791–804, 1968.

[57] Daly, H. *Beyond Growth: The Economics of Sustainable Development.* Boston: Beacon Press, 1996.

[58] Dasgupta, S., B. Laplante, H. Wang, and D. Wheeler. Confronting the environmental Kuznets curve. *Journal of Economic Perspectives*, 16(1):147–168, 2002.

[59] Dawson, D., and J.F. Shogren. Priorities and expenditures under the Endangered Species Act. *Land Economics*, 77(4):527–532, 2001.

[60] Day, K., and R.Q. Grafton. Economic growth and environmental degradation in Canada. *The Review of Economic Performance and Social Progress*, 1:293–308, 2001.

[61] de Bruyn, S.M., J.C.J.M. van den Bergh, and J.B. Opschoor. Economic growth and emissions: Reconsidering the empirical basis of environmental Kuznets' curves. *Ecological Economics*, 25:161–175, 1998.

[62] deMenocal, P.B. Cultural responses to climate change during the late Holocene. *Science*, 292(5517):667–673, 2011.

[63] Demsetz, H. *From Economic Man to Economic System: Essays on Human Behavior and the Institutions of Capitalism.* Cambridge: Cambridge University Press, 2008.

[64] Domazlicky, B., and W.L. Weber. Does environmental protection lead to slower productivity growth in the chemical industry? *Environmental and Resource Economics*, 28(3):301–324, 2004.

[65] Downs, A. Up and down with ecology: The "Issue-Attention Cycle". *The Public Interest*, pages 38–50, 1972.

[66] Duffy-Deno, K. Economic effects of endangered species preservation in the non-metropolitan west. *Growth and Change*, 28:263–288, 1997.

[67] Ehrlich, P. *The Population Bomb.* Sierra Club-Ballantine, 1968.

[68] Elmore, B.J. The American beverage industry and the development of curbside recycling programs, 1950-2000. *Business History Review*, 86:407–501, 2012.

[69] Elwell, F. W. Understanding Malthus. http://www.faculty.rsu.edu/ felwell/Theorists/Malthus/Index.htm, 2001.

[70] Energy Information Agency. Energy Information Agency: Monthly Energy Review. http://www.eia.gov/totalenergy/data/monthly, 2013.

[71] Energy Information Agency. Energy Information Agency: Short-term Energy Outlook. http://www.eia.gov/forecasts/steo/report/electricity.cfm, 2013.

[72] FAO Fisheries Department, Fishery Information, Data, and Statistics Unit. FISHSTAT Plus: Universal software for fishery statistical time series, Version 2.3, 2012.

[73] Färe, R., S. Grosskopf, and C.A.K. Lovell. *Production Frontiers.* Cambridge University Press, 1994.

[74] Färe, R., S. Grosskopf, and W.L. Weber. Shadow prices of Missouri public conservation land. *Public Finance Review*, 29(6):444–460, 2001.

[75] Färe, R., S. Grosskopf, and W.L. Weber. Shadow prices and pollution costs in U.S. agriculture. *Ecological Economics*, 1(1):89–103, 2006.

[76] Färe, R., S. Grosskopf, C. Pasurka, and W.L. Weber. Substitutability among undesirable outputs. *Applied Economics*, 44:39–47, 2012.

[77] Färe, R., S. Grosskopf, D-W. Noh, and W.L. Weber. Characteristics of a polluting technology: Theory and practice. *Journal of Econometrics*, 126:469–492, 2005.

[78] Farrow, S. Extinction and market forces: two case studies. *Ecological Economics*, 13:115–123, 1995.

[79] US Fish and Wildlife Service. Federal and state endangered and threatened species expenditures. Technical report, US Fish and Wildlife Service, 2011.

[80] Flader, S.L., and J.B. Callicott, eds. *The River of the Mother of God and Other Essays by Aldo Leopold.* Madison, WI: The University of Wisconsin Press, 1991.

[81] Foster, A.D., and M.R. Rosenzweig. Economic growth and the rise of forests. *The Quarterly Journal of Economics*, 118(2):601–637, 2003.

[82] Friedman, M. *There's No Such Thing as a Free Lunch.* Thomas Horton and Company, 1975.

[83] George, M.D. *London Life in the XVIIIth Century.* Alfred A. Knopf, 1926.

[84] Georgescu-Roegen, N. The entropy law and the economic process in retrospect. *Eastern Economic Journal*, 12(1):3–25, 1986.

[85] Glaeser, E.L. *Triumph of the City: How Our Greatest Invention Makes us Richer, Smarter, Greener, Healthier, and Happier.* New York: Penguin Books, 2011.

[86] Glaeser, E.L. The supply of environmentalism. *National Bureau of Economic Research*, 2013.

[87] Goklany, I.M. *Clearing the Air: The Real Story of the War on Air Pollution.* Washington, D.C.: Cato Institute, 1999.

[88] Goklany, I.M. Could biofuel policies increase death and disease in developing countries? *Journal of American Physicians and Surgeons*, 16(1):9–13, 2011.

[89] Goldfarb, W. *Water Law.* Chelsea, MI: Lewis Publishers, Inc., 1988.

[90] Gordon, R.J. Does the new economy measure up to the great inventions of the past? *Journal of Economic Perspectives*, 14(1):49–74, 2000.

[91] Gordon, R.J. Is U.S. economic growth over? Faltering innovation confronts the six headwinds. *National Bureau of Economic Research*, 2012.

[92] Grafton, R.Q., J. Kirkley, T. Kompas, and D. Squires. *Economics for Fisheries Management.* Ashgate, 2006.

[93] Grossman, G.M., and A.B. Krueger. Economic growth and the environment. *Quarterly Journal of Economics*, 10(2):353–377, 1995.

[94] Grossman, M., and G. C. Bryner. *U.S. Land and Natural Resources Policy.* Armenia, NY: Grey House Publishing, 2nd ed., 2012.

[95] Grubel, H.G. Economic freedom and human welfare: Some empirical findings. *Cato Journal*, 18(2):287–304, 1998.

[96] Halverson, A. *An Entirely Synthetic Fish: How Rainbow Trout Beguiled America and Overran the World.* New Haven and London: Yale University Press, 2010.

[97] Hamilton, J.T. Pollution as news: Media and stock market reactions to the Toxics Release Inventory data. *Journal of Environmental Economics and Management*, 26:98–113, 1995.

[98] Hanke, S.H., and S.J.K. Walters. Economic freedom, prosperity, and equality: A survey. *Cato Journal*, 17(3):117–146, 1997.

[99] Hanley, N., J.F. Shogren, and B. White. *Introduction to Environmental Economics.* Oxford University Press, 2001.

[100] Hanner, J. Government response to the buffalo hide trade, 1871–1883. *Journal of Law and Economics*, 24:239–271, 1981.

[101] Harbaugh, W.T., A. Levinson, and D.M. Wilson. Reexamining the empirical evidence for an environmental Kuznets' curve. *Review of Economics and Statistics*, 84(3):541–551, 2002.

[102] Hartwick, J.M., and N.D. Olewiler. *The Economics of Natural Resource Use.* Addison-Wesley, 2nd ed., 1998.

[103] Hayek, F.A. The use of knowledge in society. *The American Economic Review*, 35(4):519–530, 1945.

[104] Hemingway, E. *The Old Man and The Sea.* New York: Macmillan, 1952.

[105] Hertwich, E.G. Life cycle approaches to sustainable consumption: A critical review. *Environmental Science and Technology*, 39(13):4673–4684, 2005.

[106] Hervieux, M-S., and O. Darné. Environmental Kuznets curve and ecological footprint: A time series analysis. 2013.

[107] Hettige, H., M. Mani, and D. Wheeler. Industrial pollution in economic development: The environmental Kuznets curve revisited. *Journal of Development Economics*, 62:445–476, 2000.

[108] Hilton, F.G.H., and A.M. Levinson. Factoring the environmental Kuznets curve: Evidence from automotive lead emissions. *Journal of Environmental Economics and Management*, 35:126–141, 1998.

[109] Hirschman, A.O. *Exit, Voice, and Loyalty: Responses to Decline in Firms, Organizations, and States.* Cambridge, MA: Harvard University Press, 1970.

[110] Hobsbawm, E.J. *Industry and Empire: The Making of Modern English Society.* Pantheon Books, 1968.

[111] Hotelling, H. The economics of exhaustible resources. *Journal of Political Economy*, 39(2):137–175, 1931.

[112] Howard, D. A river reborn: A Connecticut's Naugatuck. *Trout*, pages 57–61, 2000.

[113] Hubbert, M.K. Nuclear energy and the fossil fuels. Technical report, Shell Development Company: Exploration and Production Research Division, 1956.

[114] Innes, R. and G. Frisvold. *The Economics of Endangered Species.* Volume 1 of [203], 2009.

[115] International Energy Agency. International Energy Agency: World Energy Outlook-Energy Subsidies. http://www.worldenergyoutlook.org/resources/energysubsidies/, 2013.

[116] Iudicello, S., M. Weber, and R. Wieland. *Fish, Markets, and Fishermen: The Economics of Overfishing.* Island Press, 1999.

[117] Jarrett, H., ed. *Environmental Quality in a Growing Economy.* Baltimore, MD: Johns Hopkins University Press, 1966.

[118] Jevons, W.S. *The Coal Question; An Inquiry Concerning the Progress of our Nation and the Probable Exhaustion of our Coal Mines.* London: Macmillan, 1865.

[119] Joskow, P.L., and R. Schmalensee. The political economy of market-based environmental policy: The U.S. Acid Rain Program. *Journal of Law and Economics*, XLI:37–83, 1998.

[120] Kahn, J. *The Economic Approach to Environmental and Natural Resources*. Dryden Press, 1998.

[121] Kahn, M.E. *Green Cities: Urban Growth and the Environment*. Washington, DC: Brookings Institution, 2006.

[122] Kahneman, D. *Thinking, Fast and Slow*. Farrar, Straus and Giroux, New York, 2011.

[123] Kareiva, P., and M. Marvier. *Conservation Science: Balancing the Needs of People and Nature*. Greenwood Village, CO: Roberts and Company Publishers, 2011.

[124] Kelly, T.D., and G.R. Matos. *Historical statistics for mineral and material commodities in the United States (2013 version)*. US Geological Survey Data Series 140, 2013.

[125] Kennedy, P.F. Sinning in the basement: What are the rules? The ten commandments of applied econometrics. *Journal of Economic Surveys*, 16(4):569–589, 2002.

[126] Khanna, N., and F. Plassman. The demand for environmental quality and the environmental Kuznets' curve hypothesis. *Ecological Economics*, 51(3):225–236, 2004.

[127] King, A., and M. Lenox. Exploring the locus of profitable pollution reduction. *Management Science*, 48(2):289–299, 2002.

[128] Kneese, A.V. Analysis of environmental pollution. *The Swedish Journal of Economics*, pages 59–81, 1971.

[129] Kniesner, T.J., W.K. Viscusi, C. Woock, and J.P. Ziliak. The value of a statistical life: Evidence from panel data. *Review of Economics and Statistics*, 94(1):74–87, 2011.

[130] Konar, S., and M.A. Cohen. Information as regulation: The effect of community right to know laws on toxic emissions. *Journal of Environmental Economics and Management*, 32:109–124, 1997.

[131] Krautkraemer, J.A. Nonrenewable resource scarcity. *Journal of Economic Literature*, 36(4):2065–2107, 1998.

[132] Kremer, M. Population growth and technological change: One million B.C. to 1990. *Quarterly Journal of Economics*, 108(3), 1993.

[133] Kremer, M., and C. Morcom. Elephants. *American Economic Review*, 90(1):212–234, 2000.

[134] Kriesel, W., T.J. Centner, and A.G. Keeler. Neighborhood exposure to toxic releases: Are there racial inequities? *Growth and Change*, 27:479–499, 1996.

[135] Kümmel, R. *The Second Law of Economics: Energy, Entropy, and the Origins of Wealth.* New York: Springer, 2011.

[136] Landes, D.S. *The Wealth and Poverty of Nations: Why Some Are So Rich and Some So Poor.* New York: W.W. Norton and Company, 1998.

[137] Lanoie, P., M. Patry, and R. Lajeunesse. Environmental regulation and productivity: Testing the Porter hypothesis. *Journal of Productivity Analysis*, 30(2):121–128, 2005.

[138] Lazaroff, C. On the ground: Helping ranchers, keeping wolves. *Defenders Magazine*, http://www.defenders.org/newsroom/defenders_magazine/fall_2008/, 2008.

[139] Lee, D.R., and R.B. McKenzie. How the marketplace fosters business honesty. *Business and Society Review*, pages 5–9, 1995.

[140] Lee, R. The demographic transition: Three centuries of fundamental change. *The Journal of Economic Perspectives*, 17(3):167–190, 2003.

[141] Leontief, W. Environmental repercussions and the economic structure: An input-output approach. *The Review of Economics and Statistics*, 52(3):262–271, 1970.

[142] Leopold, A. *A Sand County Almanac and Sketches Here and There.* Oxford University Press, 1949.

[143] Leopold, A. *Conservation Economics*, pages 193–202. In [80], 1991.

[144] Libecap, G.D., and R.H. Steckel, eds. *The Economics of Climate Change: Adaptations Past and Present.* 2011.

[145] Lin, J.Y. The Needham puzzle: Why the industrial revolution did not originate in China. *Economic Development and Cultural Change*, 43(2):269–292, 1995.

[146] Linn, J., and V. McConnell. How electric cars can increase greenhouse gas emissions. *Resources*, (184):33–37, 2013.

[147] Lipford, J.W., and B. Yandle. NAFTA, environmental Kuznets' curves, and Mexico's progress. *Global Economy Journal*, 10(4), 2010.

[148] Lomborg, B. *The Skeptical Environmentalist: Measuring the Real State of the World.* Cambridge: Cambridge University Press, 2001.

[149] Lueck, D. Ownership and the regulation of wildlife. *Economic Inquiry*, 29:249–260, 1991.

[150] Lueck, D., and J.A. Michael. Preemptive habitat destruction under the Endangered Species Act. *Journal of Law and Economics*, 46:27–60, 2003.

[151] Lusk, J.L., and F.B. Norwood. The locavore's dilemma: Why pineapples shouldn't be grown in North Dakota. Library of Economics and Liberty, http://www.econlib.org/index.html, 2011.

[152] Lyle, M. The Appalachian's energy development rush. *Trout*, pages 32–59, 2009.

[153] Macaulay, T.B. Southey's colloquies on society. Library of Economics and Liberty, http://www.econlib.org/library/Essays/macS1.html, 2000.

[154] Maddison, A. *The World Economy: Historical Statistics.* Development Centre of the Organisation of Economic Cooperation and Development, 2003.

[155] Maddison, A. *Contours of the World Economy, 1-2030 AD: Essays in Macro-Economic History.* Oxford University Press, 2007.

[156] Mainwaring, L. Biodiversity, biocomplexity, and the economics of genetic dissimilarity. *Land Economics*, 77(1):79–83, 2001.

[157] Mäler, K.G. A method of estimating social benefits from pollution control. *Swedish Journal of Economics*, 73(1):121–133, 1971.

[158] Mäler, K.G. *Environmental Economics: A Theoretical Inquiry.* Baltimore, MD: The Johns Hopkins University Press, 1974.

[159] Mäler, K.G., and J. R. Vincent, eds. *Handbook of Environmental Economics: Environmental Degradation and Institutional Responses*, volume 1. Amsterdam: Elsevier, 2003.

[160] Mäler, K.G., and J.R. Vincent, eds. *Handbook of Environmental Economics: Economywide and International Environmental Issues*, volume 3. Amsterdam: Elsevier, 2005.

[161] Malthus, T.R. *An Essay on the Principle of Population, as It Affects the Future Improvement of Society with Remarks on the Speculation of M. Godwin, M. Condorcet, and Other Writers.* London: J. Johnson, 1798.

[162] Managi, S., J.J. Opaluch, D. Jin, and T.A. Grigalunas. Environmental regulations and technological change in the offshore oil and gas industry. *Land Economics*, 81(2):303–319, 2005.

[163] Mann, C.C. *1491: New Revelations of the Americas before Columbus.* New York: Alfred A. Knopf, 2005.

[164] Marlow, M.L. Too much (questionable) information? *Regulation*, Winter:29–38, 2013-2014.

[165] Marquetti, A. Extended Penn World Tables: Economic growth data assembled from the Penn World Tables and other sources. https://sites.google.com/a/newschool.edu/duncan-foley-homepage/home/EPWT, 2012.

[166] McCloskey, D. *Bourgeois Dignity: Why Economics Can't Explain the Modern World.* Chicago: The University of Chicago Press, 2010.

[167] McGreggor, C.R. *Federal Land, Western Anger.* University Press of Kansas, 1996.

[168] McGuire, R.A., and P.R.P. Coelho. *Parasites, Pathogens and Progress: Diseases and Economic Development.* Cambridge, MA: The MIT Press, 2011.

[169] McKitrick, R. *Environment and Energy*, 19(5):707–711, 2007.

[170] McKitrick, R.R. A simple state contingent pricing rule for complex intertemporal externalities. *Energy Economics*, 33(1):111–120, 2011.

[171] McMillion, S. Ranching rebooted. *Nature Conservancy*, November/December 2013.

[172] McPhee, J. *Encounters with the Archdruid.* New York: Farrar, Straus, and Giroux, 1971.

[173] McPherson, M.A., and M.L. Niewsiadomy. African elephants: the effect of property rights and political stability. *Contemporary Economic Policy*, 18(1):14–26, 2000.

[174] Mealey, S.P., J.W. Thomas, H.J. Salwasser, R.E. Stewart, P.J. Balint, and P. W. Adams. *Precaution in the American Endangered Species Act as a Precursor to Environmental Decline: The Case of the Northwest Forest Plan.* In [50], 2005.

[175] Mehmood, S.R., and D. Zhang. Determinants of forest landowner participation in the Endangered Species Act Safe Harbor program. *Human Dimensions of Wildlife*, 10:249–257, 2005.

[176] Metrick, A., and M.L. Weitzman. Conflicts and choices in biodiversity preservation. *Journal of Economic Perspectives*, 12(3):21–34, 1998.

[177] Miller, R.L., D.K. Benjamin, and D.C. North. *The Economics of Public Issues.* Addison Wesley Longman, 12th ed., 2001.

[178] Mokyr, J. *The Enlightened Economy: An Economic History of Britain 1700-1850.* New Haven: Yale University Press, 2009.

[179] Muller, N.Z., and R. Mendelsohn. Weighing the value of a ton of pollution. *Regulation*, 33(2):20–24, 2010.

[180] Muller, N.Z., R. Mendelsohn, and W. Nordhaus. Environmental accounting for pollution in the United States economy. *American Economic Review*, 101:1649–1675, 2011.

[181] Nehring, K., and C. Puppe. A theory of diversity. *Econometrica*, 70(3):1155–1198, 2002.

[182] Neimark, P., and P.R. Mott, eds. *The Environmental Debate: A Documentary History with Timeline, Glossary, and Appendices.* Grey House Publishing, 2011.

[183] Nelson, R.H. Ecological science as a creation story. *The Independent Review*, 14(4):513–534, 2010.

[184] Nordhaus, W.D. Do real-output and real-wage measures capture reality? The history of lighting suggests not. *Cowles Foundation Paper*, (957):29–66, 1998.

[185] Nordhaus, W.D. Paul Samuelson and global public goods. Mimeo, 2005.

[186] Nordhaus, W.D. A review of the *Stern Review on the Economics of Climate Change*. *Journal of Economic Literature*, 45(3):686–702, 2007.

[187] OECD. *Environmental Taxes in OECD Countries.* OECD, 1996.

[188] Office of Resource Conservation and Recovery. Municipal solid waste generation, recycling, and disposal in the United States: Tables and figures for 2010. Technical report, United States Environmental Protection Agency, 2011.

[189] Olmstead, S., and C. Kousky. Unnatural disasters? *Resources*, 179:21–23, 2012.

[190] Interagency Working Group on Social Cost of Carbon. *Technical Update of the Social Cost of Carbon for Regulatory Impact Analysis Under Excecutive Order 12866.* United States Government, 2013.

[191] Ostrom, E. *Governing the Commons: The Evolution of Institutions for Collective Action.* Cambridge University Press, New York: New York, 1990.

[192] Ostrom, E. Reflections on 'Some unsettled problems of irrigation'. *American Economic Review*, 101(1):49–63, 2011.

[193] Parry, I.W.H., and K.A. Small. Does Britain or the United States have the right gasoline tax? *American Economic Review*, 95(4):1276–1289, 2005.

[194] Paulos, J. *A Mathematician Reads the Newspaper*. Anchor, 1996.

[195] Pearce, D. An intellectual history of environmental economics. *Annual Review of Energy and the Environment*, 27:57–81, 2002.

[196] Pigou, A.C. *The Economics of Welfare*. London: Macmillan, 1920.

[197] Polasky, S., C. Costello, and A. Solow. *The Economics of Biodiversity*. Volume 3 of [160], 2005.

[198] Pollock III, P. H., and M.E. Vittas. Who bears the burdens of environmental pollution? Race, ethnicity, and environmental equity in Florida. *Social Science Quarterly*, 76(2):294–310, 1995.

[199] Porter, M., and C.V.D. Linde. Toward a new conception of the environment-competitiveness relationship. *Journal of Economic Perspectives*, 9(4):97–118, 1995.

[200] Portney, P.R., I.W.H. Parry, H.K. Gruenspecht, and W. Harrington. The economics of fuel economy standards. *Journal of Economic Perspectives*, 17(4):203–217, 2003.

[201] Poterba, J.M. Global warming: A public finance perspective. *Journal of Economic Perspectives*, 7.

[202] Power, T. M. Review of *Environmental Economics for Tree Huggers and Other Skeptics*, by William K. Jaeger. *Journal of Economic Literature*, 45(4):1087–1088, 2007.

[203] Rausser, G., K. Smith, and D. Zilberman, eds. *Annual Review of Resource Economics*, volume 1. Annual Reviews, Palo Alto, CA, 2009.

[204] Roback, J. Wages, rents, and the quality of life. *Journal of Political Economy*, 90(6):1257–1278, 1982.

[205] Sabin, P. *The Bet: Paul Ehrlich, Julian Simon, and Our Gamble over Earth's Future*. Yale University Press, 2013.

[206] Samuelson, P.A. The pure theory of public expenditure. *Review of Economics and Statistics*, 36(4):387–389, 1954.

[207] Samuelson, P.A. Diagrammatic exposition of a theory of public expenditure. *Review of Economics and Statistics*, 37(4):350–356, 1955.

[208] Scarlett, L. Reshaping the Endangered Species Act: A holistic approach needed? Issue Brief 10–15, Resources for the Future, 2005.

[209] Schmalensee, R., and R. N. Stavins. The SO_2 allowance trading system: The ironic history of a grand policy experiment. *Journal of Economic Perspectives*, 25(1):103–122, 2013.

[210] Schmalensee, R., T.M. Stoker, and R.A. Judson. World carbon dioxide emissions: 1950-2050. *Review of Economics and Statistics*, 80(1):15–27, 1998.

[211] Schofield, R. The upper Green River: An ecosystem on the brink. *Trout*, pages 18–27, 2010.

[212] Schumpeter, J.A. *Capitalism, Socialism and Democracy.* Harper and Brothers, 1942.

[213] Selden, T.M., and D. Song. Environmental quality and development: Is there a Kuznets' curve for air pollution emissions? *Journal of Environmental Economics and Management*, 27(2):147–162, 1994.

[214] Sen, A. *Development As Freedom.* Alfred A. Knopf, Inc., 1999.

[215] Shafik, M. Economic development and environmental quality: An econometric analysis. *Oxford Economic Papers*, 46:753–773, 1994.

[216] Shogren, J.F., J. Tschirhart, T. Anderson, A.W. Ando, S.R. Beissinger, D. Brookshire, G.M. Brown Jr., D. Doursey, R. Innes, S.M. Meyer, and S. Polasky. Why economics matters for endangered species protection and the ESA. *Conservation Biology*, 13(6):1257–1261, 1999.

[217] Simmons, R.T. *Fixing the Endangered Species Act*, pages 1–21. In [8], 1997.

[218] Simmons, R.T. The Endangered Species Act: Who's saving what? *The Independent Review*, 3(3):309–326, 1999.

[219] Simon, J. *The Ultimate Resource 2.* Princeton, NJ: Princeton University Press, 1996.

[220] Simon, J.L. The effect of income on fertility. *Population Studies*, 23(3):327–341, 1969.

[221] Sinn, H-W. *The Green Paradox: A Supply-side Approach to Global Warming.* The MIT Press, 2012.

[222] Smith, A. *An Inquiry into the Nature and Causes of the Wealth of Nations.* 1776.

[223] Smith, W.B., and D. Darr. *U.S. Forest Resource Facts and Historical Trends.* U.S. Forest Service: Department of Agriculture, 2002.

[224] Sodhi, N.S., and P.R. Ehrlich, eds. *Conservation Biology for All.* Oxford University Press, 2010.

[225] Sparks, J., ed. *The Writings of George Washington*, volume 5. Harper and Brothers Publishers, 1847.

[226] Stavins, R.N. The problem of the commons: Still unsettled after 100 years. *The American Economic Review*, 101:81–108, 2011.

[227] Stern, N. *The Economics of Climate Change: The Stern Review*. Cambridge: Cambridge University Press, 2007.

[228] Stiglitz, J.E. Monopoly and the rate of extraction of exhaustible resources. *American Economic Review*, 66(4):655–661, 1976.

[229] Stroup, R.L. *Eco-nomics: What Everyone Should Know About Economics and the Environment*. Cato Institute, Washington, D.C., 2003.

[230] Sunstein, C. The value of a statistical life: Some clarifications and puzzles. *Journal of Benefit Cost Analysis*, 4(2):237–261, 2013.

[231] Tabarrok, A. The private provision of public goods via dominant assurance contracts. *Public Choice*, 96:345–362, 1998.

[232] Tabarrok, A. The law of unintended consequences. Marginal Revolution, http://www.marginalrevolution.com/2008/01/the-law-of-unin.html, 2008.

[233] Taylor, M.S. Buffalo hunt: International trade and the virtual extinction of the North American Bison. *The American Economic Review*, 101(2):3162–3195, 2011.

[234] The World Bank. World Development Indicators. http://data.worldbank.org/sites/product/wdi, 2012.

[235] Thompson, W. *Popular Lectures: Electrical Units of Measurement*, volume 1. 1883.

[236] Tolleffson, G. Celebrating the Clark's Fork collaborative. *Trout*, pages 12–23, 2000.

[237] Torras, M., and J.K. Boyce. Income, inequality and pollution: A reassessment of the environmental Kuznets' curve. *Ecological Economics*, 25:147–160, 1998.

[238] Tschirhart, J. *Integrated Ecological-Economic Models*. Volume 1 of [203], 2009.

[239] Tukey, J.W. *Exploratory Data Analysis*. Reading, MA: Addison-Wesley, 1977.

[240] van den Bergh, J.C.J.M., ed. *Handbook of Environmental and Resource Economics*. Edward Elgar, 1999.

[241] Vitousek, P.M., H.A. Mooney, J. Lubchenco, and J.M. Melillo. Human domination of Earth's ecosystems. *Science*, 227(5325):494–499, 1997.

[242] Voigtländer, N., and H-J. Voth. Gifts of Mars: Warfare and Europe's early rise to riches. *Journal of Economic Perspectives*, 27(4):165–186, 2013.

[243] Voigtländer, N., and H-J. Voth. How the West "invented" fertility restriction. *The American Economic Review*, 103(6):2227–2264, 2013.

[244] Warde, P. *Energy Consumption in England and Wales 1560-2000*. ISSM-CNR, Napoli, 2007.

[245] Weber, C.L., and H.S. Matthews. Food-miles and the relative climate impact of food choices in the United States. *Environmental Science and Technology*, 42(1):3508–3513, 2008.

[246] Weber, W.L., and B. Domazlicky. Productivity growth and pollution in state manufacturing. *Review of Economics and Statistics*, 83(1):195–199, 2000.

[247] Weiss, H., and R.S. Bradley. What drives societal collapse? *Science*, 291(5504):609–610, 2001.

[248] Weitzman, M.L. Prices versus quantities. *Review of Economic Studies*, 41:447–491, 1974.

[249] Weitzman, M.L. The Noah's ark problem. *Econometrica*, 66(6):1279–1298, 1998.

[250] Weitzman, M.L. A review of *The Stern Review on the Economics of Climate Change*. *Journal of Economic Literature*, 45(3):703–724, 2007.

[251] Wessel, D. From Greenspan, a (truly) weighty idea. *Wall Street Journal*, page B1, May 20, 1999.

[252] West, J.J., S.J. Smith, R.A. Silva, V. Naik, Y. Zhang, Z. Adelman, M.M. Fry, S. Anenberg, L.W, Horowitz, and J-F. Lamarque. Co-benefits of mitigating global greenhouse gas emissions for future air quality and human health. *Nature Climate Change*, 3:885–889, 2013.

[253] Wheeler, D. Racing to the bottom? Foreign investment and air pollution in developing countries. *Journal of Environment and Development*, 10(3):225–245, 2001.

[254] Wilcove, D.S. *Endangered Species Management: The US Experience*. In [224], 2010.

[255] Wilson, R., and E.A.C. Crouch. Risk assessment and comparisons: An introduction. *Science*, 236:267–270, 1987.

[256] Wingo, P.A., C.J. Cardinez, S.H. Landis, R.T. Greenlee, L.A.G. Ries, R.N. Anderson, and M.J. Thun. Long-term trends in cancer mortality in the United States, 1930-1998. *Cancer*, 97(12):3133–3275.

[257] Wolf, C., Jr. A theory of non-market failure: Framework for implementation analysis. *Journal of Law and Economics*, 22(1):107–139, 1979.

[258] Wolf, C., Jr. *Markets for Governments: Choosing between Imperfect Alternatives.* The MIT Press, 1988.

[259] World Coal Association. Coal facts 2012. World Coal Association, http://www.worldcoalorg/resources/coal-statistics, 2012.

[260] World Health Organization. 10 facts on malaria. http:www.who.int/features/factfiles/malaria/en/, 2013.

[261] Worm, B., E.B. Barbier, N. Beaumont, J.E. Duffy, C. Folke, B.S. Halpern, J.B.C. Jackson, H.K. Lotze, F. Micheli, S.R. Palumbi, E. Sala, K.A. Selkoe, J.J. Stachowicz, and R. Watson. Impacts of biodiversity loss on ecosystem services. *Science*, 314(5800):787–790, 2006.

[262] Wrigley, E.A. *Energy and the English Industrial Revolution.* Cambridge University Press, 2010.

[263] Yandle, B., M. Vijayaraghavan, and M. Bhattarai. The environmental Kuznets curve: A primer. *PERC Research Study*, 2(1):1–24.

[264] Zhu, J. *Quantitative models for Performance Evaluation and Benchmarking: Data Envelopment Analysis with Spreadsheets and DEA Excel Solver.* New York: Springer Science+Business Media, 2003.

Index

For Product Safety Concerns and Information please contact our EU representative GPSR@taylorandfrancis.com
Taylor & Francis Verlag GmbH, Kaufingerstraße 24, 80331 München, Germany

www.ingramcontent.com/pod-product-compliance
Ingram Content Group UK Ltd.
Pitfield, Milton Keynes, MK11 3LW, UK
UKHW020021200926
475702UK00014B/1994
9781138075092